ENVIRONMENTAL MODELING

A Practical Introduction

T0179294

ENVIRONMENTAL MODELING

A Practical Introduction

Michael J. Barnsley

CRC Press
Taylor & Francis Group
Boca Raton London New York

CRC Press is an imprint of the
Taylor & Francis Group, an **informa** business

CRC Press
Taylor & Francis Group
6000 Broken Sound Parkway NW, Suite 300
Boca Raton, FL 33487-2742

First issued in paperback 2019

© 2007 by Taylor & Francis Group, LLC
CRC Press is an imprint of Taylor & Francis Group, an Informa business

No claim to original U.S. Government works

ISBN-13: 978-0-415-30054-4 (hbk)
ISBN-13: 978-0-367-38947-5 (pbk)

Library of Congress Cataloging-in-Publication Data

Barnsley, Michael J. (Michael John)
    Environmental modeling : a practical introduction / Michael John Barnsley.
        p. cm.
    Includes bibliographical references and index.
    ISBN-13: 978-0-415-30054-4 (alk. paper)
    ISBN-10: 0-415-30054-1 (alk. paper)
    1. Environmental sciences--Mathematical models. I. Title.

GE45.M37B37 2007
577.01'5118--dc22   2006030555

Visit the Taylor & Francis Web site at
http://www.taylorandfrancis.com

and the CRC Press Web site at
http://www.crcpress.com

Contents

List of Figures

List of Tables

List of Programs

Preface

The initial motivation to write this book stemmed from the perception of a growing divide between the increasingly sophisticated computer-based models that are being developed to represent various aspects of Earth's environmental systems (including those pertaining to its climate, ecosystems, biogeochemical cycles and hydrological processes) and the ability of many undergraduate students, and even some graduate students, of the environmental sciences to engage constructively with these models. The aim of this book, therefore, is to provide a practical introduction to the various methods, techniques and skills involved in computerized environmental modeling, including (i) representing an environmental problem in conceptual terms (i.e., developing a conceptual model), (ii) formalizing the conceptual model using mathematical expressions (i.e., formulating a mathematical model), (iii) converting the mathematical model into a program that can be run on a desktop or a laptop computer (i.e., implementing a computational model) and (iv) examining the results produced by the computational model (i.e., visualizing the output from a model and checking the model's validity in comparison with observations of the target system).

The contents of this book are based on a course that I have taught for many years to honors degree undergraduate students at Swansea University. The objectives of the course, and of this book, are to introduce the student to the broad arena of environmental modeling, to show how computational models can be used to represent environmental systems, and to illustrate how such models can improve our understanding of the ways in which environmental systems function. Equally important, the book also aims to impart a set of associated analytical and practical skills, which will allow the reader to develop, implement and experiment with a range of computerized environmental models. The emphasis is, therefore, on active engagement in the modeling process rather than on passive learning about a suite of well-established models. A practical approach is adopted throughout, one that tries not to get bogged down in the details of the underlying mathematics and that encourages learning through "hands on" experimentation. To this end, a set of software tools and data sets are provided free-of-charge under the General Public License (GPL) and Gnuplot License so that the reader can work through the various examples and exercises presented in each chapter.

Most of the data sets used in this book relate to an area immediately south of Llyn Efyrnwy, Powys in mid-Wales, UK. Apart from the fact that this is a particularly beautiful part of the world, I chose this site because it is one of a relatively small number of locations in the UK at which the solar irradiance measurements used in Chapter 5 are routinely recorded. I should also confess to deriving a certain amount of innocent amusement thinking about the additional challenge that the pronunciation of this particular Welsh place name will present to many readers. If nothing else, it will help to take the reader's mind off the demands of environmental modeling, every now and then.

I am deeply grateful to the UK Ordnance Survey, and in particular Ed Parson, for making available the digital elevation data used in Chapters 2 and 10. Thanks are also due to the UK MetOffice for permission to use the various meteorological data sets pertaining to Llyn Efyrnwy, and also to the staff at the British Atmospheric Data Centre (BADC), which is operated by the UK's Natural Environmental Research Council (NERC), for providing the excellent service through which I was able to access these data.

This book was put together using a range of "open source" software, including the GNU/Linux operating system. Most of the figures were produced using gnuplot (http://www.gnuplot.info/); the majority of the remainder were created using the PSTricks class in LaTeX. LyX (http://www.lyx.org/) and LaTeX (http://www.latex-project.org/) were used to produce the camera-ready copy, and the Beamer class in LaTeX was used to create the presentation files. I should like to thank the developers of all of these software packages.

I should also acknowledge the many cohorts of undergraduate students at Swansea University who have acted as a test bed for much of the material presented in this book: a sea of blank faces is undoubtedly the most immediate and effective signal that the material being presented is inadequately explained or otherwise confusing, and I hope that the salutary lessons that my students have taught me along the way have resulted in a clearer exposition in this book. On a more positive note, I am deeply gratified by those students who, having been introduced to computer-based environmental modeling for the first time, have honed their newly acquired skills and gone on to greater things. I hope that this is also the case for the readers of this book.

Finally, I should like to acknowledge the support of various friends and colleagues who have offered help and advice, and above all provided much-needed injections of humor at numerous points during the production of this book: to Mat Disney, Philip Lewis ("Lewis") and Tristan Quaife at University College London, to Tim Fearnside, Sietse Los, Adrian Luckman, Peter North and Rory Walsh at Swansea, and to Paul Mather at the University of Nottingham, *diolch yn fawr iawn* (thank you very much). Paul Mather deserves special mention for kindly reading through the drafts of this book and for providing many detailed comments and helpful suggestions. As it has become conventional to say at this point, though, any oversights, omissions or errors that remain are mine alone. I understand that in the world of "closed source" computer software such things are often described as "features"; I hope, however, that "features" of this type are few and far between herein. *Pob lwc!* (Good luck!)

Mike Barnsley
School of the Environment and Society
Swansea University, UK

Chapter 1

Models and Modeling

Topics

- Why model?

- The modeling process

- A typology of models

- Systems analysis and systems dynamics

1.1 WHY MODEL?

A model is a simplified representation of a more complex phenomenon, process or
system; an environmental model is one that pertains to a specific aspect of either
the natural or the built environment. Environmental models have been developed
to represent, among other things, elements of Earth's climate system, hydrological
processes, ecosystems and biogeochemical cycles. The principal purposes of these
models are threefold: to increase knowledge, and hence to reduce uncertainty, of the
phenomenon, process or system that the model purports to represent (i.e., to improve
understanding); to provide a tool with which to estimate the state of the phenomenon,
process or system at times and locations other than those for which observations are
presently available (i.e., to facilitate prediction); and to provide a framework within
which "what if" questions can be asked about possible changes to the state and operation of the phenomenon, process or system under specified conditions (i.e., to perform
simulations).

The production of an environmental model involves a process of abstraction:
in the sense that most environmental models deal with abstract concepts and ideas
(i.e., mathematical formulae and computational code) rather than physical objects

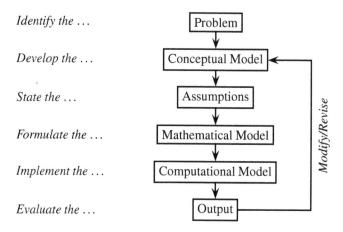

Figure 1.1: Schematic representation of the modeling process.

and events (i.e., conceptualization); in terms of identifying and extracting the most important elements of the phenomenon, process or system and discarding the least significant ones (i.e., selection); and in the sense of summarizing the essence of the phenomenon, process or system (i.e., encapsulation). Building a well-designed model therefore forces one to examine carefully, analytically and in detail the component elements of an environmental system, the processes and structures that govern the relationships and interactions between them, and the spatial and temporal scales over which they operate.

1.2   THE MODELING PROCESS

In principle, the process of designing, building and using an environmental model can be divided into a series of discrete stages. These stages are shown schematically in Figure 1.1 and are described in detail below. In practice, the boundaries between the different stages are not always well defined and progression from one stage to the next is seldom as straightforward or as linear as Figure 1.1 implies. Nevertheless, this diagram provides a useful framework within which to introduce the basic concepts.

1.2.1   Identifying the Nature and Scope of the Problem

The first step is to identify the specific science question, or problem, that is to be addressed and then to establish both whether and how a model will help to answer this question (Wainwright and Mulligan 2004). The problem should be sufficiently well defined and focused so that it is amenable to solution using the knowledge, skills and resources at hand (these factors influence the tractability of the problem), but it should also be sufficiently generic so that it is of more than just parochial interest (this implies that a compromise is negotiated between the specificity and the generality of the model). If the problem is poorly defined at the outset, the model-building process

Table 1.1: Four main phases of systems analysis (after Huggett 1980).

Phase	Actions
Lexical	Define the system boundaries (closure). Choose the system components, i.e., state variables (entitation). Estimate the values (i.e., the state) of the state variables (quantitation).
Parsing	Define verbally, statistically or analytically the relationships between the state variables.
Modeling	Model construction. Model operationalization (i.e., running the model).
Analysis	Model validation and verification (i.e., compare the results of the model with observations of the target system).

is likely to be more difficult, more time-consuming and more complex. Worse still, the resultant model may not be appropriate to the task for which it was originally intended.

A related consideration is the scope of the model, in terms of those elements of the science question that the model is, and is not, intended to address. The scope of the model may have to be limited in various ways to produce a tractable solution. For instance, the model may need to be designed so that it represents a selected part of the target environmental system, a particular spatial domain, a specified period of time, or perhaps a combination of all three.

1.2.2 Developing the Conceptual Model

After specifying the science question, the next step is to develop a conceptual model of the problem. The term *conceptual model* is used here to refer to a model that is expressed verbally or in written or diagrammatic form (i.e., concepts), as distinct from one that is represented in terms of mathematical formulae (i.e., a mathematical model) or one that is constructed from physical materials (i.e., a physical model).

The development of a conceptual model necessarily involves a comprehensive analysis of the target phenomenon, process or system with the aim of identifying its component parts, their respective inputs and outputs, the relationships between them, and the processes and structures that govern their interaction. This stage in the model-building process is therefore closely related to the lexical and parsing phases of systems analysis, a branch of science concerned with the study of complex systems, including their composition, structure, function and operation (Huggett 1980, Table 1.1). In each case, it is assumed that the "real world" can be divided into a number of more or less discrete systems, which can be further sub-divided into their component parts and processes, identified by careful analysis and detailed observation (Hardisty *et al.* 1993).

Table 1.2: Important definitions in environmental modeling.

Element	Definition	Example
Constant	Quantity whose value does not vary in the target system.	Speed of light.
Parameter	Quantity whose value is constant in the case considered, but may vary in different cases.	Total solar radiation at the top of Earth's atmosphere.
Variable	Quantity whose value may change freely in response to the functioning of the system.	Amount of precipitation.
Relation	Functional connection or correspondence between two or more system elements.	Rainfall, run-off and soil erosion.
Relationship	State of being related.	—
Process	Operation or event, operating over time (temporal process) or space (spatial process) or both, which changes a quantity in the target system.	Evapotranspiration.
Scale	Relative dimension, in space and time, over which processes operate and measurements are made.	Local, regional, global; diurnal, seasonal, annual.
Structure	Manner in which component parts of a system are organized.	—
System	Set of related elements (e.g., constants, parameters and variables), the relations between them, the functions or processes that govern these relations and the structure by which they are organized.	Forest ecosystem, drainage basin, global carbon cycle, Earth's climate.

The component elements of an environmental system typically include inputs, outputs, constants, parameters, variables (also known as stocks, stores, pools and reservoirs), processes (flows), relations (links or connectors) and structures (Edwards and Hamson 1989); see Table 1.2 for definitions. The boundaries, or limits, of the target environmental system must also be specified. In this context, environmental systems are sometimes classified in terms of their degree of openness: open systems, also known as forced systems, have exogenous (or forcing) variables; closed systems, also known as unforced systems, have no exogenous variables (i.e., all of the variables are endogenous to the system) (Hardisty *et al.* 1993).

The process of developing a conceptual model is often aided by using diagrams to represent the component parts of the target system and the connections between them (e.g., Figure 1.2). These diagrams vary from the simple to the complex, from the

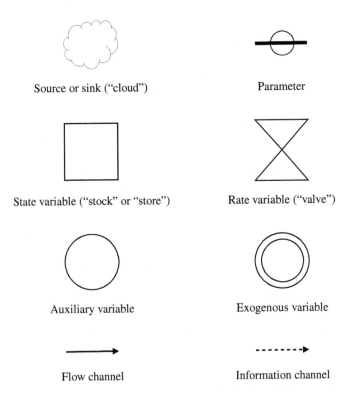

Source or sink ("cloud")

Parameter

State variable ("stock" or "store")

Rate variable ("valve")

Auxiliary variable

Exogenous variable

Flow channel

Information channel

Figure 1.2: Examples of some of the symbols used in Forrester diagrams.

schematic to the formalized. The symbols employed for this purpose differ somewhat between studies, although Forrester diagrams (Forrester 1973) and their derivatives are used quite widely, particularly in the field of ecosystem dynamics (Ford 1999, Deaton and Winebrake 2000).

1.2.3 Stating the Assumptions

Every environmental model is founded on a set of assumptions. These assumptions may be made so that a complex environmental system can be simplified sufficiently to produce a working model (e.g., when deciding which elements of the target system should be included in the model and which should be omitted) or they may reflect the limits to current knowledge of the target environmental system (e.g., concerning the nature and form of the relations between its component parts). The validity and the scope of any such assumptions ultimately determine the value of the resultant model (Edwards and Hamson 1989, Wainwright and Mulligan 2004).

Some of the assumptions that are made when a model is first created may later be found to be incorrect, in which case it is often possible to revise the assumptions in subsequent versions of the model. Some may be known to be wrong at the outset, but nevertheless they may be retained because they have a relatively insignificant effect on the output of the model and are necessary for reasons of simplicity or efficiency (Wainwright and Mulligan 2004). What is important is that each of the assumptions inherent in a model is recognized, understood and stated explicitly (Edwards and Hamson 1989, Wainwright and Mulligan 2004). The primary reasons for this are twofold: first, it clarifies the nature, purpose and limitations of the model, not least in the mind of the modeler; second, it helps potential users of the model to understand its scope (i.e., the range of conditions over which the model is known, or thought, to be valid), to challenge the assumptions on which it is based and, hence, to develop improved versions of the model in the future. For both the modeler and the user, it is particularly important to consider the consequences, whether they are intended or not, of the assumptions made in the model and to identify any assumptions that may have been made implicitly (i.e., without recognizing that this is the case) (Edwards and Hamson 1989).

Although it is not always mentioned in this context, the modeler should make clear the spatial and temporal scales over which the relations and processes that are being modeled operate. Many environmental processes operate, or are dominant, over a restricted range of spatial and temporal scales, and different processes operate at different scales; this phenomenon is sometimes referred to as "domains of scale" in the field of landscape ecology (Wiens 1989). It is inadvisable, therefore, to apply a model outside the range of scales for which it is designed; that is to say, most models are scale dependent.

1.2.4   Formulating the Mathematical Model

The next stage is to represent the conceptual model in mathematical terms; that is, using mathematical tools and concepts, such as variables, functions and equations. This process can be described as one of formulating the mathematical model, since it involves translating the conceptual model into mathematical formulae (Edwards and Hamson 1989). It is frequently the most challenging stage in the development of a model. Sometimes this is because the solution to the problem demands the use of advanced mathematical techniques, although each of the models considered in this book requires only basic skills in algebra and trigonometry. More often it is because there is more than one way in which the system can be represented mathematically and it is not immediately apparent which approach is best. Therefore, deriving a suitable mathematical formulation of a model is often a trial-and-error process, but it is also a skill that improves with practice.

The range and the diversity of mathematical models that have been developed to study environmental systems are considerable, and various different schemes have been proposed to group them by type. While these classification schemes differ at the level of detail, most are founded on a common set of principles that include the following considerations: the extent to which the model is derived from theory or from observations (i.e., empirical models versus theoretically informed models); the degree to which random events and effects play a major role in the target system and, hence, in the model (i.e., deterministic versus stochastic, or probabilistic, models); the level of knowledge or understanding of the target system that the model purports to represent (i.e., black box versus white box models); whether the model is operated predominantly in forward or inverse mode (i.e., forward-mode versus inverse-mode models); whether the model deals with environmental processes that are static or dynamic with respect to time and space (i.e., static versus dynamic models); whether the model deals with environmental processes that can be considered to operate in a discrete or continuous manner with respect to time and space (i.e., discrete versus continuous models); and whether the model parameters are lumped or distributed (i.e., lumped-parameter versus distributed-parameter models). Each of these considerations is examined in greater detail in the sections that follow.

Empirical and Theoretical Models

An empirical model is one that is primarily or solely based on observations, as distinct from one that is derived from theory. In models of this type, the relationships between the component elements of an environmental system are established by examining measurements of the variables concerned. The form of each of these relationships is then defined by a mathematical function (Edwards and Hamson 1989). The decision as to which one of a set of candidate mathematical functions (e.g., Figure 1.3) should be used typically involves a compromise between how well the candidate functions fit the data and the relative simplicity of their mathematical form; the decision, however, is not informed by theory (Edwards and Hamson 1989). Regression analysis is widely used in this context to fit the function to the data.

While empirical models are frequently valuable in terms of making predictions in the cases for which they are developed, they typically lack generality (i.e., they are specific to a particular circumstance or set of data). It is often difficult, therefore, to employ them at other spatial locations or different points in time.

Deterministic and Stochastic (Probabilistic) Models

Loosely defined, a deterministic model is one in which the outputs (i.e., the results) are uniquely and consistently determined by the inputs (i.e., the values used to drive the model) (Edwards and Hamson 1989). Deterministic models therefore function in the same way, and produce exactly the same output, each time they are run using a particular set of input values. This definition can be recast a little more precisely as follows: the state (i.e., the value) of a variable in a deterministic model is uniquely and consistently determined by the initial conditions of the model (i.e., the values of the constants and parameters input to the model) and, subsequently, the previous states of the variable itself. In contrast to empirical models, the implication is that deterministic models are based on assumptions, theory or knowledge of the nature and form of the relations between the variables in the target system. Deterministic models tend, therefore, to offer a much greater degree of generality than empirical models. Consequently, if they are properly configured and suitably implemented, deterministic models can usually be applied to different spatial locations and points in time from those for which they were originally developed and tested.

In contrast, stochastic (or probabilistic) models are ones in which random events and effects play an important role (Edwards and Hamson 1989). The state of the

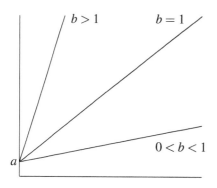

a) Positive linear, $y = a + bx$

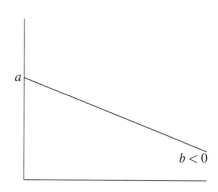

b) Negative linear, $y = a + bx$

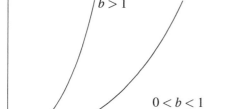

c) Positive exponential, $y = a\exp(bx)$

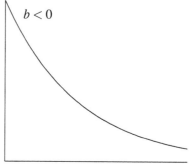

d) Negative exponential, $y = a\exp(bx)$

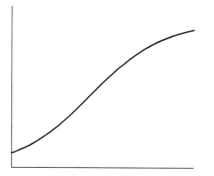

e) Logistic, $y = \dfrac{K}{\left(1 + \left(\frac{K-y_0}{y_o}\right)\exp(-rx)\right)}$

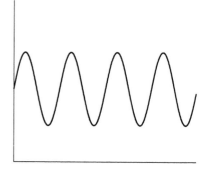

f) Periodic, $y = \sin(x)$

Figure 1.3: Forms of various mathematical functions.

model variables is therefore described by probability distributions, rather than single values. As a result, the output from such a model will vary from run to run even when the input values used are the same. This class of model is therefore suitable in those circumstances where apparently random fluctuations in the system processes, and hence in the system variables, render deterministic models inappropriate (Kirkby *et al.* 1993). The random fluctuations that are observed in the system, and hence in the model, may be due to environmental processes and events that are truly random in nature or they may be pseudo-random; that is, where knowledge of a potentially deterministic process is inadequate or incomplete, such that it has to be treated as though it is random.

Black Box and White Box Models

Mathematical models can also be classified according to the degree to which the composition, structure and operation of the target system is known and, hence, is represented in the model. The two extreme cases in this respect are usually referred to as white box (or clear box) models and black box models. In white box models, the internal workings of the target system are known, completely understood and clearly stated. This *a priori* information may be based on detailed observation of the target system or it may be derived from theory (Kirkby *et al.* 1993). By contrast, in black box models the target system is treated as a sealed unit, with no attempt made to understand the variables of which it is composed or the relations between them. Whereas white box models are based on knowledge or theory of the target system, black box models are usually defined empirically (Kirkby *et al.* 1993). In practice, most models fall somewhere between these two extremes; as such, they might be described as gray box models.

Forward and Inverse Modeling

Mathematical models are typically specified in terms of functional relations between two or more variables of the target system. For example, all other things being equal, the growth and abundance of a particular plant species in a given environment is functionally related to the ambient light conditions, the air temperature, the amount of precipitation and the availability of nutrients in the soil. A functional relation, such as this, can be expressed mathematically as follows:

$$growth = f(light, temperature, precipitation, nutrients) \qquad (1.1)$$

Equation 1.1 indicates that the variable *growth* is a function f of four other system variables: *light, temperature, precipitation,* and *nutrients.* Thus, the value of *growth* is related to, and is in some way dependent on, the values of these four variables. If the form of the function f is known, or can be derived from theory or else be obtained by induction from a set of observations (i.e., empirically), it can be expressed as a rule, which allows the value of the dependent variable (i.e., *growth*) to be determined for any value of the independent variables (i.e., *light, temperature, precipitation,* and *nutrients*). For instance, if $y = f(x)$ describes a functional relation between two system variables x and y, and the function f is defined in terms of the rule $f(x) = x^2$, then $y = f(3) = 3^2 = 9$ when $x = 3$. This will be referred to as forward modeling, for reasons that will become clear once inverse modeling is introduced, below.

Sometimes the modeler may be more interested in the inverse functional relation between two or more variables in the target environmental system. For instance, the modeler may wish to use knowledge of the relation between temperature and plant growth to reconstruct past climatic conditions (e.g., temperature) from an analysis of the abundance of a given plant species (e.g., inferred from pollen counts in cores taken from lake sediments). Inverse functional relations, such as this, can be expressed mathematically as follows:

$$x = f^{-1}(y) \tag{1.2}$$

Equation 1.2 states that there is a function f^{-1}, which is the inverse of function f, that relates the value of variable y to that of variable x (Biggs 1989, Piff 1992). Note that this does not imply that x is dependent on y (e.g., ambient temperature is not controlled in a direct way by the abundance of a particular plant species), merely that they are functionally related in some way. The rule for f^{-1} is obtained by reversing the rule for f (Grossman 1995). For example, if $y = f(x)$ and $f(x) = x^2$, then $x = f^{-1}(y)$ where $f^{-1}(y) = \sqrt{y}$. This is referred to as the inverse model. For simple empirical models, obtaining the rule for the inverse function is relatively straightforward: the values of variable y are regressed on the corresponding values of variable x, rather than *vice versa*. For more complex deterministic and stochastic models, however, the problem is much more challenging and involves the use of analytical and numerical solutions that are beyond the scope of this book.

One area in which the application of inverse models is increasingly common is the field of Earth observation, also known as terrestrial remote sensing. In part, this area of science is concerned with the reflectivity, at different wavelengths, of Earth's surface materials (e.g., vegetation, soil, rocks, snow, ice and water). The reflectivity of these materials is known to be a function of their respective physical, chemical and biological properties. Various models have been developed to represent these functional relations. These models can be used to predict the amount of solar radiation (i.e., sunlight) reflected by a terrestrial surface, based on information about its physical, chemical and biological properties. The inverse functional relation, however, is typically of greater interest for practical purposes; that is, the ability to estimate the properties of Earth's surface materials from measurements of spectral reflectance made by sensors mounted onboard aircraft or Earth-orbiting satellites. This requires the use of sophisticated analytical and numerical techniques to invert the forward model against the measured data (Goel 1989, Kuusk 1995, Verstraete *et al.* 1996).

Static Models, Dynamic Models, Equilibrium, Stability and Feedback

Mathematical models can be divided into two further categories: static and dynamic (Ford 1999). The former deal with systems that do not change, or at least are thought not to change, appreciably with respect to time. This type of model focuses on the

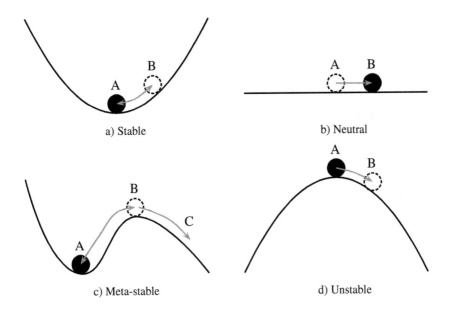

a) Stable

b) Neutral

c) Meta-stable

d) Unstable

Figure 1.4: Graphical representation of different system states. The system state is represented by the position of a ball on the topographical surface.

processes, or forces, that keep the system in a state of equilibrium. Dynamic models, in contrast, deal with systems that change over time, which is much more common in environmental systems. Dynamic mathematical models are typically constructed from difference equations or differential equations (see Chapter 8).

The need to understand the behavior of an environmental system with respect to time leads the modeler to consider the important issue of system stability. A number of generic system states can be envisaged in this context; these states are known as stable, neutral, meta-stable and unstable (Huggett 1980, Figure 1.4). In a stable system, the system returns to its original state (position A, Figure 1.4a) having been perturbed (i.e., moved to position B, Figure 1.4a) by an external force or process. In a neutral system (Figure 1.4b), the action of an external force or process causes no change in the state of the system variables. In a meta-stable system, the system is, initially, in a weak stable state (position A, Figure 1.4c), but the effect of an external force or process may be to move it to an unstable transitional state (position B, Figure 1.4c) from which it may either return to its original state or change to a different one (position C, Figure 1.4c). Finally, an unstable system is one in which the system state changes, perhaps irrevocably, as a result of an external force or process (Figure 1.4d).

The relative stability of an environmental system is partly controlled by feedback mechanisms. Feedback is the process by which a fraction of the output from a given system, or part thereof (i.e., a sub-system), is returned (i.e., is fed back) as input to the same system or sub-system. This process is represented schematically in Figure 1.5. Feedback is an important feature of many environmental systems, not least because

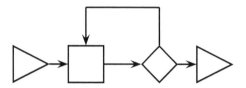

Figure 1.5: Schematic representation of a feedback relation.

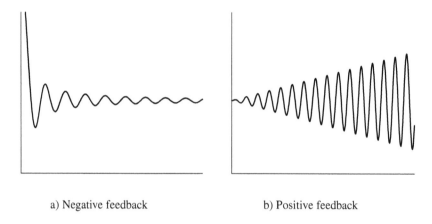

a) Negative feedback   b) Positive feedback

Figure 1.6: Negative and positive feedback relations.

it makes their operation more complex and their behavior more difficult to predict. Feedback is said to be either negative or positive (Figure 1.6). Negative feedback is produced by deviation-damping processes that counteract the effect of a perturbation to a system and, hence, tend to maintain the stability of the system: a system that displays a propensity for this type of behavior is said to exhibit homeostasis. In contrast, positive feedback is the result of processes that amplify an initial perturbation to a system and that tend to keep the system changing toward a new state: this tendency is sometimes referred to as homeorhesis. A notable example of a positive feedback loop in Earth's climate system is one that involves sea ice and albedo (i.e., the average reflectivity of a surface). In this feedback loop, warmer atmospheric conditions (the initial perturbation) result in increased melting of the Arctic and Antarctic ice sheets. This reveals more of the relatively dark (i.e., lower albedo) polar oceans. These, in turn, absorb more solar radiation, which causes the oceans to warm further, so that still more of the polar ice sheets melt, and so on.

Before proceeding further, it is worth noting that many environmental systems not only exhibit feedback mechanisms but also branching or splitting events (Figure 1.7). These occur where two or more inputs feed into a single process, or where two or more outcomes are possible depending on certain conditions (Huggett 1980). As is the case with feedback mechanisms, branching events make the behavior of the system more complex and, hence, more difficult to predict.

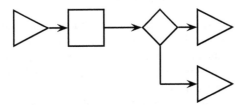

Figure 1.7: Schematic representation of a branching or splitting relation.

Feedback mechanisms and branching events are examined in greater detail in the context of the Daisyworld model, a simple biospheric feedback model, which is analyzed in Chapter 9.

Continuous and Discrete Models

Mathematical models can also be divided into two further categories: discrete and continuous. The distinction between the two relates to the way in which each treats time (or spatial location or both) as a variable. In discrete models, time proceeds in a series of finite steps that are usually, although not necessarily, of equal length; each step relates to a specific interval, such as an hour, a day, a month or a year. The state of the system is updated at the end of each time step, taking into account changes in the system variables resulting from the processes operating on the system over that time step. In continuous models, by contrast, the system state is updated instantaneously in response to continuous changes in the forcing factors. Discrete and continuous models also differ in terms of the mathematical equations on which they are founded: the former are expressed in terms of difference equations; the latter employ differential equations. Discrete and continuous mathematical models are explored in detail, in the context of studies of population growth, in Chapter 8.

Distributed-Parameter and Lumped-Parameter Models

Environmental systems often exhibit considerable spatial heterogeneity in the value of their variables and in the processes that control them. Some environmental models account for this spatial variation explicitly in their mathematical formulation. This is achieved by dividing the geographical domain of the model into a number of discrete spatial units, such as tessellating polygons (particularly square cells, which are then known as rasters), triangular irregular networks (TINs) or irregularly shaped spatial objects (Wainwright and Mulligan 2004). Models that assign different values of the system parameters and variables to each of these spatial units are known as distributed-parameter models. This approach is not always feasible, however. Sometimes this is because of computational constraints; on other occasions it may be due to a lack of data on the spatial variability of the model parameters and variables. In such circumstances, it may be necessary to assign a single "lumped" value across the whole of the model's spatial domain; models such as this are often referred to as lumped-parameter models.

Analytical and Numerical Solutions to Mathematical Models

It is possible, in many cases, to construct a mathematical model of an environmental system by analytical means; that is, by formulating the model so that it is expressed concisely in terms of mathematical equations, functions, variables and constants. This approach is sometimes referred to as an analytical or closed-form solution. Not all environmental models are amenable to this approach, however; some must be solved numerically. The numerical approach relies heavily on the power of modern computers to solve the model by performing a sequence of operations rapidly over and over again, each time applying the operations to the result of the previous iteration (British Computer Society 1998). Sometimes an estimate, or guess, is made regarding the correct solution, to which the iterative procedure is initially applied. Provided that the mathematical model has been formulated and implemented correctly, the expectation is that the procedure will move ever closer to the correct solution with each successive iteration. Depending on the nature of the model, and in some cases the accuracy of the initial estimate, the iterative procedure will eventually reach the correct solution or else a very close approximation to it. The best known example of this approach is Newton's iterative method for calculating $\sqrt{x}$ (Harris and Stocker 1998). Examples of the analytical and numerical approaches to modeling are examined in detail in Chapter 7.

1.2.5 Implementing the Computational Model

General Considerations

Once an appropriate formulation has been derived for the mathematical model, the next step is to convert the equations and formulae into some form of software that can be run on a computer; that is, to implement the computational model. There is an enormous range of options available to the modeler in this context, including spreadsheet packages, specialized modeling software, high-level computer programming and scripting languages, and integrated modeling environments (Table 1.3). Each has its advantages and disadvantages. Ultimately, the decision regarding which software the modeler chooses to use is likely to be conditioned by issues of availability, cost, personal preference and prior experience, in addition to the ease with which the software is learned and its suitability for the task at hand.

It has been argued that it is easier to learn how to use spreadsheet packages than formal programming languages (Hardisty *et al.* 1993). Spreadsheets are certainly ubiquitous and, hence, are familiar to many computer users. Most spreadsheet packages offer the added advantage of built-in facilities for data visualization, statistical analysis and function-fitting. Spreadsheets have been used quite widely, therefore, in environmental modeling; for example, in conservation biology and landscape ecology (Donovan and Welden 2001, 2002). Nevertheless, spreadsheets have intrinsic limitations that present problems when producing more sophisticated environmental models, especially those requiring iteration to reach the solution. Most importantly, perhaps, spreadsheets do not encourage a structured approach to model building, and the potential for sharing and reusing code is limited.

Table 1.3: Selection of tools suitable for implementing computer-based environmental models.

Category	Name	Reference
Spreadsheets	Calc	http://www.openoffice.org/
	Excel	http://office.microsoft.com/
	Gnumeric	http://www.gnome.org/projects/
	Kspread	http://www.koffice.org/
	Quattro Pro	http://www.corel.com/
Specialized modeling environments	MODELMAKER	http://www.modelkinetix.com/
	POWERSIM	http://www.powersim.com/
	SIMILE	http://simulistics.com/
	STELLA	http://www.iseesystems.com/
	VENSIM	http://www.vensim.com/
High-level computer programming languages	BASIC	http://www.freebasic.net/
	C	Kernighan and Ritchie (1988)
	C++	Oualline (1995)
	FORTRAN	Hahn (1994)
	JAVA	Niemeyer and Knudsen (2005)
	PASCAL	Buchanan (1989)
	VB	Willis and Newsome (2005)
Scripting languages	awk	Aho *et al.* (1988)
	JavaScript	Flanagan (2001)
	Perl	Wall and Schwartz (1993)
	Python	Lutz (1996)
	PHP	Atkinson (2000)
	Rexx	Cowlishaw (1990)
	Ruby	Thomas and Hunt (2000)
	Tcl/Tk	Welch *et al.* (2003)
Integrated modeling environments	IDL	http://www.ittvis.com/
	MATLAB	http://www.mathworks.com/
	OCTAVE	http://www.gnu.org/software/octave/
	R	http://www.r-project.org/
	S	http://www.insightful.com/

Specialized modeling software also offers many important benefits, including the ability to construct models diagrammatically using graphical interfaces, as opposed to typing lines of computer code, and the provision of an extensive range of built-in analytical and numerical functions and data visualization routines. This type of software tends to be made available on a commercial basis, however; that is, at some financial cost to the modeler. It may also be difficult to share models developed in one software package with users who employ a different modeling suite. As a result, there is a danger that the modeler becomes "locked in" to a product offered by a particular vendor. Finally, the modeling software may restrict the user to a particular set of modeling techniques, depending on the functions and the range of expressions offered by the software.

High-level computer programming and scripting languages represent yet another option for the modeler. In many senses this is the most flexible option, albeit one that requires the modeler to learn the commands, grammar and syntax of the chosen language. Despite suggestions to the contrary, however, this is not a major challenge and, importantly, the reward is a highly transferable skill that can be applied in a wide range of contexts. The choice of which programming language to use is largely a matter of personal preference. Although there are differences between programming languages, most can be used effectively to implement a computer-based environmental model. Moreover, many programming languages share a set of common features so that once the user has learned one language it is relatively easy to master another.

Irrespective of the specific software that is used for the purpose, the process of implementing a computational model involves a translation between the vocabulary, grammar and syntax of one language (i.e., the functions and equations of mathematics) and those of another (i.e., the instructions of a spreadsheet or high-level computer programming language). Just as in the translation between natural languages, such as English and French, there is rarely a single "correct" solution to this process, although some solutions might be regarded as being intrinsically better than others. The flexibility and versatility of computer software means that the implementation can often be realized in a number of ways. In some instances, the differences may amount to little more than a matter of programming style, broadly analogous to the variations in style of written prose among literary authors; on other occasions, however, the differences may have a significant impact in terms of the efficiency of the resulting code, including the speed with which the result is computed and the demands that the program places on various aspects of the computer's resources, such as memory usage and hard-disk access. Thus, some implementations may be considered to be more elegant or otherwise preferable to others.

Computer Software Used in This Book

A simple, but powerful, high-level scripting language, known as awk, is used in this book. The acronym awk is derived from the surnames of the individuals who initially developed the awk language and utility: Aho, Weinberger and Kernighan (Aho *et al.* 1988). There are several versions of awk. The version used here is the one produced by the GNU Project (http://www.gnu.org/), which is known as gawk

(Dougherty 1996, Robbins 2001). gawk has two main advantages in this context: it is an interpreted language, meaning that the computer interprets each line of the program (or script) as it runs, so that the modeler does not have to compile the code before it is executed by the computer; and it is data-driven, in as much as it provides a simple mechanism for reading data from files structured in terms of fields (columns) and records (lines or rows). The second of these two features greatly simplifies the process of incorporating data to models. More generally, gawk provides a powerful and flexible tool for data manipulation and a convenient framework within which to develop a range of environmental simulation models. It is also available free-of-charge for use on a wide range of computer platforms and operating systems.

While the gawk programming language is used to manipulate data stored in files and to implement computational models, a separate utility is required to visualize the inputs to, and outputs from, models. A software package known as gnuplot is used for this purpose. gnuplot is primarily, but not exclusively, an interactive, command-driven, function and data plotting program (Williams and Kelly 1998). It can be used to generate two-dimensional (2D) scatterplots and line diagrams in either rectangular or polar coordinates, as well as plots of points, lines, vectors and surfaces in three dimensions (3D). It is able to fit user-defined functions to a data set and to output the results to a wide range of graphical file formats, as well as to the computer screen and printer. Apart from its analytical and plotting capabilities, gnuplot has two other important advantages: it is available free-of-charge under the terms of its license (Appendix C) and versions of the software are available for use on a wide variety of computer platforms and operating systems.

Model Parameterization

Before proceeding further, it is worth noting that many computational models require what is sometimes known as parameterization (also referred to as parametrization or parameter estimation); that is, the values of the parameters and the initial values of the variables used in the model need to be determined so that the model produces sensible results (i.e., the model output matches observations of the target system under the conditions being modeled). This process is also known as tuning or calibrating the model. Parameterization can be performed graphically (i.e., visually), statistically (e.g., using least-squares estimation, often known as regression) or analytically (e.g., by solving a set of simultaneous linear equations).

1.2.6 Evaluating the Model

Verification and Validation

Once the computational model has been implemented, the next step is to run the model to check, first, that it works and, second, that it produces acceptable results. If the result of the first of these tests is unsatisfactory, the implementation should be checked and, where appropriate, revised. The second test is addressed by a process known as model verification and validation (or V&V). The difference between *verification* and *validation* is a subtle one: verification involves an assessment that the computational model satisfies the criteria set out in the model specification (has the modeler built the model correctly?); validation is concerned with assessing whether the computational model is suitable for its intended purpose (has the modeler built the correct model?).

The obvious way to validate a model is to examine how it performs over a range of conditions compared to observations of the target system. This usually involves an evaluation of the goodness-of-fit between the modeled and observed values for a given set of conditions, typically using an index of association or correlation, such as the coefficient of determination, R^2, or the chi-squared value, χ^2. The closer the fit, the more accurate the model.

There are several reasons why the values output from a model may differ from observations of the target system made under the same set of conditions. The reasons include incorrect assumptions made about the target system in the specification of the conceptual model, errors or inappropriate methods employed in the construction of the mathematical model, mistakes or incorrect techniques applied in the implementation of the computational model, inexact arithmetic (i.e., rounding errors) performed in the computational model, uncertainty in the data used to parameterize the model, and errors in the data used to test the model (i.e., measurement error). If the output from a model differs significantly from observations of the target system, the sources of error need to be investigated. This investigation may then lead to the model being revised and some or all of the model-development cycle (Figure 1.1) being revisited.

In some instances it may only be possible to measure (or to infer) the values of some of the variables used in the model. This situation might be described as a partial validation. In the extreme case, it may be necessary to validate a model by comparing the results that it produces with those of another model designed to represent the same system. If one compares a simple model to a more complex one, the conditions of the more complex model should be set to replicate those of the simpler one. This form of validation is generally less satisfactory, in that it only highlights differences between the two models, unless the model used as the standard has been previously validated rigorously (e.g., against observations of the target system). Sometimes, however, it is the only practical solution.

Accuracy, Error and Precision

It is important, in the context of model verification and validation, to be aware of the distinction between the terms accuracy, error and precision. Broadly speaking, the term accuracy refers to the fidelity with which a model represents the processes and relations of the target environmental system, but it is also used in a somewhat narrower sense to indicate the degree to which the model output conforms to the actual (or true) values for the corresponding system. Interpreted in the latter way, accuracy is the complement of error (i.e., 95% accuracy implies 5% error).

Error is often assessed by comparing the output from a model against a set of independent observations (measurements) made on the variables incorporated in the model, and is usually expressed in terms of the average (mean) difference between the observed and modeled values for those variables (Hardisty *et al.* 1993). The root mean square error (RMSE), which is given as follows,

$$\text{RMSE} = \sqrt{\frac{1}{N} \sum_{i=1}^{N} (x_{\text{observed}} - x_{\text{model}})^2} \qquad (1.3)$$

is a commonly used measure of error. Note that the value so obtained is only an estimate of the error of the model because it is based on a sample set of observations, typically acquired over a limited range of the total set of possible conditions that the model variables can take.

Precision has two related meanings: the first indicates the degree of agreement among a group of related observations or model outputs; the second refers to the units of the least significant digit of an observation or model output. Note that it is possible for the output of a model to be precise but inaccurate, or to be accurate but imprecise. The difference between error and precision is perhaps best illustrated by the following example: if the value of some property predicted by our model is -0.4515, while the expected value is 1905.25, then the modeled value is very precise (it is provided to four decimal places) but very inaccurate (there is a large error).

Sensitivity Analysis

A further component of model evaluation is the investigation of a model's sensitivity to the values with which it is initialized. If a small change in the value of an input parameter produces a large change in the model's output, the model is said to be sensitive to that parameter and the parameter is said to have a high influence on the model (Ford 1999). Conversely, if a large change in a parameter produces a small change in the model's output, the model is said to be insensitive to the parameter and the parameter is said to have a low influence on the model.

The process of establishing the sensitivity of a model to its parameters is known as sensitivity analysis (Saltelli *et al.* 2000, Saltelli *et al.* 2004). Knowledge gained by performing sensitivity analysis on a model can help to elucidate the way in which the modeled system functions and to identify those parameters of the model whose values need to be specified most accurately. The ultimate aim of sensitivity analysis is to focus attention on the critical parts of the model and the environmental system that it purports to represent.

In practice, sensitivity analysis is usually performed by perturbing the values of the model parameters by known amounts, measuring the effect that these variations have on the model outputs. The simplest method is to vary the value of one parameter at a time, while the values of the other parameters are held constant. This approach, known as one-at-a-time (OAT) or univariate sensitivity analysis, is used to quantify the influence that each parameter exerts on the model. This method has a significant limitation, however, in that it does not account for the sensitivity of the model to two or more parameters, which may have limited influence when considered in isolation, that interact to produce major changes in the model output in combination. This problem requires the use of more sophisticated multivariate sensitivity analysis techniques, such as Monte Carlo simulation with simple random or Latin Hypercube sampling (Saltelli *et al.* 2000, Saltelli *et al.* 2004).

Figure 1.8: Photograph of Llyn Efyrnwy.

1.3   Llyn Efyrnwy

Many of the data sets and some of the models examined in this book relate to the area around Llyn Efyrnwy in the Berwyn Mountains (*Mynydd y Berwyn*), Powys, Wales, UK (52°46′58″ N, 3°30′46″ W; Ordnance Survey (OS) grid reference SJ008178; Figure 1.8). Llyn Efyrnwy is a man-made reservoir, which was created in 1888 by the construction of a large masonry dam, supplying water to the city of Liverpool approximately 70 miles away. Covering an area of $8.24\,km^2$, Llyn Efyrnwy is the largest lake in Wales. Much of the surrounding area is a dedicated wildlife reserve, which is owned by Severn Trent Water (http://www.stwater.co.uk/) and managed by the Royal Society for the Protection of Birds (RSPB) (http://www.rspb.org.uk/). The nature reserve covers an area of roughly 6475 hectares (16,000 acres). Various species of bird can be seen around the lake and in the sessile oak woodlands, heather moorlands and coniferous plantations beyond (Table 1.4). The meteorological data sets that are used in Chapters 2 through 5 and Chapter 10 relate to the MetOffice station located close to Llyn Efyrnwy (52°44′55″ N, 3°28′16″ W; OS grid reference SJ 008178; elevation 235 m above Ordnance Datum Newlyn (ODN)).

1.4   Structure and Objectives of the Book

In the chapters that follow, the reader is taken through the various stages of model development outlined above. One of the challenges in this context is that the reader

Table 1.4: Some of the bird species found in the area surrounding Llyn Efyrnwy.

Habitat	Common name	Latin name
Lake	Common Sandpiper	*Actitis hypoleucos*
	Goosander	*Mergus merganser*
	Great Crested Grebe	*Podiceps cristatus*
	Peregrine	*Falco peregrinus*
Oak woodlands	Nuthatch	*Sitta europaea*
	Pied Flycatcher	*Ficedula hypoleuca*
	Redstart	*Phoenicurus phoenicurus*
	Siskin	*Carduelis spinus*
	Wood Warbler	*Phylloscopus sibilatrix*
Heather moorlands	Black Grouse	*Tetrao tetrix*
	Brambling	*Fringilla montifringilla*
	Buzzard	*Buteo buteo*
	Hen Harrier	*Circus cyaneus*
	Merlin or Pigeon Hawk	*Falco columbarius*
	Red Grouse	*Lagopus lagopus*
Coniferous plantations	Coal Tit	*Parus ater*
	Common or Red Crossbill	*Loxia curvirostra*
	Goldcrest	*Regulus regulus*
	Goshawk	*Accipiter gentilis*
	Nightjar	*Caprimulgus europaeus*
	Raven	*Corvus corax*

must be introduced to three distinct themes: the mathematics underpinning individual models, the computer software or programming language in which the mathematical model is implemented, and the computer software that is used to visualize the inputs to, and outputs from, the computational model. One approach is to cover each of these topics separately, introducing them in sequence. This implies, however, that the reader must learn about a range of environmental models before finding out how these models can be implemented in computer code, or else learn how to write computer programs before being introduced to the models themselves. Either way, the reader is likely to become discouraged before he or she has fully implemented and tested a single environmental model. This book therefore adopts a different approach; one in which, as far as possible, elements of all three themes (the mathematical foundations of a model, the aspects of the computer programming language that are required to implement it, and the components of the data visualization software that are needed to explore its output) are introduced in combination. The material in each chapter therefore progresses incrementally from topics that require fairly basic data visualization and data manipulation techniques through to ones that demand somewhat more sophisticated mathematical modeling and computer programming procedures.

Chapter 2 provides an introduction to the computer software (gnuplot) used to visualize a range of environmental data sets, including the output from environmental models, examined throughout this book. The ability to present data graphically is central to the model-building process and, more generally, to understanding the operation and the behavior of environmental systems. Data visualization allows the modeler to explore the form and the strength of relationships between the system variables prior to building a model and, subsequently, to examine its output. This chapter therefore provides a tutorial on how to use gnuplot to handle a number of different types of data and to create a range of different plot styles. Other features and facilities of gnuplot are introduced, as they are required, in later chapters.

Chapter 3 provides an introduction to gawk, the scripting language that is used in this book to process a range of environmental data sets and to implement several environmental models. Rather than diving straight into the intricacies of model implementation, however, this chapter aims to familiarize the reader with the basics of gawk by showing how it can be used to process a small example data set. The data set contains measurements of precipitation made at Llyn Efyrnwy every 12 hours throughout 1998. The data are given in the format provided by the MetOffice. gawk is used to reformat the data so that they can be visualized more readily in gnuplot.

In Chapter 4, further elements of the gawk scripting language are introduced in the context of measuring and modeling wind speed, wind energy and wind power at Llyn Efyrnwy. Wind speed measurements made at hourly intervals are processed using gawk to calculate the annual mean wind speed and the relative frequency distribution of different wind speeds at this site. gnuplot is used to fit a probability density function (PDF) model of wind speed to the measured data. This model is used to determine the likelihood with which wind speeds capable of driving a small wind energy conversion system (WECS) are observed at Llyn Efyrnwy. gawk is then used to implement a model of the potential for electricity generation by wind power at Llyn Efyrnwy, based on established mathematical formulae.

The transition from measurements to models continues in Chapter 5, in which the amount of solar radiation reaching Earth's surface, known as solar irradiance, is examined. A mathematical model is constructed from equations and formulae published in the scientific literature. This model is implemented in gawk. The resulting computational model is used to predict the amount of solar radiation incident at Llyn Efyrnwy at different times of the year. The model is validated by comparing its output to hourly measurements of total solar irradiance made at the local meteorological station.

Chapter 6 continues the solar radiation theme by modeling its interaction with plant canopies on Earth's surface. Rather than returning to the extensive literature in this area for existing mathematical models, however, this chapter takes a different approach, developing a simple model from scratch. The intention is to demonstrate the process of model development, from specification of the conceptual model and its assumptions, through formulation of the mathematical model and its implementation in gawk code, to testing the model against observations of the target system. Several iterations of the model development cycle (Figure 1.1) are required before the model replicates adequately the observations.

The final model developed in Chapter 6 is reformulated a number of times in Chapter 7 to demonstrate the difference between analytical and numerical solutions to environmental models. In the process, various additional features of the gawk scripting language, such as arrays, iterative methods and control-flow constructs, are introduced.

Chapter 8 covers a range of discrete and continuous models of population growth. It provides a basic introduction to the mathematics of difference equations and differential equations, in addition to constrained (or density-dependent) and unconstrained (or density-independent) models of population growth. This chapter also shows how chaotic (i.e., unpredictable) behavior can be produced, in certain circumstances, by deterministic systems. The initial models are developed further so that they take into account competition for resources among individuals of a single species (intraspecific competition) and between individuals of two or more species (inter-specific competition), in addition to the effect of predator-prey relationships. This requires the introduction of some basic techniques for numerical integration, namely the methods of Euler and Runge-Kutta, and further aspects of control-flow structures in the gawk scripting language.

Many of the elements introduced in the first eight chapters are brought together in Chapter 9, which examines a model feedback mechanism between the biota (the living organisms) and the abiotic environment on an imaginary planet, known as Daisyworld. For instance, the Daisyworld model explores the growth of two species of daisy over time (Chapter 8), as a function of planetary temperature. The planetary temperature is, in turn, partly controlled by changes in the amount of solar radiation that is incident on the planet's surface (Chapter 5). The two species of daisy differ solely with respect to color: one is dark, the other is light, compared to the soil substrate in which they grow. Thus, the model is also concerned with the interaction between solar radiation and plant canopies, or, more specifically, the fractions of incident radiation that are reflected or absorbed by the daisies (Chapters 6 and 7). While the focus of this chapter is the operation and implications of the Daisyworld model, a number of technical aspects are also covered. These aspects include the application of user-defined functions in gawk to produce modular, and hence more manageable, computer code and a consideration of sensitivity analysis as a tool for understanding the operation of the model and for exploring its computational implementation.

In the final chapter, Chapter 10, the model of incident solar radiation at Earth's surface that was introduced in Chapter 5 is extended to account for the effects of sloping terrain. The revised model is applied to data on terrain gradient and aspect derived from a digital elevation model (DEM) covering an area immediately to the south of Llyn Efyrnwy. This necessitates an introduction to handling 2D arrays in gawk. In the second part of the chapter, the information on terrain slope is used to predict the local drainage direction (LDD) network of the study area. This network is compared to the "blue line" features (i.e., rivers and streams) extracted from the corresponding OS digital topographic map. The LDD network and "blue line" features are visualized together using gnuplot's vector plotting capabilities.

A series of additional exercises are set throughout the book, which the reader is encouraged to try. The recommended solutions to these exercises are presented in Appendix E. It is worth noting, however, that there are often a number of different ways of solving a specific problem, so that the reader may find alternative solutions to the ones suggested in Appendix E.

1.5   Resources on the CD-ROM

The CD-ROM that accompanies this book contains copies of the computer software (i.e., gawk and gnuplot), data sets, gawk programs and gnuplot scripts used in the following chapters. Copies of gawk and gnuplot can be found in the utils directory on the CD-ROM. Instructions on how to install this software on desktop computers running a version of either the GNU/Linux or the Microsoft Windows operating systems are given in Appendix A, which also provides details on where to find and how to download copies of the latest versions of these packages. Note that both gawk and gnuplot are provided free-of-charge and can therefore be installed on as many computers as required within the fairly broad constraints of their respective licenses (Appendix B and Appendix C, respectively).

As previously noted, the gawk scripting language is used throughout this book to process a range of environmental data sets and to develop computer implementations of various environmental models. Both the data sets and the gawk programs are stored in ASCII (American Standard Code for Information Interchange) text files. These files can therefore be created and edited using, among other things, basic text editor software; however, it is recommended that standard word-processing packages not be used for this purpose. The resources associated with each chapter (i.e., data sets, gawk programs and gnuplot scripts) are stored in separate directories on the CD-ROM, labeled chapter2, chapter3, chapter4 and so on. Updates and errata will be made available via http://stress.swan.ac.uk/~mbarnsle/envmod/.

Files containing material suitable for lecture presentations are also included on the CD-ROM. These files can be found in the present sub-directory and are provided in Portable Document Format (PDF), which can be displayed from a laptop or desktop computer, via a data projector, using suitable software, such as Adobe Acrobat Reader. This software can be downloaded free-of-charge from the following web site: http://www.adobe.com/products/acrobat/readstep2.html. Once a file has been loaded into Acrobat Reader, the presentation can be set to full screen mode by typing $\boxed{\text{Ctrl}}$ + $\boxed{\text{L}}$.

1.6 Typographical Conventions

The following typographical conventions are used throughout the book. When the reader is required to type an instruction on the command line (i.e., in a GNU/Linux shell or at the Microsoft Windows command prompt), the instruction is presented in a rectangular box as shown below.

```
gawk -f myprog.awk mydata.dat
```
1

The number on the right-hand side of the box is used to refer to different instructions within each chapter. Sometimes the command line will be too long to fit on a single line of the book. In this case, the ⇨ symbol is used to indicate that the command line continues on subsequent lines, as shown below.

```
gawk -f myprog.awk -v variable1=3.1415 -v variable2=0.125 ⇨
    ⇨-v variable3=23.67 mydata.dat > result.dat
```
2

Note that this command should be entered on one line and that the blank space before the first of the ⇨ symbols is significant and should be respected. The command line shown above redirects the output from the program into a data file (`result.dat` in this instance). The contents of a data file are presented in a box with rounded corners, such as that illustrated below.

```
1  199801010900     1.80    -0.30
2  199801012100     7.70     1.60
3  199801020900     6.80     3.90
4  199801022100     6.50     3.80
5  199801030900     9.70     4.80
```

The individual lines of data (i.e., records) in the file may or may not be numbered. Where they are numbered, the line numbers appear of the right-hand side of the page.

Instructions to be typed in gnuplot are shown in a rectangular box with a gray background, as illustrated below.

```
reset
set style data lines
plot 'mydata.dat' using 1:2
```
1
2
3

The line numbers on the right-hand side of this box are used to refer to individual gnuplot instructions. Finally, the gawk code for the computational models developed in this book is presented in the form of listings between horizontal lines, such as that shown in Program 1.1. Once again, for the sake of convenience, line numbers

Program 1.1: introduction.awk

```
(NR >1 && $10 != -999) {                                                  1
  wind_speed=$10;                                                         2
  sum_speed=sum_speed+wind_speed;                                         3
  num_obs=num_obs+1;                                                      4
}                                                                         5
                                                                          6
END {                                                                     7
  mean_wind_speed =0.515*sum_speed/num_obs;                               8
  printf("Mean wind speed=%3.1f m/s\n", mean_wind_speed);                 9
}                                                                        10
```

are appended to the right-hand side of the listing; these numbers are not part of the computer code and should not be included when entering the program.

Chapter 2

Visualizing Environmental Data

Topics

- Scientific data visualization

Methods and techniques

- Generating 2D plots in gnuplot

- Handling time-series data in gnuplot

- Visualizing 3D data sets in gnuplot

- Exporting graphics from gnuplot

- Saving gnuplot commands in script files

2.1 INTRODUCTION

The ability to present data graphically is central to the model-building process and, more generally, to understanding the operation and the behavior of environmental systems. When designing a model, for instance, graphical representations of data can help the modeler identify the major trends and anomalies that are characteristic of a particular environmental system. They can also reveal the degree of influence that specific components of a system exert on the state of the system as a whole. This information can, in turn, be used to decide which parameters and variables should be built into a model (i.e., those that have a major influence on the system state) and which ones may safely be omitted (i.e., those that have a negligible impact). Simi-

Table 2.1: Examples of "open source" software for scientific data visualization.

Software	Description and URL
DX	IBM Open Visualization Data Explorer
	URL: http://www.opendx.org/
GMT	Generic Mapping Tools
	URL: http://gmt.soest.hawaii.edu/
gnuplot	A scientific plotting package
	URL: http://www.gnuplot.info/
Grace	WYSIWYG 2D plotting tool
	URL: http://plasma-gate.weizmann.ac.il/Grace
Gri	Interpreted language for scientific graphics
	URL: http://gri.sourceforge.net/
XGobi	Data visualization system for multidimensional data
	URL: http://www.research.att.com/areas/stat/xgobi

larly, graphical representations of data can help the modeler understand the ways in which individual elements of the system interact and, hence, elucidate the environmental processes that govern their interaction. Later, during the model construction and testing phases, the validity and the accuracy of a model can be evaluated by visually comparing the data that it generates with *in situ* measurements of the corresponding environmental system. Where this procedure highlights errors in the output from the model, particularly systematic ones, it may initiate a further cycle of analysis, explanation and model enhancement. The ultimate objective of this process is to reduce uncertainty in the output from the model and, through this, to understand better the environmental system that the model purports to represent.

Environmental data sets typically contain very large numbers of values. They are also frequently multidimensional, in the sense that the data often represent measurements made on more than one variable (e.g., incident solar radiation, precipitation, temperature and wind speed) at a particular location and point in time. When the data describe measurements made at more than one spatial location and at different points in time, the dimensionality of the data increase further still. These characteristics impose special demands on the computer software used to present environmental data graphically, some of which (e.g., the ability to handle 3D surfaces, contours and vectors) are not well supported by standard business graphics software, such as that associated with many spreadsheet packages. Consequently, a range of specialist software has been developed to visualize scientific data, including those relating to environmental systems (Table 2.1).

The software of choice here is gnuplot. As is the case with many "open source" software packages, gnuplot is available free-of-charge for use on a variety of computer platforms, including GNU/Linux, MacOS X and Microsoft Windows (Williams and Kelly 1998, DiBona *et al.* 1999, Raymond 1999). gnuplot is used throughout this book to generate everything from simple *x-y* scatter plots to time-series diagrams, contour maps and vector-flow plots, and to visualize 3D surfaces. The purpose of

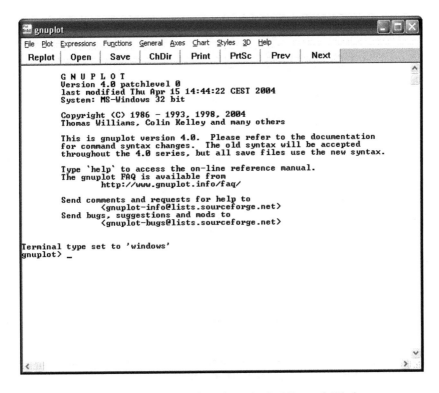

Figure 2.1: GUI for the version of gnuplot for Microsoft Windows.

this chapter is, therefore, to introduce this software through a structured tutorial. This is intended to demonstrate some of gnuplot's graphical and analytical capabilities. It also illustrates a number of scientific data visualization techniques that can be applied to environmental data sets (Tufte 2001). The reader is strongly encouraged to run the examples presented in this chapter on his or her own computer, and to experiment more freely with this software. Further aspects of gnuplot's plotting capabilities are examined in subsequent chapters.

Two versions of gnuplot are included on the CD-ROM that accompanies this book (henceforth referred to simply as "the CD-ROM"): one is for use with GNU/Linux; the other is for use with Microsoft Windows. They are virtually identical in terms of functionality, although the version for Microsoft Windows has a built-in graphical user interface (GUI) that allows the user to interact with the software by means of point-and-click techniques (Figure 2.1). In this book, however, the instructions used to generate plots are entered via gnuplot's command-line interface; that is, they are typed-in beside the gnuplot> prompt. One reason for doing this is to standardize the presentation of the GNU/Linux and Microsoft Windows versions of the software. More importantly, though, the command-line approach confers significant advantages in the long run, including the ability to maintain a record of how individual data sets were plotted and the potential to automate the visualization of several related data sets. Moreover, the learning curve associated with the command-line method is about

```
      1  19980101T0900     1.80    -0.30                                    1
      2  19980101T2100     7.70     1.60                                    2
      3  19980102T0900     6.80     3.90                                    3
      4  19980102T2100     6.50     3.80                                    4
      5  19980103T0900     9.70     4.80                                    5
      6  19980103T2100     6.50     0.20                                    6
      7  19980104T0900     4.20     2.30                                    7
      8  19980104T2100     4.80     0.20                                    8
      9  19980105T0900     3.20     0.60                                    9
     10  19980105T2100     2.30    -0.20                                   10

    706  19981227T0900     5.50     1.80                                  706
    707  19981227T2100     5.70     2.70                                  707
    708  19981228T0900     3.90     0.30                                  708
    709  19981228T2100     3.80     0.30                                  709
    710  19981229T0900     3.00    -0.30                                  710
    711  19981229T2100     6.20     3.00                                  711
    712  19981230T0900     9.00     6.10                                  712
    713  19981230T2100     7.40     4.10                                  713
    714  19981231T0900     5.40     3.70                                  714
    715  19981231T2100     7.00     4.80                                  715
```

Figure 2.2: Extract from the data on air temperature at Llyn Efyrnwy throughout 1998 (le98temp.dat; first and last 10 lines only).

the same as that of the point-and-click approach, largely because a small set of core commands is used to generate many different types of plot. Detailed instructions on how to install and run gnuplot are given in Appendix A.

2.2   CREATING 2D PLOTS

2.2.1   Creating a Simple *x-y* Plot

The file le98temp.dat, which can be found on the CD-ROM, contains information on air temperatures at Llyn Efyrnwy recorded by an automatic weather station (AWS) operated by the UK MetOffice. The AWS records, among other things, the minimum and maximum air temperature at this site every 12 hours. The file le98temp.dat is derived from the AWS data set for 1998 (Figure 2.2). It consists of four columns of data: the first contains the sequential identifying number of the measurement, the second indicates the date and time at which that measurement was made, and the third and fourth refer to the maximum and minimum air temperature (°C), respectively, recorded over the preceding 12 hours. Note that the date and time information, presented in the second column, is given in the format *YYYYMMDD*T*hhmm*, where *YYYY* denotes the calendar year (e.g., 1998), *MM* is the month of the year (i.e., 01 to 12, for January through to December), *DD* is the day of the month (i.e., 01 to 31), *hh* is the number of complete hours that have passed since midnight (i.e., 00 to 11)

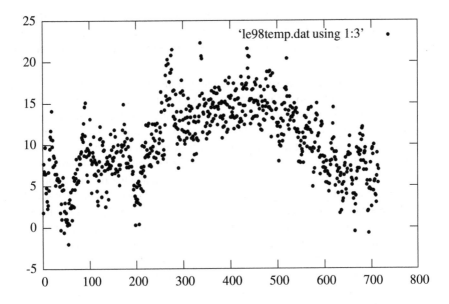

Figure 2.3: Simple *x-y* plot of the maximum air temperature measured at Llyn Efyrnwy every 12 hours throughout 1998.

and *mm* is the number of complete minutes that have elapsed since the start of the hour (i.e., 00 to 59). The letter T, known as the time designator, is used to separate the date and time components. This format conforms to the "basic" version of the international standard for date and time notation (International Organization for Standardization (ISO) 8610:2004; Appendix D.1).

Suppose that one wished to examine how the maximum air temperature at Llyn Efyrnwy varied throughout 1998. A simple way to do this would be to present the data in the form of a 2D scatter-plot in which maximum air temperature (*y*-axis) is plotted against measurement number (*x*-axis). This can be achieved in gnuplot by typing the following instruction next to the gnuplot> command-line prompt and then pressing the ⏎ or Enter key to submit the instruction (i.e., to run the command).

```
plot 'le98temp.dat' using 1:3                                          1
```

This instruction produces a plot resembling the one presented in Figure 2.3, which should appear in a separate pop-up window on the computer screen.

In the example given above, the plot command generates a simple *x-y* plot of the data contained in the named file. Note that the file name must be surrounded by single quotation marks. If the file is not located in the working directory, its full or relative path name must be given (e.g., /mnt/cdrom/chapter2/le98temp.dat under GNU/Linux or d:\chapter2\le98temp.dat in Microsoft Windows; Appendix A). The rest of the plot command (i.e., using 1:3) informs gnuplot which columns of the data file should be used to generate the plot. In this example, the data in columns 1 (measurement number) and 3 (maximum air temperature) of the file are plotted along the *x*-axis and *y*-axis, respectively. The general syntax is using x:y. Thus, the same data could be plotted with the measurement number on the *y*-axis and the maximum air temperature on the *x*-axis, if so desired, by altering the last part of the command line so that it reads using 3:1.

It is possible to obtain the same result using point-and-click methods in the version of gnuplot for Microsoft Windows, although the command-line approach is preferred here. As was noted earlier, the command-line approach confers significant benefits in the long run. For example, it is possible to store a sequence of gnuplot commands in an ASCII text file, sometimes known as a "script" or a "macro", which can be run in a "batch" whenever the user requires (see Section 2.7). There are at least two reasons why this is good practice. First, it ensures that a permanent record is kept of the commands employed to generate a plot and the order in which they should be applied. This can be used, for instance, to reproduce a plot quickly should it be accidentally deleted, or if disk space (or bandwidth) are limited and it is not possible to store (or transmit) the resulting graphics file. Second, it introduces the possibility of automating the production of standard plots from a number of related data sets. Thus, the same commands can be applied, subject perhaps to only minor modification, to a collection of data files. As an example, one might wish to generate separate plots of the maximum air temperatures recorded at Llyn Efyrnwy over several years, where the data for each year are contained in separate files.

It is acknowledged, however, that some readers may have little or no experience with command-line interfaces because the point-and-click approach is now standard in many elements of human–computer interaction. Apart from the need to remember the syntax of specific commands, one aspect of the command-line approach that new users generally find challenging is the requirement, on occasion, to type in relatively long commands. This process can be tiresome, especially if a simple error is made when entering a command, so that the whole command line must be typed in again. The impact of this problem can be reduced by storing the commands in script files, as outlined above, but gnuplot also helps in this respect by maintaining a list of the most recently issued commands. This list is known as the command-line history. It is possible to step back through the command-line history by pressing the ⬆ key in the gnuplot command-line window; similarly, the ⬇ key can be used to step forward through the list. Thus, a previously entered command can be issued again without modification by pressing the ⬅ or Enter key when the appropriate command appears on the screen. It is also possible to edit previously submitted commands, using the ⬅ and ➡ keys to move left or right, respectively, along the command line and by deleting or inserting text as required.

Exercise 2.1: Use the ⬆ key to recall the last plot command. Edit this command line so that it reads plot 'le98temp.dat' using 3:1 and then press the ⬅ or Enter key to submit the revised command. How does the resulting plot differ from the output presented in Figure 2.3? Use the ⬆ key again to recall and run the initial plot command once more.

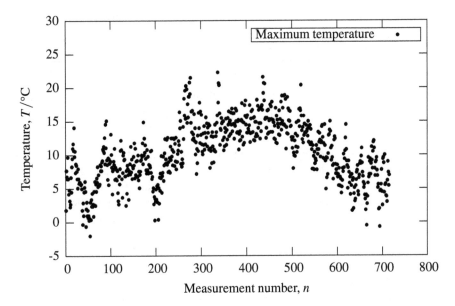

Figure 2.4: Simple *x-y* plot of the maximum air temperature measured at Llyn Efyrnwy every 12 hours throughout 1998, with labeled axes and a boxed key.

2.2.2   Labeling the Axes of a Plot

The plot presented in Figure 2.3 is rather bald. It portrays the seasonal variation in maximum air temperature at Llyn Efyrnwy, with the lowest values toward the start and end of the measurement sequence (i.e., January and December) and the highest values toward the middle (i.e., July and August), which is typical of this mid-latitude northern hemisphere site. Neither of the axes has been labeled, so that the contents of the plot can only be guessed at by the casual reader. Moreover, the automatically generated key, which identifies the source of the data, employs the arguments of the plot command to label the data series, instead of using something more instructive. These deficiencies can be overcome by entering a few extra instructions, listed below, via the **gnuplot** command line, which produce the plot shown in Figure 2.4.

```
set xlabel "Measurement number, n"                          2
set ylabel "Temperature, T/degree Celsius"                  3
set key top right box                                        4
set yrange [-5:30]                                           5
plot 'le98temp.dat' using 1:3 title "Maximum temperature"   6
```

The set command used on lines 2 through 5, above, assigns values to various parameters that control the appearance of plots in gnuplot. In this particular example, lines 2 and 3 specify the labels that should be placed beside the *x*- and *y*-axes, respectively. Note that the text used to define these labels must be enclosed by double quotation marks, so that **gnuplot** knows where the labels begin and end. Line 4 indicates that a key should be placed in the top right-hand corner of the plot, and that

Table 2.2: Selected options of the `plot` command in gnuplot.

Option	Abbreviation	Interpretation
using	u	Use the data in these columns to create the plot.
with	w	Plot the data in the following style (see Table 2.3).
title	t	Give this data series the following title in the key.

this should be surrounded by a box. Line 5 sets the range of values that should be displayed on the y-axis, here using a lower limit of $-5\,°C$ and an upper limit of $30\,°C$. These changes do not take effect, however, until the data have been plotted once more (line 6). The `plot` command issued on line 6 also gives an explicit title to the data series (`title "Maximum temperature"`); this information is used to produce the key.

2.2.3 Plotting Multiple Data Series

As has already been noted, the file `le98temp.dat` contains information on both the maximum and minimum air temperatures recorded at Llyn Efyrnwy. Typically, one might wish to plot both series in the same figure, perhaps to gauge the diurnal temperature range at this site. This can be achieved by modifying the previous `plot` command (line 6) so that it reads as follows:

```
plot 'le98temp.dat' u 1:3 t "Maximum temperature", \        7
     'le98temp.dat' u 1:4 t "Minimum temperature"           8
```

Note that the \ ("backslash") character on the right-hand side of line 7, above, which is known as the line continuation symbol, is used to split a single long gnuplot command over two or more lines so that it is easier to enter and read. Note that the line continuation symbol must be the last character on the line. In this example, the line continuation symbol is used to continue the `plot` command from line 7 to line 8. Thus, lines 7 and 8 represent a single gnuplot command. When entered manually on the gnuplot command line, the line continuation symbol causes the command prompt to change from `gnuplot>` on the first line to `>` on the continuation lines. Also, note that the keywords `using` and `title` have been abbreviated to u and t, respectively, on lines 7 and 8, to reduce the amount of typing required. gnuplot allows many of its basic keywords to be abbreviated in this way (see, for example, Table 2.2 and Table 2.3). The output of the revised `plot` command is presented in Figure 2.5.

The syntax of the command on lines 7 and 8 is relatively straightforward. The `plot` keyword is followed by a comma-separated list of the file (or files) containing the data series that are to be plotted, together with instructions on how each series should be presented. In this example, the first series is contained in column 3 of the file `le98temp.dat` and is plotted against the data in column 1 of that file (i.e., maximum temperature on the y-axis versus measurement number on the x-axis). The second series is stored in column 4 of the same file and is also plotted against the data in column 1 of that file (i.e., minimum temperature on the y-axis versus measurement number on the x-axis). Different titles are used to label each data series in the key.

Table 2.3: Selected data style options of the `plot` command in gnuplot.

Option	Abbreviation	Interpretation
`lines`	`l`	Connect consecutive data points (x,y) with straight line segments. Requires two columns of data.
`points`	`p`	Plot a symbol at each data point (x,y). Requires two columns of data.
`linespoints`	`lp`	Plot a symbol at each data point (x,y) and connect consecutive data points with straight line segments. Requires two columns of data.
`impulses`	`i`	Draw a vertical line from the x-axis ($y=0$) to each data point (x,y). Requires two columns of data.
`dots`	`d`	Plot a small dot at each data point (x,y). Requires two columns of data.
`boxes`	-	Draw a vertical bar (box) centered on the x-value from the x-axis ($y=0$) to each data point (x,y). Requires two columns of data.
`xerrorbars`	`xe`	Plot a dot at each data point (x,y) and a horizontal error bar from $(x-\Delta x,y)$ to $(x+\Delta x,y)$, or from $(x_{\min},y)$ to $(x_{\max},y)$. Requires three $(x,y,\Delta x)$ or four $(x,y,x_{\min},x_{\max})$ columns of data.
`yerrorbars`	`ye`	As above, but with vertical error bars. Requires three $(x,y,\Delta y)$ or four $(x,y,y_{\min},y_{\max})$ columns of data.
`xyerrorbars`	`xye`	As above, but with both horizontal and vertical error bars. Requires four $(x,y,\Delta x,\Delta y)$ or six $(x,y,x_{\min},x_{\max},y_{\min},y_{\max})$ columns of data.
`xerrorlines`	`xerrorl`	As `xerrorbars`, but with consecutive data points (x,y) connected by straight line segments. Requires three $(x,y,\Delta x)$ or four $(x,y,x_{\min},x_{\max})$ columns of data.
`filledcurves`	`filledc`	Plots either the current curve closed and filled, or the region between the current curve and a given axis, horizontal or vertical line, or a point, filled with the current drawing color.
`vectors`	`v`	Draws vectors from (x,y) to $(x+\Delta x,y+\Delta y)$. Requires four columns of data $(x,y,\Delta x,\Delta y)$.

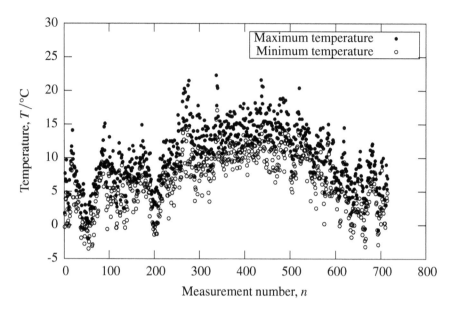

Figure 2.5: Simple *x-y* plot of the maximum and minimum air temperatures measured at Llyn Efyrnwy every 12 hours throughout 1998.

2.2.4 Plotting with Different Data Styles

In Figure 2.5, individual measurements from both data series are represented by point symbols, although **gnuplot** offers a large number of other data styles (Table 2.3). The default data style can be changed using the set style data command, as outlined below:

```
set style data lines                                       9
set xrange [200:300]                                      10
replot                                                    11
```

Thus, line 9 sets the default data style to be lines, so that consecutive data points in each series are connected by straight line segments. The command on line 10 is used to restrict the range of values displayed along the *x*-axis to measurement numbers 200 to 300, which correspond to data acquired between March 12, 1998 and July 1, 1998. This instruction demonstrates how **gnuplot** can be used to visualize a chosen subset of a data file, allowing the user to examine those data in greater detail. These changes do not take effect, however, until the data have been plotted once more, which is performed on line 11. The replot command used here submits the preceding plot command (lines 7 and 8 in this example) again, taking into account any changes that have been made to the axes labels, plotting style, and such, since that command was last issued. The result of these commands is shown in Figure 2.6.

The style of each data series can be controlled separately using the keyword with, which abbreviates to w (Table 2.2). For instance, the following **gnuplot** command uses a combination of straight lines and point symbols (linespoints or lp) to represent

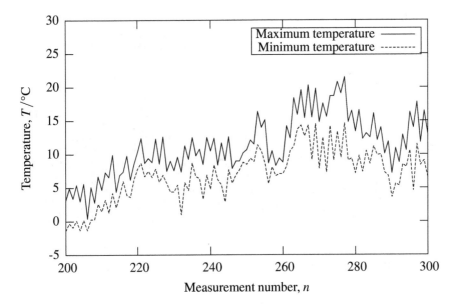

Figure 2.6: Plot of the maximum and minimum air temperatures measured at Llyn Efyrnwy every 12 hours between March 12, 1998 and July 1, 1998, illustrating the lines data style.

the data from the first series (i.e., maximum temperature) and vertical bars (boxes) to represent the data from the second series (i.e., minimum temperature; see Figure 2.7).

```
plot 'le98temp.dat' u 1:3 t "Maximum temperature" w lp,\       12
     'le98temp.dat' u 1:4 t "Minimum temperature" w boxes      13
```

The second of these two data styles is probably not the most appropriate choice in these circumstances, but it helps to illustrate the different styles that are available.

2.3   PLOTTING TIME-SERIES DATA

Although the file le98temp.dat contains time-series data, the values have not been treated explicitly as such thus far. Instead, the preceding examples have plotted the maximum and minimum air temperature data against the sequential measurement number. This section therefore investigates how gnuplot can be used to visualize explicit time-series data. Only a few extra commands are involved. These commands indicate which axis should be employed to represent the time dimension, how the date and time information is formatted (so that the input data can be correctly interpreted), and how the tic-marks along the time axis should be labeled. An example of each of these commands is shown on lines 14 to 20, below:

```
set xdata time                    14
set timefmt "%Y%m%dT%H%M"         15
set format x "%d/%m"              16
```

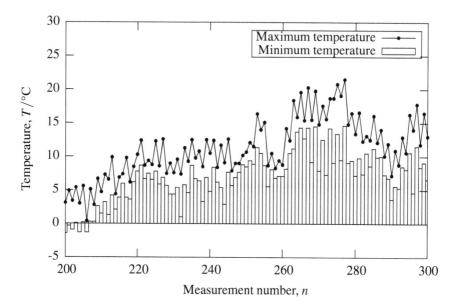

Figure 2.7: Plot of the maximum and minimum air temperatures measured at Llyn Efyrnwy
every 12 hours between March 12, 1998 and July 1, 1998, illustrating the
linespoints and boxes data styles.

```
set xlabel "Day/Month (1998)"                                          17
set xrange ["19980401T0000":"19980701T0000"]                           18
plot 'le98temp.dat' u 2:3 t "Maximum temperature" w lp, \              19
     'le98temp.dat' u 2:4 t "Minimum temperature" w lp                 20
```

Line 14, for instance, instructs gnuplot to treat the data plotted on the x-axis as
date and time values. Line 15 specifies the format in which these values are stored
(Table 2.4). This is given by the format string, "%Y%m%dT%H%M", which indicates that
the date and time values comprise a four-digit number denoting the year (%Y), a two-
digit number representing the month (%m), a two-digit number for the day of month
(%d), the letter T, which is used to separate the date and time components, a two-digit
number for the hour of day (%H) and a two-digit number for the minutes (%M). So,
for example, 9 am on January 5, 1998 is represented as 19980105T0900. This format
corresponds to the "basic" version of the ISO 8610:2004 notation for times and dates
(Appendix D.1) and, hence, the structure of the data in column two of temp981e.dat.

Line 16 instructs gnuplot to use a combination of the day of month (%d) and the
month of year (%m), separated by a forward-slash (or "solidus") symbol (/), to label
the tic-marks along the x-axis. Line 17 alters the text string used to label this axis.
Line 18 specifies the range of values to plot along the x-axis, in this case starting at
"19980401T0000" (i.e., midnight on March 1, 1998) and ending at "19980701T0000"
(i.e., midnight on July 1, 1998). Finally, the plot command, which is split over two
lines (19 and 20) using the line continuation symbol (\), displays the data using the
linespoints (lp) style for both time series (Figure 2.8).

Table 2.4: Selected time and date format specifiers in gnuplot.

Specifier	Interpretation
%a	Abbreviated name of the day of the week (Sun, Mon, ... , Sat)
%A	Full name of the day of the week (Sunday, Monday, ... , Saturday)
%b	Abbreviated name of the month (Jan, Feb, ... , Dec)
%B	Full name of the month (January, February, ... , December)
%d	Day of the month (1–31)
%H	Hour (00–23)
%j	Day of the year (1–366)
%m	Number of the month (01–12)
%M	Minute (00–59)
%S	Second (00–59)
%y	Year (two digits; 00–99)
%Y	Year (four digits; e.g., 2006)

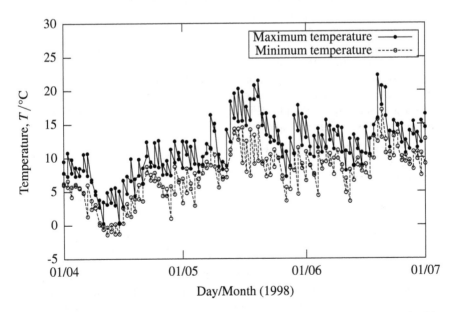

Figure 2.8: Time-series plot of the maximum and minimum air temperatures measured at Llyn Efyrnwy every 12 hours between March 12, 1998 and July 1, 1998.

```
 1 19980101T0900     0.00                                                    1
 2 19980101T2100    14.60                                                    2
 3 19980102T0900     5.20                                                    3
 4 19980102T2100     1.60                                                    4
 5 19980103T0900    35.80                                                    5
 6 19980103T2100     5.60                                                    6
 7 19980104T0900     6.60                                                    7
 8 19980104T2100    20.80                                                    8
 9 19980105T0900     4.40                                                    9
10 19980105T2100     3.00                                                   10
11 19980106T0900     4.00                                                   11
13 19980107T0900     2.40                                                   12
```

Figure 2.9: Extract from the data on precipitation at Llyn Efyrnwy during 1998 (le98rain.dat; first 12 lines only).

2.3.1 Plotting Multiple Time-Series

It is possible to display two or more time-series, drawn from different data files, in a single plot. Suppose, for example, that one wished to present information on both air temperature and precipitation at Llyn Efyrnwy. The latter is contained in a file called le98rain.dat (Figure 2.9), which can be found on the CD-ROM. The general structure of this file is similar to that of le98temp.dat. More specifically, it consists of three columns of data denoting the sequential measurement number, the date and time at which the measurements were made (every 12 hours) and the total precipitation (in mm) that accumulated over the preceding 12 hours.

Since temperature and precipitation are measured in different units ($°C$ and mm, respectively), and may also exhibit a very different range of values, it is not always possible to plot them effectively on the same scale on the y-axis. To overcome this problem, gnuplot provides a second y-axis, known as the $y2$-axis, which is drawn up the right-hand side of the plot (Figure 2.10). The following gnuplot commands illustrate how this facility is employed.

```
set y2label "Precipitation, P/mm"                                          21
set ytics nomirror                                                         22
set y2tics nomirror                                                        23
set xlabel "Month (1998)"                                                  24
set format x "%b"                                                          25
set xrange ["19980101T0000":"19990101T0000"]                               26
plot 'le98temp.dat' u 2:3 t "Maximum temperature" w lp, \                  27
     'le98rain.dat' u 2:3 axes x1y2 t "Precipitation" w i                  28
```

Line 21 sets the label to be drawn along the $y2$-axis. Line 22 instructs gnuplot that the tic-marks drawn on the y-axis (the left-hand side of the plot) should not be mirrored on the $y2$-axis (the right-hand side of the plot), and *vice versa* on line 23. Thus, a separate plotting scale is used for each of the two y-axes. Line 24 resets the label to be drawn along the x-axis. Line 25 instructs gnuplot to use the three-letter abbreviation for the name of the month (i.e., Jan, Feb, ... , Dec) to label the tic-marks

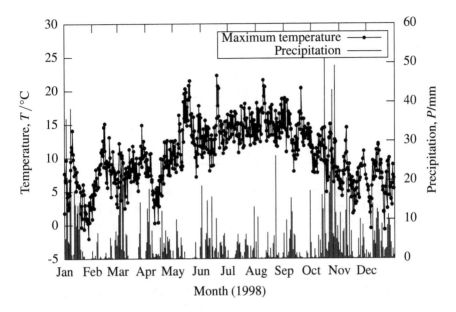

Figure 2.10: A plot of two time-series (maximum air temperature and precipitation at Llyn Efyrnwy) contained in separate data files.

along the x-axis (see Table 2.4). Line 26 specifies the range of values to plot along the x-axis, in this case starting at "19980101T0000" (i.e., midnight on January 1, 1998) and ending at "19990101T0000" (i.e., midnight on January 1, 1999). Lines 27 and 28 plot the two data series. Note that the data are taken from separate files (le98temp.dat and le98rain.dat) and that the temperature data are plotted with the linespoints (w lp) style, while the precipitation data are plotted with the impulses (w i) style. The latter draws vertical bars rising from the x-axis to each data point. gnuplot is given an explicit instruction (line 28) to plot the second data series on the first x-axis (x1; i.e., the one drawn along the base of the plot) and the second y-axis (y2; i.e., the one drawn up the right-hand side of the plot), axes x1y2. Unless specified in this way, gnuplot assumes that all data are plotted using the x1y1 (bottom and left-hand side) axes. For completeness sake it is worth noting that the $x2$-axis refers to the axis along the top of the plot.

2.3.2 Further Control over Plotting Styles

The preceding sections provide a brief introduction to a selection of gnuplot's 2D plotting capabilities, including its ability to handle time-series data. Although the possibilities are not explored further here, it is worth noting that gnuplot also offers considerable control over both line and point styles, including their size (for points), width (for lines) and color (for both lines and points). Some of the command-line options that control these properties are listed in Table 2.5. The nature of the line and point styles available depends on the graphical device, known as a "terminal" in gnuplot parlance, to which the plot is sent. Thus, the default point and line styles

Table 2.5: Selected command-line options to control the appearance of data series in gnuplot.

Option	Abbreviation	Interpretation
pointtype	pt	Symbol type used to denote data points.
pointsize	ps	Relative size of point symbol.
linetype	lt	Line type used to connect consecutive data points.
linewidth	lw	Relative line width.

differ between the versions of gnuplot for Microsoft Windows and GNU/Linux. This issue is revisited later in this chapter (see Table 2.6 in Section 2.5). For the moment it is sufficient to note that the available styles can be inspected using the test command.

```
test                                                                    29
```

Figure 2.11, for example, was generated using the test command and gnuplot's epslatex "terminal". The epslatex terminal produces a special type of Encapsulated PostScript (EPS) file, and is used to generate most of the figures presented in this book. In the epslatex terminal, line type three (lt 3) corresponds to a fine dashed line and point type seven (pt 7) corresponds to a filled black circle (Figure 2.11). A significant advantage of the epslatex terminal is that it allows various mathematical symbols (e.g., α, β, γ, W·m^{-2} and °C), as well as different typographical fonts, to be used in the resulting plot. Similar capabilities are also available in the jpeg, pdf, png, postscript, svg and x11 (i.e., GNU/Linux) terminal types (see Table 2.6 in Section 2.5), but not in the default terminal for Microsoft Windows (windows). In the latter case, the required symbols must be emulated; for instance, W·m^{-2} can be represented as W.m^{-2}, °C as degree Celsius, ρ as rho, and so on. The reader is referred to the file visual.gp on the CD-ROM to inspect the commands used to generate the figures presented in this chapter.

2.4   PLOTTING IN THREE DIMENSIONS

It was noted earlier that one of the characteristic features of environmental data sets is that they tend to be multidimensional, often highly so; that is, they frequently consist of co-located measurements made simultaneously (or, at least, quasi-simultaneously) on several different properties (e.g., temperature, precipitation and wind speed) of a particular environmental system. It is possible, of course, to explore the inter-relationships between these properties by plotting them in pairs using, for example, simple *x-y* scatter diagrams. This is not a particularly efficient or effective approach, though, because it may require the interpretation of tens or even hundreds of such plots, depending on the number of independent variables (i.e., dimensions) involved. Consequently, various alternative methods have been developed, including Chernoff faces (Chernoff 1973), star glyphs (Fienberg 1979) and parallel coordinates (Inselberg 1985), each of which attempts to represent multidimensional data in a 2D plot.

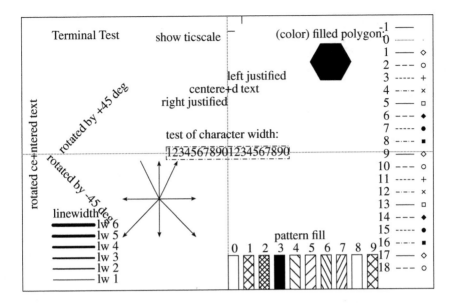

Figure 2.11: Output from gnuplot's `test` command for the `epslatex` terminal type.

Here, however, attention is focused on the visualization of 3D data sets, especially those in which two of the three dimensions describe the spatial location of the measurement (e.g., the Easting and Northing, or latitude and longitude, derived from a map) and the third relates to the value of an environmental property recorded at that location. Some of the techniques that are examined can also be applied to aspatial data sets in which the dimensions are defined by co-located or contemporaneous measurements of three different environmental properties.

2.4.1   Description of the Digital Elevation Data Set

The data used in this section are contained in the file `efyrnwy.dem` (Figure 2.12), which can be found on the CD-ROM. These data are derived from a single tile of the UK Ordnance Survey Land-Form PROFILE® digital elevation model (DEM), and are reproduced here by kind permission of Ordnance Survey. The original data comprise a grid of terrain elevation values posted at 10 m spatial intervals, covering a 5 km × 5 km area southwest of Llyn Efyrnwy between Eastings 295000 and 300000 and Northings 315000 and 320000 in the UK National Grid coordinate system. These data have been re-sampled onto an array of 51 × 51 elevation values, with a 100 m spatial posting, to reduce the volume of data that has to be handled here. The resulting file, `efyrnwy.dem`, consists of three columns of data that describe, respectively, the Easting, Northing and elevation of the re-sampled data points (Figure 2.12). The Eastings and Northings are measured in meters relative to a fixed reference point in the UK National Grid, known as the false origin; the elevation values are measured in meters relative to the datum at Newlyn, Cornwall, known as the Ordnance Datum Newlyn (ODN).

```
295000   315000     407                                                   1
295000   315100     410                                                   2
295000   315200     419                                                   3
295000   315300     422                                                   4
295000   315400     421                                                   5
295000   315500     415                                                   6
295000   315600     398                                                   7
295000   315700     385                                                   8
295000   315800     367                                                   9
295000   315900     390                                                  10

300000   319100     347                                                2592
300000   319200     348                                                2593
300000   319300     363                                                2594
300000   319400     381                                                2595
300000   319500     391                                                2596
300000   319600     393                                                2597
300000   319700     382                                                2598
300000   319800     364                                                2599
300000   319900     351                                                2600
300000   320000     353                                                2601
```

Figure 2.12: First 10 lines and last 10 lines of the Llyn Efyrnwy DEM data file, efyrnwy.dem
(© Crown Copyright License Number NC/03/11298).

2.4.2   Visualizing 3D Data in gnuplot

gnuplot provides a command, splot, which can be used to plot 3D data, such as those contained in efyrnwy.dem. The syntax of the splot command is very similar to that of its 2D counterpart, plot. The splot keyword is followed by the name of the data file, the columns of data in this file that should be plotted on the *x*-, *y*- and *z*-axes, respectively, and any instructions relating to the way in which the data should be presented. For example, Figure 2.13 was generated as follows:

```
reset                                                                    30
unset key                                                                31
set style data points                                                    32
set xlabel "Easting/m"                                                   33
set ylabel "Northing/m"                                                  34
set zlabel "Elevation/m"                                                 35
set xtics 295000, 1000, 300000                                          36
set ytics 315000, 1000, 320000                                          37
set ztics 250, 100, 550                                                 38
splot 'efyrnwy.dem' u 1:2:3                                             39
```

The reset command (line 30) causes all of the parameters that were modified by the set commands issued in the preceding section (axis labels, plotting ranges and so on) to be restored to their default values. Line 31 instructs gnuplot not to add a key to subsequent plots. More generally, the unset command is used to "turn off"

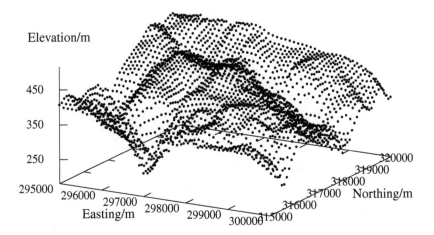

Figure 2.13: Visualization of the Llyn Efyrnwy DEM. Data reproduced by kind permission of Ordnance Survey (© Crown Copyright License Number NC/03/11298).

or reset specific features of a plot. Line 32 selects point symbols as the default data style. Lines 33 through 35 specify the labels that are to be placed beside the x-, y- and z-axes, respectively. Lines 36 through 38 control the position of the tic-marks that are to be placed along the axes. The three values associated with each of these commands specify the position of the first tic-mark, the interval between the tic-marks, and the position of the final tic-mark, respectively, on the given axis. Thus, line 36 indicates that the tic-marks on the x-axis should commence at a value of 295000 and be plotted at 1000 unit intervals, ending at 300000. Finally, line 39 generates a 3D plot of the data in the file `efyrnwy.dem`, plotting the data in columns 1, 2 and 3 on the x-, y- and z-axes, respectively (Figure 2.13).

Strictly speaking, Figure 2.13 is not a 3D plot but a visualization of a 3D data set projected onto a 2D plane (i.e., a flat piece of paper), sometimes known as a 2.5D (two-and-a-half dimensional) plot. To avoid having to use this sort of convoluted linguistics, however, the term 3D plot will be used as shorthand to describe figures of this type throughout the rest of the book.

2.4.3 Altering the View Direction

While Figure 2.13 provides an indication of the topography near Llyn Efyrnwy, it is impossible to obtain a complete appreciation of the nature of the terrain by observing it from just one direction. The problem is that data points located in the foreground partially obscure those in the background. Moreover, it is often difficult to distinguish between the two because we do not have a proper sense of perspective. This problem can be overcome, to a certain extent, by rotating the plot so that the terrain is viewed from a number of directions (Figure 2.14). In gnuplot, this can be controlled using the `set view` command, as follows:

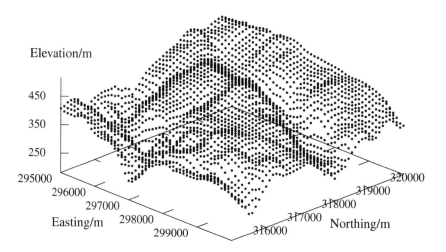

Figure 2.14: Viewing the Llyn Efyrnwy DEM from a different direction. Data reproduced by kind permission of Ordnance Survey (© Crown Copyright License Number NC/03/11298).

```
set view 45, 45, 1, 1                                                    40
replot                                                                   41
```

Line 40 sets the view direction to be 45° with respect to the *y*-axis (effectively tilting the plot toward the viewer) and 45° with respect to the *z*-axis (rotating the plot around that axis). By comparison, the default values used to produce Figure 2.13 are 60° and 30°, respectively. The final two arguments on line 40 control, in turn, the scaling of the plot in both the *x*- and *y*-axes, and in the *z*-axis. A value greater than 1 stretches that axis; a value less than 1 compresses it. In this case, the plot has not been re-scaled along any axis. Line 41 plots the data again, taking into account the revised view direction (Figure 2.14).

> **Exercise 2.2**: Plot the Llyn Efyrnwy DEM data set, viewing it from several directions. Experiment with the scaling of the *x*-, *y*- and *z*-axes.

2.4.4 Generating a 3D Surface Plot

It is often more appropriate to visualize 3D data as a quasi-continuous surface rather than a cloud of data points. One solution is to connect adjacent data points using straight line segments so that they form a wire-frame model (Figure 2.15). Various methods can be used for this purpose depending on the distribution of the original data and the intended application of the derived surface model (Burrough and Mc-Donnell 1998). If the data points are distributed irregularly across the measurement space, two approaches are generally possible. The first involves the construction of a

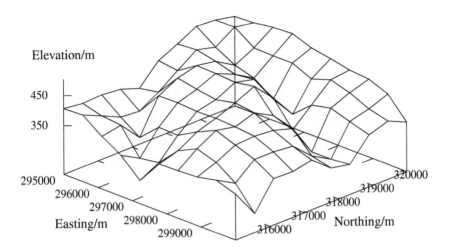

Figure 2.15: Wire-frame surface (10 × 10 element grid) generated from the Llyn Efyrnwy DEM. Data reproduced by kind permission of Ordnance Survey (© Crown Copyright License Number NC/03/11298).

triangulated irregular network (TIN), in which the data points are connected so that they form a network of triangular surface facets of irregular size and shape. This approach is frequently applied to digital elevation data, often using a method known as Delaunay Triangulation (Watson 1992, Bonham-Carter 1994). The second approach involves interpolating the original data onto a regular rectangular grid. Several methods can be used to achieve this, including trend-surface analysis, inverse-distance weighting and kriging. The relative merits of TIN and grid-based approaches are discussed extensively by Burrough and McDonnell (1998); both methods are commonly implemented in current GIS.

Simple wire-frame surface models can be generated in gnuplot using the dgrid3d command. This performs an interpolation of the original z-axis values onto a regular rectangular grid using the inverse distance weighting (IDW) method. The interpolated z-value for a particular intersection point on the resulting grid is computed from the weighted average of the original z-values in the local neighborhood, with the weights being inversely proportional to the distance between the grid point and the data points according to Equation 2.1:

$$\text{weight} = \frac{1}{\Delta x^w + \Delta y^w} \tag{2.1}$$

where Δx and Δy are the distances along the x- and y-axes between the original data point and the interpolated grid point, and w is a weighting factor (Williams and Kelly 1998). Thus, the closer the original data point is to an interpolated grid point, the greater the influence it has on the interpolated z value. Moreover, as the value of the weighting factor, w, is increased, data points further from the grid point have a decreasing effect on the interpolated z value.

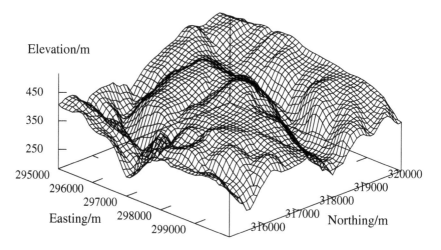

Figure 2.16: Wire-frame surface (51 × 51 element grid) generated from the Llyn Efyrnwy
           DEM. Data reproduced by kind permission of Ordnance Survey (© Crown Copy-
           right License Number NC/03/11298).

The dgrid3d command is used to construct a wire-frame model of the terrain
around Llyn Efyrnwy even though the digital elevation data are provided in the form
of a regular grid of values. If the dgrid3d command is not used, the resulting straight
line segments connect the data points in the sequence in which they occur in the
data file, in the manner of a child's join-the-dots diagram, instead of their immediate
spatial neighbors in the grid. Thus, Figure 2.15 is generated as follows:

```
set style data lines                                                      42
set dgrid3d 10, 10, 16                                                     43
replot                                                                     44
```

Line 42 sets the default data style to be lines, ensuring that points on the interpolated
grid are connected by straight line segments. Line 43 instructs gnuplot to treat the
data as scattered point values and to generate from these a gridded data set. The first
two arguments to this command determine the number of rows and columns in the
output grid. In this case, the grid is specified in terms of 10 equally spaced samples
along both the x- and y-axes (i.e., a 10×10 grid). The third argument is the weighting
factor, w, in Equation 2.1. Finally, line 44 plots the digital elevation data taking these
commands into account.

    The density of the interpolated grid can be varied by altering the arguments to the
dgrid3d command. For instance, Figure 2.16 was produced as follows:

```
set dgrid3d 51, 51, 16                                                     45
replot                                                                     46
```

The use of a 51×51 element grid makes sense in this case because it matches the
number of points in the original data set. The resulting plot reveals much more in-

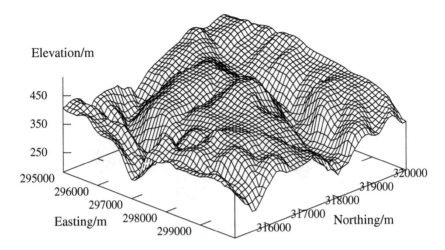

Figure 2.17: Wire-frame surface (51 × 51 element grid) generated from the Llyn Efyrnwy DEM, with hidden line removal. Data reproduced by kind permission of Ordnance Survey (© Crown Copyright License Number NC/03/11298).

formation about the surface topography in this area, including the main ridges and valleys, as well as more subtle geomorphological features. In other cases, selecting the correct density for the grid may be a matter of trial and error. If the grid is too dense, it may take a long time to generate the wire-frame surface even on a fast computer. Interpretation of the resulting patterns may also be difficult. If the grid is too sparse, important detail may be lost, as in Figure 2.15.

It is worth noting that wire-frame surface plots can be rotated and viewed from different directions in gnuplot either by using the set view command, described in Section 2.4.3, or by moving the mouse to the left or right within the plot window while holding down the left-hand mouse button. Similarly, diagonal movements of the mouse within the plot window will rotate the surface plot around a different axis, while upward and downward movements of the mouse will increase or decrease the vertical (z-axis) scaling of the plot, respectively.

2.4.5 Hidden-Line Removal

Although the dense grid used to generate Figure 2.16 means that a greater amount of detail is evident in the resulting wire-frame model, visual interpretation is made difficult by the mass of intersecting lines in the plot. This problem can be alleviated by removing those lines that would normally be hidden from view if the surface was solid (i.e., so that vertical protrusions obscure the features behind them; Figure 2.17). This process is known as hidden-line removal and the relevant gnuplot commands are

```
set hidden3d                                                           47
replot                                                                 48
```

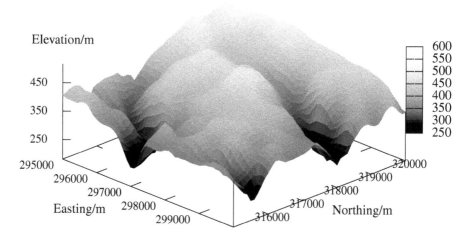

Figure 2.18: Grayscale rendered surface model of the terrain elevation close to Llyn Efyrnwy. Data reproduced by kind permission of Ordnance Survey (© Crown Copyright License Number NC/03/11298).

Thus, line 47 enables hidden-line removal and line 48 plots the data accordingly. Note that it is easier to identify the main geomorphological features of the terrain around Llyn Efyrnwy (e.g., ridges, valleys and breaks of slope), and that parts of the x- and y-axes are hidden, in the resulting plot.

2.4.6   Producing Solid Surface Models

The visualization of 3D data can be enhanced further still by generating a solid (cf. wire-frame) surface model, in which each facet of the surface is rendered with a different graytone or color depending on its elevation, gradient, aspect or other environmental property (Figure 2.18). The benefit of this approach is that the human interpreter is provided with two visual cues with which to comprehend the data, the general shape of the surface and the gray tone or color of each surface facet. This can be achieved in gnuplot by issuing the following commands.

```
set style line 9                                    49
set pm3d at s hidden3d 9                             50
unset hidden3d                                       51
splot 'efyrnwy.dem' u 1:2:3 with pm3d                52
```

Thus, line 49 determines the line style used to draw the wire-frame grid (a solid black line in this example). Line 50 instructs gnuplot to produce a grayscale or color visualization of the elevation data and to render this onto the wire-frame surface (pm3d at surface, which abbreviates to pm3d at s). The rendering can also be placed at the base of the plot (pm3d at b), on top of the plot (pm3d at t), or some combination of these (e.g., pm3d at bst). Here, the pm3d style is asked to perform hidden-line removal on the wire-frame grid and the surface model (hidden3d; line 50), although it uses a different algorithm from the one employed in the preceding examples. Consequently, line 51 "unsets" the standard hidden-line removal algorithm. Lastly, line 52 plots the data once more, this time using the pm3d style.

> **Exercise 2.3**: Using the mouse, rotate the solid 3D surface plot of the Llyn Efyrnwy DEM about its *x*-, *y*- and *z*-axes. Experiment with the set pm3d command, altering the position of the rendered surface with respect to the plot (e.g., pm3d at b). Remember to replot the data each time.

2.4.7 Contouring 3D Surface Plots

Another way of visualizing 3D data is to produce a contour plot. This is perhaps most commonly applied to environmental data sets for which the *x*- and *y*-axes refer to the spatial location of the measurements made on the *z*-axis variable. An obvious example is the topographic map, where the contours denote lines of equal terrain elevation (Watson 1992). Other environmental examples include plots of barometric air pressure (isobars), precipitation (isohyets), temperature (isotherms), water depth (isobaths) and salinity (isohalines). Contour plots can also be used to visualize data sets in which the contours portray variation in the value of the *z*-axis property as a function of two other aspatial properties, which define the *x*- and *y*-axes.

Contours can be added to a wire-frame surface model in gnuplot as follows:

```
unset pm3d                                                    53
set hidden3d                                                  54
set contour base                                              55
set cntrparam levels discrete 250, 300, 350, 400, 450, 500    56
splot 'efyrnwy.dem' u 1:2:3                                   57
```

Line 53 instructs gnuplot to cease plotting the data as a solid surface model. As a result, subsequent 3D plots are produced as standard wire-frame models. Line 54 invokes the standard algorithm for hidden line removal. Line 55 indicates that the contours should be drawn on the base of the plot: the other options are to draw them on the wire-frame surface itself (set contour surface) or on both the base and the surface of the plot (set contour both). Rather than accepting the default values, line 56 instructs gnuplot to modify the contour parameters (cntrparam) so that the contour lines are drawn at pre-defined intervals in elevation; in this example they are plotted at 50 m intervals from 250 m to 500 m. Finally, line 57 plots the data as a wire-frame model and a set of contours (Figure 2.19).

Sometimes it is helpful to remove the 3D surface altogether, leaving only the contour plot. The simplest way to do this in gnuplot is to enter the following commands:

```
unset surface                                                 58
set view map                                                  59
set size square                                               60
replot                                                        61
```

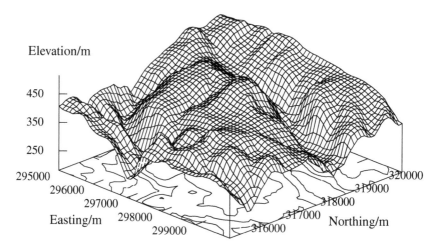

Figure 2.19: Visualization of the Llyn Efyrnwy DEM, with a contour map added to the base of the plot. Reproduced by kind permission of Ordnance Survey (© Crown Copyright License Number NC/03/11298).

where line 58 removes the 3D surface and line 59 sets the view direction so that one is effectively looking vertically down onto the plot. This command is equivalent to set view 0,0,1,1. It gives the plot the appearance of a planimetric map even though it is generated using the splot command. Line 60 causes gnuplot to produce a plot with an aspect ratio (i.e., the ratio of the length of the y-axis to the length of the x-axis) of 1. This produces a square plot, appropriate to the presentation of a conventional planimetric map, unless the window size is altered manually using the mouse. Line 61 plots the data taking these changes into account. The result is not shown here.

It is possible to gain greater control over the appearance of the resulting contour plot, and hence to produce a better quality figure, by saving the data that define the positions of the contour lines in a separate data file. These can then be read back into gnuplot and drawn using the plot command, as shown below.

```
set terminal push                                          62
set terminal table                                         63
set output 'contours.dat'                                  64
splot 'efyrnwy.dem' using 1:2:3                             65
set output                                                 66
set terminal pop                                           67
set grid                                                   68
plot 'contours.dat' index 0 u 1:2, \                       69
     'contours.dat' index 1 u 1:2, \                       70
     'contours.dat' index 2 u 1:2, \                       71
     'contours.dat' index 3 u 1:2, \                       72
     'contours.dat' index 4 u 1:2, \                       73
     'contours.dat' index 5 u 1:2                          74
```

The set terminal push command (line 62) stores information concerning the current terminal type and its settings, which can subsequently be recalled using the set terminal pop command (line 67). This pair of commands provides a convenient way of saving the default terminal type (e.g., the one employed when gnuplot starts up), before the terminal type is changed for whatever reason, and of restoring it later. Significantly, this approach is independent of the operating system on which gnuplot is run. Thus, the same commands can be employed without modification on different computers; for example, on one running GNU/Linux and another running Microsoft Windows. These commands are therefore said to be "portable" between operating systems.

In the example given above, after storing information on the default terminal type (line 62), gnuplot is instructed to use the table terminal type until further notice (line 63). Instead of producing graphical output, this terminal type generates several columns of data in ASCII text format (Figure 2.20). Line 64 ensures that these data are output to a file called contours.dat located in the working directory, as opposed to displaying them on the screen. If a file by this name already exists in that directory, its contents are over-written; otherwise, a new file is created. Line 65 then plots the contours, storing the x,y,z values that denote their locations in the named data file. Note that the data are organized so that contour lines are listed in descending order of elevation. Moreover, gnuplot separates the data for contour lines at the same elevation (e.g., 500 m) by a single blank line, and data from contours at different elevations (e.g., 300 m and 250 m) by two blank lines (e.g., lines 851 and 852 in Figure 2.20). Line 66 closes the output file.

Line 67 restores the default terminal type so that subsequent plots are directed to the computer screen rather than to a data file. Line 68 draws a grid across the resulting plot. Lines 69 through 74 plot the data contained in the newly created file, contours.dat. Note that this is a single plot command, which is split over several lines using the line continuation symbol (\). This command plots the data contained in the first two columns of the data file (i.e., u 1:2). The index keyword provides control over which section of the data file to plot. Thus, index 0 tells gnuplot to read in the data from the start of the file up to the first pair of consecutive blank lines (lines 58 and 59 in Figure 2.20), index 1 refers to the subsequent block of data up to the next pair of blank lines, and so on. Six contour levels were specified (250 m, 300 m, 350 m, 400 m, 450 m and 500 m), so contours.dat contains six blocks of data (blocks 0 to 5). Note that lines in the data file that start with a hash symbol (#) are treated as comments and are ignored by gnuplot (they are not plotted).

The result of the gnuplot commands described above is presented in Figure 2.21. One feature evident in this figure is the valley that runs roughly WNW–ESE, roughly along Northing 318000. This valley contains the Afon Conwy, a large stream that drains into the Afon Efyrnwy below Llyn Efyrnwy.

gnuplot offers further control over the production of contour plots, including a range of contour interpolation routines, such as piecewise linear, cubic splines and B-splines. These are controlled via options to the set cntrparam command. For more professional-quality output, however, it is generally necessary to use specialist software for digital mapping (Jones 1997, Burrough and McDonnell 1998).

```
#Surface 0 of 1 surfaces                               1
                                                       2
# Contour 0, label:      500                           3
 295000   319420   500                                 4
 295100   319456   500                                 5
 295200   319500   500                                 6
 295300   319500   500                                 7
 295400   319420   500                                 8
 295500   319400   500                                 9
                                                      10

 296380   319900   500                                55
 296400   319975   500                                56
 296500   320000   500                                57
                                                      58
                                                      59
# Contour 1, label:      450                          60
 298400   319400   450                                61
                                                      62
 298500   319700   450                                63
                                                      64
 296794   317600   450                                65
 296800   317590   450                                66
 296843   317500   450                                67
 296816   317400   450                                68
 296800   317362   450                                69
 296729   317300   450                                70

 299800   315317   300                               840
 299700   315369   300                               841
 299650   315400   300                               842
 299700   315450   300                               843
 299729   315500   300                               844
 299777   315600   300                               845
 299800   315630   300                               846
 299833   315700   300                               847
 299854   315800   300                               848
 299900   315843   300                               849
 300000   315850   300                               850
                                                     851
                                                     852
# Contour 5, label:      250                         853
 297100   315000   250                               854
 297000   315067   250                               855
 296990   315000   250                               856
```

Figure 2.20: Extracts from Llyn Efyrnwy contour line data file (contours.dat; lines 1–10, 55–70 and 840–856).

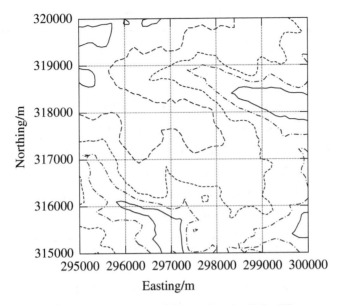

Figure 2.21: Contour map of the terrain around Llyn Efyrnwy.

Exercise 2.4: Using gnuplot, produce a contour plot similar to that shown
in Figure 2.21 based on the Llyn Efyrnwy DEM. The resulting plot should
display elevation contours at 25 m intervals from 300 m to 500 m.

2.5   PRINTING PLOTS

The method of printing a plot in the version of gnuplot for Microsoft Windows is to
click the right-hand button of the mouse over the title bar of the plot window, select
"Options" and then "Print ... ". This sequence should reveal a standard printer dialog
box with options to select the appropriate printer, paper size and orientation. Clicking
on the OK button will send the current plot to the selected printer.

The method for printing plots in the version of gnuplot for GNU/Linux is slightly
different and is driven from the command-line interface as follows:

```
set output '|lpr'                                                    75
set terminal push                                                    76
set terminal postscript                                              77
replot                                                               78
set output                                                           79
set terminal pop                                                     80
```

Line 75 sets the destination to which the plot should be output. In this example, the
plot is "piped" (|) through to the printer (lpr). Information about the current (default)
terminal type is stored (line 76) before the terminal type is changed to postscript,

Table 2.6: Selected output file formats supported by gnuplot.

Format	Description
aqua	For display on Mac OS X
emf	Enhanced Metafile Format
pdf	Adobe Portable Document Format (PDF)
png	Portable Network Graphics file format
postscript	PostScript and Encapsulated PostScript file format
svg	Scalable Vector Graphics file format
table	Dump an ASCII table of x y [z] values to output
windows	For use with Microsoft Windows
x11	For use with X servers (e.g., under GNU/Linux and UNIX)

on the assumption that the printer is PostScript-compatible (line 77). Line 78 plots the current figure, sending the resulting PostScript commands directly to the printer instead of the screen. Lines 79 and 80, respectively, close the pipe to the printer and send subsequent plots to the previously stored default terminal (i.e., to the screen).

2.6 EXPORTING GRAPHICS FILES

As well as printing directly from gnuplot, there is often a need to incorporate plots into reports and other types of document. This book, for example, has been produced using LyX (http://www.lyx.org), a document preparation system based on the LaTeX typesetting package (Goossens *et al.* 1994, Lamport 1994). Among other things, this software allows EPS figures, such as the ones presented in this chapter, to be included within a document (Merz 1996, Goossens *et al.* 1997). Standard word-processing packages, such as OpenOffice (http://www.openoffice.org/), Corel WordPerfect and Microsoft Word, have similar functionality. Each is able to import figures stored in a number of different formats supported by gnuplot, including vector formats, such as Enhanced Metafile Format (EMF) and Scalable Vector Graphics (SVG), and bitmap images, such as Portable Network Graphics (PNG) (Table 2.6). In general, vector formats are preferable because the size of the plot can be altered subsequently in the word-processing package without loss of resolution.

Output from the version of gnuplot for Microsoft Windows can be saved as an EMF file as follows:

```
set terminal push                          81
set terminal emf "Times Roman" 12          82
set output 'myplot.emf'                     83
replot                                      84
set output                                  85
set terminal pop                            86
```

Line 82 instructs gnuplot to create the next plot in the form of an EMF file, using 12 pt Times Roman font for the text. Line 83 ensures that the plot is output to a file called myplot.emf in the working directory. Line 84 re-issues the previous plot or splot command, sending the results to the named file. Line 85 closes this file and line 86 restores the previous terminal type. The resulting graphics file, myplot.emf, can be imported directly into most standard word-processing or graphics packages. The exact method differs from one package to another, but the relevant functions can often be found under the "Insert" option on the main menu bar.

In the version of gnuplot for Microsoft Windows it is also possible to copy the current plot to the clipboard and from there paste it directly into a document (e.g., a Word file). This is done by clicking the right-hand button of the mouse over the title bar of the plot window, selecting "Options" and then "Copy to Clipboard". In most Microsoft Windows packages the contents of the clipboard can then be pasted into the current document by typing Ctrl + v .

2.7 COMMAND-LINE SCRIPTS

In Section 2.2.1, one of the justifications given for using command-line entry, rather than point-and-click methods, was the ability to store the gnuplot commands used to create a plot (or plots) in a simple ASCII text file. This file can then be employed to process similar data sets using the same set of commands. Files of this type are sometimes known as scripts or macros. In the context of gnuplot, a script is simply an ASCII text file that contains a series of valid gnuplot commands. For example, the commands listed in lines 1 through 86 throughout this chapter could be saved in an ASCII text file called visual.plt using a standard text editor. The commands contained in this file can then be invoked ("run") by typing

```
load 'visual.plt'                                          87
```

at the gnuplot> command-line prompt, assuming that visual.plt is located in the working directory. If the script file is located somewhere else on the computer, the full or relative path name of the file must be given (e.g., /mnt/cdrom/chapter2/visual.plt for GNU/Linux or d:\chapter2\visual.plt for Microsoft Windows). Thus, the load command (line 87) reads in and executes each line of the named script file as though it had been typed interactively on the command line (Williams and Kelly 1998).

2.8 SUMMARY

This chapter provides a general introduction to the graphical and analytical functions of gnuplot, which is given as an example of the type of computer software that can be used to visualize a range of environmental data sets. The coverage is by no means exhaustive either in terms of what can be achieved with gnuplot or, indeed, the full scope of scientific data visualization. gnuplot is, however, used throughout the rest of the book to examine key data sets and to plot the output from various environmental models. In doing so, other features of the software are explored. In the meantime, it is worth noting that gnuplot comes with extensive built-in documentation, which can be accessed by typing `help`, or `help <command>` for help on a specific command (e.g., `help plot`).

> **Exercise 2.5**: Use a text editor to create and save a file that contains the gnuplot commands required to generate the plot shown in Figure 2.10. The script file should output the resulting plot as an EMF or PNG file. Use the gnuplot command line to load and run the script file. Paste the graphics file that it generates into a word-processor document to check that it has been created properly. N.B. If you use word-processing software instead of a text editor to create and save the gnuplot script file, you must ensure that the file is saved in ASCII text format. This can often be achieved by using the word-processor's "Save as" facility.

SUPPORTING RESOURCES

The following resources are provided in the `chapter2` sub-directory on the CD-ROM:

`le98temp.dat`	Air temperature data set for Llyn Efyrnwy, 1998
`le98rain.dat`	Precipitation data set for Llyn Efyrnwy, 1998
`efyrnwy.dem`	Digital elevation data set for Llyn Efyrnwy (© Crown Copyright License Number NC/03/11298)
`visual.plt`	gnuplot commands listed in this chapter
`visual.gp`	gnuplot commands used to generate the EPS figures presented in this chapter

Chapter 3

Processing Environmental Data

Topics

- Data pre-processing and post-processing

- Records and fields

Methods and techniques

- Creating and running a simple gawk program

- Selecting and manipulating specific fields and records of data

- Command-line switches

- Formatting the output from a gawk program

- Output redirection

- Handling "missing data"

3.1 INTRODUCTION

The ability to manipulate data is an important part of the environmental modeling process, and not only within the model itself. Where a model requires input values, for instance, the data must be presented in a form that the model can handle. This may differ, however, from the format in which the data are originally provided. In such circumstances the data must be reformatted so that they are compatible with the model; this task is sometimes referred to as data pre-processing. Similarly, the data output from a model may have to be post-processed so that they can be compared to, or combined with, other data sets. These might include *in situ* measurements of the target environmental system or the output from another model. Moreover, it may be necessary to post-process the output from a model so that it can be visualized effectively or presented in summary form as part of a report.

The data sets examined in the preceding chapter represent a case in point. The files used to explore the temporal variations of air temperature (le98temp.dat) and precipitation (le98rain.dat) at Llyn Efyrnwy are derived from the AWS data sets recorded by the UK MetOffice. The original data were pre-processed by the author, however, to simplify the introduction to gnuplot. Specifically, several columns of data were removed to reduce the size of the data sets, and the date and time information was reformatted to conform to the ISO 8610:2004 standard. Figure 3.1, for example, shows part of the original precipitation data set (rain98le.dat) used to generate the file le98rain.dat, which was examined in the preceding chapter (Figure 2.10).

Many different tools can be employed to process environmental data sets in the various ways outlined above. These include spreadsheets and databases, as well as bespoke software written by the modeler in a computer programming language of his or her choice. Each has its advantages and disadvantages, some of which are discussed in Chapter 1. A simple, but powerful, computer programming language, known as awk, is used in this book. The acronym awk is derived from the surnames of the individuals who developed the language, specifically Aho, Weinberger and Kernighan (Aho *et al.* 1988). There are several versions of awk; the one used here is the GNU Project (http://www.gnu.org/) implementation, which is known as gawk (Dougherty 1996, Robbins 2001). gawk has a number of important advantages in this context, some of which are outlined in Table 3.1. Significantly, gawk not only provides a powerful tool for data manipulation, but also a convenient framework in which to develop a range of environmental simulation models. These properties are demonstrated in subsequent chapters. By way of introduction, though, the current chapter explores how gawk can be used to reformat a particular data set so that it can be visualized in gnuplot.

3.2   STRUCTURE OF THE LLYN EFYRNWY PRECIPITATION DATA

The exercises covered in this chapter make use of the file rain98le.dat, which is included on the CD-ROM. This file contains information on the precipitation that accumulated at Llyn Efyrnwy over the preceding 12 hours recorded twice daily (at 09:00 and 21:00 GMT) throughout 1998 by an AWS (Figure 3.1). Unlike the files used in the preceding chapter, the data are given in the original format supplied by the UK MetOffice, as downloaded from the BADC web site (http://badc.nerc.ac.uk/).

The data in rain98le.dat can be thought of as being divided into rows and columns, similar to the structure of a spreadsheet. The rows and columns are known formally as records and fields, respectively. There are 717 records in rain98le.dat, each of which contains 10 fields of data (Figure 3.1). In this example, the fields are separated by white space (i.e., blank spaces or tab stops) and every record is terminated by a newline character, which is not displayed but ensures that each record is placed on a separate line. In other files, commas are sometimes used to separate fields; these are known as comma-separated value (CSV) files.

ID	IDTYPE	MET_DOM	YEAR	MONTH	DAY	END_HOUR	COUNT	AMT	DUR
425000	RAIN	NCM	1998	1	1	900	12	0	-999
425000	RAIN	NCM	1998	1	1	2100	12	146	-999
425000	RAIN	NCM	1998	1	2	900	12	52	-999
425000	RAIN	NCM	1998	1	2	2100	12	16	-999
425000	RAIN	NCM	1998	1	3	900	12	358	-999
425000	RAIN	NCM	1998	1	3	2100	12	56	-999
425000	RAIN	NCM	1998	1	4	900	12	66	-999
425000	RAIN	NCM	1998	1	4	2100	12	208	-999
425000	RAIN	NCM	1998	1	5	900	12	44	-999
425000	RAIN	NCM	1998	1	5	2100	12	30	-999
425000	RAIN	NCM	1998	1	6	900	12	40	-999
425000	RAIN	NCM	1998	1	6	2100	12	-999	-999
425000	RAIN	NCM	1998	1	7	900	12	24	-999
425000	RAIN	NCM	1998	1	7	900	24	148	-999
425000	RAIN	NCM	1998	1	7	2100	12	66	-999
425000	RAIN	NCM	1998	1	8	900	12	136	-999
425000	RAIN	NCM	1998	1	8	2100	12	384	-999
425000	RAIN	NCM	1998	1	9	900	12	10	-999
425000	RAIN	NCM	1998	1	9	2100	12	2	-999
425000	RAIN	NCM	1998	1	10	900	12	0	-999
425000	RAIN	NCM	1998	1	10	2100	12	0	-999
425000	RAIN	NCM	1998	1	11	900	12	0	-999
425000	RAIN	NCM	1998	1	11	2100	12	12	-999
425000	RAIN	NCM	1998	1	12	900	12	2	-999

Figure 3.1: Partial contents of the file rain981e.dat (first 25 records).

Table 3.1: Selected properties of the gawk programming language.

Data driven	Reading data from files, selecting the required elements from these data, manipulating the chosen elements and writing the results to files are very much simpler than in many other computer programming languages.
Powerful	Extensive built-in control-flow constructs and functions. Can be extended through user-defined functions.
C-like syntax	Syntax similar to the C programming language (Kernighan and Ritchie 1988), which is widely used in industry, commerce and academia. Consequently, it is relatively easy to migrate to C having learned gawk.
Interpreted	Interprets each line of the program at run-time. Removes the additional distractions and complexities associated with languages in which the code must be compiled (i.e., converted into a binary format that can be executed by the computer).
Platform independent	Available for a host of different computer platforms, including GNU/Linux, UNIX, MacOS X and Microsoft Windows/DOS.
Free	Covered by the General Public License (GPL; see Appendix B), sometimes referred to as the *Copyleft* agreement. Can be freely copied and redistributed. Legal to place a copy of gawk on any number of computers without infringing copyright.

The first record of rain98le.dat consists of brief textual descriptions of the data contained in the corresponding fields of the remaining records; these are, in effect, the field names (Table 3.2). For example, the first field contains the number used by the MetOffice to identify (ID) the rain gauge at this meteorological station, the second field specifies the type of measurement made at this site (RAIN), and the third field indicates that the data are reported in the National Climate Message (NCM) format (see http://badc.nerc.ac.uk/data/surface/ukmo_guide.html for further details).

It is important to note that the precipitation values given in the ninth field of Figure 3.1 (AMT) are reported in tenths of a millimeter (0.1 mm). Thus, a value of 18 indicates that 1.8 mm of precipitation accumulated over the preceding 12 hours. Furthermore, the value −999 is used to indicate "missing data". This may refer to data that have been lost for some reason (e.g., owing to instrument failure) or that are not collected at this site.

3.3   CREATING AND RUNNING A SIMPLE gawk PROGRAM

The first thing that one might wish to do with the file rain98le.dat is to examine its contents on the computer screen. Various command-line tools and software packages have been developed for this purpose, including text editors and spreadsheets. Here, though, gawk is used to achieve the same result, primarily because it serves

Table 3.2: Interpretation of the data fields in the file `rain981e.dat`.

Field	Field name	Data contained in this field
1	ID	Meteorological station ID number
2	IDTYPE	Meteorological station type
3	MET_DOM	Meteorological domain
4	YEAR	Year
5	MON	Month (1 to 12)
6	DAY	Day of the month (1 to 31)
7	END_HOUR	Hour of observation (09:00 or 21:00 GMT)
8	COUNT	Period (hours) over which the observations were recorded
9	AMT	Precipitation amount (in units of 0.1 mm)
10	DUR	Precipitation duration (minutes)

as a convenient introduction to the structure of a basic **gawk** program. Thus, the following instruction must be entered on the command line to inspect the contents of `rain981e.dat` using **gawk**, assuming that the file `rain981e.dat` is located in the working directory:

```
gawk '{print}' rain981e.dat
```
1

If the file is located elsewhere on the computer, its full or relative pathname must be used (e.g., `/mnt/cdrom/chapter3/rain981e.dat` in GNU/Linux or `d:\chapter3\rain981e.dat` in Microsoft Windows).

In the example given above, the keyword `gawk` invokes the **gawk** utility (Robbins 2001). This interprets the **gawk** program specified by the user. Here, the program `{print}` is entered in full via the command line. The **gawk** utility applies the "rules" specified in the program to the data contained in the file named on the command line `rain981e.dat`. Thus, the general syntax is `gawk 'program' datafile`. The program consists of a pair of curly braces, `{}`, and a single **gawk** statement, `print`. The curly braces denote the beginning and end of this particular program, and the `print` statement instructs the computer to print out each record of data exactly as it appears in the named file. The output is sent to the computer screen unless otherwise specified. This process, and the resulting output, is illustrated schematically in Figure 3.2. Note that a pair of single quotation marks is placed around the **gawk** program to prevent it from being interpreted as an operating system command; this procedure is known as *quoting*. The quotation marks are not part of the program and are only required when the program is typed in full on the command line (cf. supplied from a program file; see Section 3.5).

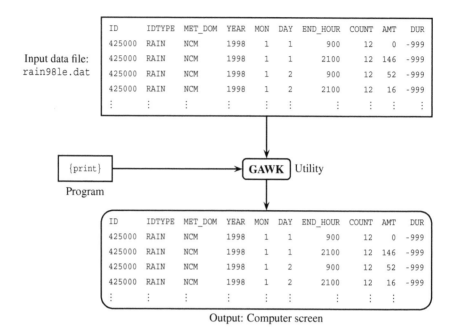

Figure 3.2: Operation of a gawk program. The arrows show the flow of data through the gawk utility and the supply of instructions from the gawk program, which controls the way the data are processed.

3.4   USING gawk TO PROCESS SELECTED FIELDS

If gawk could only echo the contents of a data file to the computer screen, it would be a very limited tool indeed. It is, in contrast, a remarkably powerful and flexible utility, one that can be used to select particular items of data from a file and to process them according to the user's instructions (Robbins 2001). gawk facilitates this by breaking up a data file into its constituent records and by splitting each record into its component fields; moreover, it performs this task automatically. By default a record equates to a single line of data in a file and a field corresponds to a single item of data in a record, where individual fields are separated by white space. This behavior can be altered, however, as required. For instance, gawk can also be instructed how to handle CSV files by setting the field separator to be a comma, although this possibility is not explored here.

Suppose, for example, that one wished to extract from the file rain98le.dat those data pertaining to the 12-hourly accumulation of precipitation, expressed in units of millimeters (cf. tenths of a millimeter), together with the year, month, day and time at which the observations were made; that is, to recreate the file le98rain.dat used in the preceding chapter. This process involves selecting the appropriate fields from each record of the input data file, and in the case of the precipitation values converting them into millimeters by dividing by 10.0. This can be achieved using gawk by typing the instruction shown on the following command line:

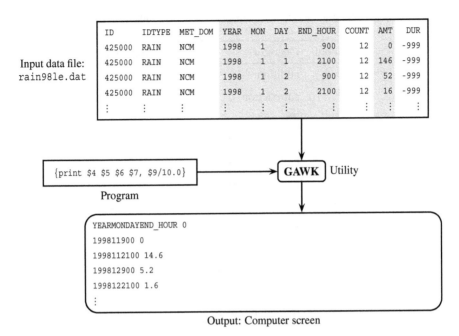

ID	IDTYPE	MET_DOM	YEAR	MON	DAY	END_HOUR	COUNT	AMT	DUR
425000	RAIN	NCM	1998	1	1	900	12	0	-999
425000	RAIN	NCM	1998	1	1	2100	12	146	-999
425000	RAIN	NCM	1998	1	2	900	12	52	-999
425000	RAIN	NCM	1998	1	2	2100	12	16	-999
⋮	⋮	⋮	⋮	⋮	⋮	⋮	⋮	⋮	⋮

Input data file: rain98le.dat

{print $4 $5 $6 $7, $9/10.0}

Program

GAWK Utility

```
YEARMONDAYEND_HOUR 0
199811900 0
1998112100 14.6
199812900 5.2
1998122100 1.6
⋮
```

Output: Computer screen

Figure 3.3: Operation of a **gawk** program designed to manipulate and print selected fields of an input data file. The areas of the input data file highlighted in gray indicate the fields selected for processing by the **gawk** program.

```
gawk '{print $4 $5 $6 $7, $9/10.0}' rain98le.dat
```
2

The operation and output of this program is shown schematically in Figure 3.3.

Although the revised command line is slightly longer than the one employed in the preceding example, the general syntax is the same; only the program differs. More specifically, the print statement is now followed by six entities, known as arguments, which instruct **gawk** exactly what to print and in which order. If no arguments are supplied, **gawk** prints the entire record. A more detailed explanation of the revised program is given below.

gawk uses the dollar sign ($) followed by a number to refer to a specific field within a record (Robbins 2001). So, for example, $1 refers to the first field, $2 refers to the second, and so on. Thus, the revised program instructs **gawk** to print out the values in the fourth ($4), fifth ($5), sixth ($6), seventh ($7) and ninth ($9) fields of each record in the input data file, rain98le.dat. The data in the other fields are discarded. The selected fields contain data on the year, month, day, time and the accumulated precipitation, respectively. The values in the ninth field ($9) are divided by 10.0 before they are printed (i.e., $9/10.0) to convert them into millimeters. Note that the comma symbol (,) used in this program instructs **gawk** to leave a space between the corresponding arguments in the output; where the comma is omitted the arguments are joined together (concatenated). Here, the comma symbol is used to insert a space between the date and time information and the precipitation values.

Table 3.3: Selected mathematical operators available in gawk.

Symbol	Operator	Example
+	addition	`$2+$3`
-	subtraction	`$2-$3`
/	division	`$3/$2`
*	multiplication	`$3*6.5`
**	exponentiation	`$2**2`
^	exponentiation	`$2^2`
sqrt	square root	`sqrt($2)`
sin	sine	`sin(0.5)`
cos	cosine	`cos($3)`
log	natural logarithm	`log($4)`
exp	exponential of *x*	`exp($2/$3)`
rand	random number generator	`rand()`

It may be helpful to think of the gawk utility as a filter, which can be configured to allow some or all of the input data file to pass through and, in this case, to be displayed on the computer screen (Figure 3.3). It also allows the input data to be modified as it is filtered (e.g., dividing the contents of the ninth field by 10.0). In this context, gawk provides a wide range of mathematical operators that can be used to manipulate data (Table 3.3). These operators are explored in subsequent chapters.

> **Exercise 3.1**: Modify the program specified in the command line given on page 65 so that it reads `'{print $4, $5, $6, $7, $9/10.0}'`. Run the revised program. How does the output differ from that shown in Figure 3.3? What effect do the additional comma symbols have? What happens if the commas are removed?

3.5   STORING THE gawk PROGRAM IN A FILE

In the examples studied thus far, the gawk programs are entered via the command line. This method is convenient if the program is short and it is unlikely that it will be needed again (sometimes described as "throw away" code). Indeed, the facility to run simple one-line programs in this way is one of the main advantages of gawk. For longer programs, however, and where there is a need to apply the same set of instructions to several data files, this approach can be time-consuming, inefficient and error-prone. The preferred alternative, therefore, is to store the program in a separate file.

Program 3.1: selcols.awk

```
{print $4 $5 $6 $7, $9/10.0}                                                1
```

Program 3.1 shows the contents of a file, selcols.awk, such as might be created using a standard text editor. The program is identical to the one employed in the preceding example (see page 65): the only difference is that it is stored in a file, rather than entered directly via the command line. The program stored in selcols.awk can be applied to the data contained in rain98le.dat as follows:

```
gawk -f selcols.awk rain98le.dat                                            3
```

assuming that both files are located in the working directory; if not, either the full or the relative pathname of each file must be provided. Note that the -f symbol, which is known as a switch, instructs the **gawk** utility to read the program from a named file, in this case selcols.awk. The rest of the command line indicates that the rule contained in selcols.awk should be applied to the data in rain98le.dat. Also note that the quotation marks, which were employed to demarcate the **gawk** program in each of the preceding examples, are no longer required because the program is stored in a file. The output produced by the revised command line is identical to that presented in Figure 3.3.

3.6 USING gawk TO PROCESS SELECTED RECORDS

Close inspection of Figure 3.3 reveals something unintended in the first record (line) of the output, which reads YEARMONDAYEND_HOUR 0. This result is produced because the program instructs the **gawk** utility to apply the print $4 $5 $6 $7, $9/10.0; statement to every record in the input data file, even though it is not really appropriate to the first of these because it contains the text strings that specify the field names. In the circumstances, **gawk** has done its best to apply the print statement to these data, including "dividing" one of the text strings (i.e., AMT) by 10.0. The result in this case is reported as 0, but this is clearly not what is intended.

One way to overcome this problem is to edit the input data file by deleting the first record. This is unsatisfactory, though, because the original (unmodified) version of the file may be needed at some point in the future. Equally, keeping two or more subtly different versions of the same file (i.e., modified and unmodified) consumes additional disk space on the computer and, more importantly, is a recipe for confusion in the future. A much better approach is to alter the **gawk** program so that it skips the first record of the input data file (i.e., the text containing the field names), but processes subsequent records in the prescribed way. Program 3.2, selcols2.awk, demonstrates one way in which this can be achieved. This program can be run from the command line as follows:

```
gawk -f selcols2.awk rain98le.dat                                           4
```

Program 3.2: selcols2.awk

```
(NR >1){print $4 $5 $6 $7, $9/10.0}
```
1

A new feature, (NR>1), known as a pattern, is introduced in Program 3.2. The gawk utility reads each record of the data file in turn and checks whether the current record matches this pattern. If it does, gawk performs the corresponding action, which in this case consists of printing the specified fields, before proceeding to read the next record of data from the named file. If the current record does not match the pattern, however, the action is not performed, and gawk moves on to the next record. This sequence is repeated until every record in the data file has been read and processed.

The pattern used in this program employs one of gawk's special features, the built-in variable NR (Robbins 2001). The term *variable* refers to a named entity that can be used to store a specific value (i.e., a number or a text string) within a program (British Computer Society 1998). Variables are so-called because the value that each holds can change (i.e., vary) in response to the actions performed by the program. The term *built-in* indicates that this particular variable is integral to (i.e., is automatically provided by) the gawk programming language (cf. user-defined variables, which are created by the computer programmer as he or she requires; see Chapter 4).

The built-in variable NR is used by gawk to keep track of the total number of records that have been read from the input data file thus far (Robbins 2001). Thus, NR has the value 1 when the program reads the first record from the input data file, it has the value 2 for the second record, and so on. The pattern (NR > 1) checks to see whether the value of NR is greater than 1. This pattern matches the second and subsequent records of the data file, but not the first. As a result, the corresponding action, {print $4 $5 $6 $7, $9/10.0}, is performed on all of the records in the input data file with the exception of the first. This process is illustrated schematically in Figure 3.4.

The preceding example illustrates the general syntax of a gawk program, namely (*pattern*){*action*}. The pairing of a pattern and an action is sometimes known as a rule (Robbins 2001) because it describes what action should be taken if the current record matches the specified pattern. A simple analogy from everyday life would be "if it is raining (*pattern*), take an umbrella {*action*}".

It is worth noting that the pattern (NR>1) involves a comparison between two numbers (i.e., the value of the built-in variable NR for the current data record and the numerical constant 1). The outcome of this comparison is either true or false. If the value of NR is greater than 1, the pattern matches the current record (i.e., the pattern evaluates as true) and, hence, the corresponding action is performed; otherwise, if the value of NR is less than or equal to 1, the pattern does not match the current record (i.e., the pattern evaluates as false) and the action is skipped. gawk provides other numerical comparison operators, which are listed in Table 3.4. These operators can be used individually or in combination to construct sophisticated patterns that allow specific parts of a data set to be selected for processing, as is demonstrated later in this chapter.

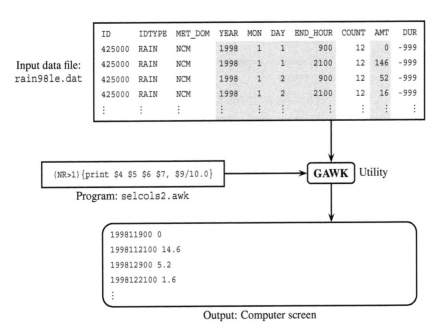

Figure 3.4: Operation of a gawk program designed to print selected records and fields of an input data file. The areas of the input data file highlighted in gray correspond to the records and fields selected by the gawk program.

Table 3.4: Numerical comparison operators in gawk.

Operator	Example	Evaluates as ...
==	x==y	True if *x* and *y* are equal; otherwise false.
!=	x!=y	True if *x* and *y* are *not* equal; otherwise false.
>	x>y	True if *x* is greater than *y*; otherwise false.
<	x<y	True if *x* is less than *y*; otherwise false.
>=	x>=y	True if *x* is greater than or equal to *y*; otherwise false.
<=	x<=y	True if *x* is less than or equal to *y*; otherwise false.

Table 3.5: Selected `printf` format control specifiers.

Format control	Interpretation	Examples
`%i`	Integer	`50, -347267`
`%e`	Scientific (exponential) notation	`1.950e+03`
`%f`	Floating-point number	`3.1415927, -29.621`
`%g`	Scientific or floating-point notation, whichever uses fewer characters	`1.950e+03` or `1950.00`
`%s`	Character string	`AMT, DUR`
`\t`	Tab space character	
`\n`	Newline character	

Program 3.3: selcols3.awk

```
(NR>1){printf("%i%i%i%i %f\n", $4, $5, $6, $7, $9/10.0)}          1
```

3.7   CONTROLLING THE FORMAT OF THE OUTPUT

The `print` statement, used in each of the preceding programs, provides basic control over the way in which the output is presented. A related command, `printf`, offers much greater flexibility, allowing one to specify exactly how the output should be formatted (Robbins 2001). In general terms, the syntax of the `printf` statement is as follows:

```
printf("format-string", arg1, arg2, ..., argN)
```

In addition to the `printf` keyword, the `printf` statement contains a comma-separated list of arguments (arg1, arg2, ..., argN) and a format string (`"format-string"`): the former specifies what is to be printed and in which order; the latter comprises a string of characters that specifies exactly how the output should be formatted, by means of format specifiers (Table 3.5). Each argument has a corresponding format specifier.

   Program 3.3, which is functionally identical to Program 3.2, illustrates how the `printf` statement is used. There are five arguments in this case: the first four are the contents of fields 4 (year), 5 (month), 6 (day) and 7 (hour) of the input data file, respectively; the fifth is the value in the ninth field of the input data file divided by the numerical constant 10.0 (i.e., accumulated precipitation in millimeters (mm)). The corresponding format string (`"%i%i%i%i %f\n"`) indicates that each of the first four arguments should be printed as an integer value (`%i`); that is, a number without a fractional part or, more formally, a member of the infinite set of positive and negative natural numbers ($\ldots -3, -2, -1, 0, 1, 2, 3\ldots$). By contrast, the format string indicates that the fifth argument should be printed as a floating-point number (`%f`);

Table 3.6: Selected `printf` format modifiers. Note that ␣ marks a blank space in the output.

Modifier	Interpretation	Examples		
		Value	Modifier	Result
width	Minimum width in characters	62	`%9i`	␣␣␣␣␣␣␣62
		62	`%2i`	62
		3.141 5927	`%1i`	3
		3.141 5927	`%9f`	␣3.141593
		"string"	`%9s`	␣␣␣string
		"string"	`%2s`	string
.prec	Precision	3.1415927	`%9.2f`	␣␣␣␣␣3.14
			`%9.4f`	␣␣␣3.1416
—	Left-justify	3.141 5927	`%-9.2f`	3.14␣␣␣␣␣
leading zero	Pad output with leading zeros	1	`%02s`	01
		1	`%03s`	001

that is, a number with a fractional part or, more formally, a member of the infinite set of positive and negative real numbers, such as 3.1415927. The format string also indicates that the fifth argument should be separated from the other four arguments by a single blank space, and that a newline character (\n) should be inserted at the end of each line in the output. The effect of the `printf` format specifiers, therefore, is to control how particular values are presented in the output. For instance, applying the format specifier `%i` to the numerical constant 3.1415927 causes it to be printed out as the integer value 3, by truncating the fractional part. Similarly, the format specifier `%f` converts the integer constant 1 to a floating-point number, which is printed out as 1.000000.

> **Exercise 3.2**: Run Program 3.3, `selcols3.awk`, on the data contained in `rain98le.dat` and confirm that the output is the same as that shown in Figure 3.4. Edit the program by removing the newline character (\n) from the format string. What effect does this have when you re-run the program on the same data set? What is the role of the newline character?

It is possible to achieve still greater control over the format of the output from the `printf` statement by using format modifiers (Table 3.6). Among other things, these can be employed to allocate different field widths in the output for each argument, to specify the precision (i.e., the number of decimal places) with which floating-point numbers are reported, to left-justify or right-justify within a field, and to pad-out the values in particular fields with leading zeros.

Program 3.4: selcols4.awk

```
(NR >1) {                                                                  1
   printf ("%4i%02i%02iT%04i  %5.1f\n", $4, $5, $6, $7, $9/10.0) ;2
}                                                                           3
```

Program 3.4 shows how format modifiers are used with the `printf` statement. In this example, the value of the first argument (i.e., the year number) is printed as an integer and is allocated a space at least four characters wide in the output (`%4i`; e.g., `1998`). The second and third arguments (i.e., the month and the day) are also printed as integers, each being allocated a space at least two characters wide in the output. If the value of either of these two arguments consists of just one character (i.e., a number in the range 0 to 9) it is padded out by a leading zero (`%02i`); for example, the value 1 is output as 01. Similarly, the fourth argument, also printed as an integer, is allocated a space at least four characters wide, padded with leading zeros where necessary (`%04i`; e.g., the value 900 is output as 0900). The format string also specifies that the third and fourth arguments should be separated by the letter "T", so that the output conforms to the ISO 8610:2004 standard for the representation of date and time information in numerical format (see Appendix D). Lastly, the fifth argument is printed out as a floating-point number, is allocated a space of at least five characters wide in the output, is reported to one decimal place (1 d.p.; `%5.1f`) and is separated from the other arguments by a single blank space.

In addition to the changes that have been made to the format string, Program 3.4 (`selcols4.awk`) is also laid out slightly differently from Program 3.3 (`selcols3.awk`). More specifically, the action part of Program 3.4 is spread over three lines of code, and the second line is indented by one tab stop, whereas it is presented on a single line in Program 3.3. This makes no difference to the functioning of the code, but it helps to improve its readability. Note that line 2 of Program 3.4 is terminated by a semi-colon (`;`). Strictly speaking, this is not required by the **gawk** utility, but it serves two useful purposes: first, it confirms that the line of code is meant to end at this point; second, the convention of terminating a statement with a semi-colon eases the transition to coding in other programming languages, such as C, where it is obligatory. Matters of programming style such as these are developed further in subsequent chapters, but it is important to understand that they are guidelines rather than rules and that, ultimately, consistency is as important as programming style (Oualline 1997).

Program 3.4 can be applied to the data in `rain98le.dat` by typing the following instruction on the command line:

```
gawk  -f  selcols4.awk  rain98le.dat                                      6
```

The first 12 lines of the resulting output are shown in Figure 3.5. Notice how the time and date values have been padded with leading zeroes where required; thus, 9 am on January 1, 1998 is reported as `19980101T0900` (cf. `199811T900`). As a result, the two fields of data output are properly formatted and aligned. Note, also, the spurious negative precipitation value reported at 9 pm on January 6, 1998 (i.e., the final record

```
19980101T0900    0.0
19980101T2100   14.6
19980102T0900    5.2
19980102T2100    1.6
19980103T0900   35.8
19980103T2100    5.6
19980104T0900    6.6
19980104T2100   20.8
19980105T0900    4.4
19980105T2100    3.0
19980106T0900    4.0
19980106T2100  -99.9
```

Figure 3.5: First 12 lines of output produced by Program 3.4 applied to the data in
rain98le.dat. Note the spurious negative precipitation reported at 9 pm on January 6, 1998 (last record shown; see text for details).

shown in Figure 3.5). This has arisen because, for whatever reason, a precipitation
value was not recorded by the AWS on this occasion. Instead, the "missing data"
value (−999) was inserted in this record of the input data file (rain98le.dat). The
gawk program, selcols4.awk, has faithfully divided this number by 10.0 to produce
a value of −99.9 in the output, although this is clearly not what was intended. The
following sections explore two possible solutions to this problem: one of which uses
gnuplot; the other, gawk. First, though, a method for saving the output from a gawk
program to a file is demonstrated.

3.8 REDIRECTING THE OUTPUT TO A FILE

In each of the examples examined thus far, the results of data processing are directed
to the computer screen for immediate inspection. It is sometimes necessary, though,
to save the results for future reference so that they can be combined with other data
sets, incorporated within a technical report, or simply output to a printer. This can
be achieved by sending the output to a file, rather than to the screen. This process is
known as redirection and is demonstrated below.

```
gawk -f selcols4.awk rain98le.dat > rain98le.out
```
7

The request for redirection is indicated by the > symbol on the command line, which
sends the results of the gawk program selcols4.awk (which is, in turn, applied to
the input file rain98le.dat) to the output data file rain98le.out. Note that the redirection symbol (>) forms part of the command line under GNU/Linux and Microsoft
Windows, and should not be confused with the "greater than" logical operator (>),
which is part of the gawk programming language. It may also help to think of the
redirection symbol as an arrow that points in the direction in which the data flows,
i.e., from the gawk program (on the left) to the output data file (on the right).

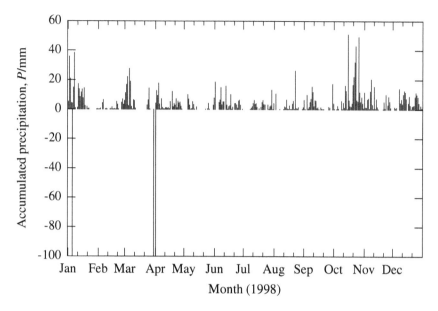

Figure 3.6: Hourly precipitation accumulation at Llyn Efyrnwy throughout 1998 illustrating
the "missing data" problem.

3.9 VISUALIZING THE OUTPUT DATA

The data file, `rain98le.out`, produced in the preceding section can be visualized in
gnuplot using the following commands, which should be familiar from Chapter 2:

```
reset                                                            1
unset key                                                        2
set xdata time                                                   3
set timefmt "%Y%m%dT%H%M"                                        4
set format x "%b"                                                5
set xlabel "Month (1998)"                                        6
set ylabel "Accumulated precipitation, P/mm"                     7
set xrange ["19980101T0000":"19990101T0000"]                     8
plot 'rain98le.out' u 1:2 w i                                    9
```

These commands produce the plot shown in Figure 3.6. The problem with Figure
3.6 is immediately apparent: "missing data" values are plotted as spurious negative precipitation readings. These points badly distort the plot and, in any case,
are highly misleading. A cosmetic solution to this problem is to limit the range
of data values displayed on the *y*-axis to numbers greater than zero (i.e., by typing
`set yrange [0:*]` in gnuplot) because it is not possible to have negative precipitation
values. This merely hides the problem, though; it does not solve it. A much better
approach is to use the "missing data" facility in gnuplot, which is demonstrated below.

```
set datafile missing "-99.9"                                    10
replot                                                          11
```

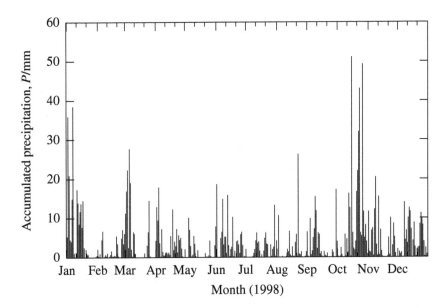

Figure 3.7: Hourly precipitation accumulation at Llyn Efyrnwy throughout 1998 with the "missing data" values removed by gnuplot.

Line 10 tells gnuplot which character string is used to denote "missing data" in the data file. In the original precipitation data set (rain98le.dat) the missing data value is −999, but in the output data file (rain98le.out) it is −99.9 because the raw precipitation values are divided by 10.0 to convert them into millimeters. Thus, line 10 tells gnuplot to treat any data item that has the value −99.9 on the y-axis as missing data. As a result, these data records are not plotted. Line 11 plots the data once again, this time taking the missing data value into account. The result is shown in Figure 3.7, which indicates clearly that Llyn Efyrnwy received rainfall every month during 1998, although it peaked in January, March and, in particular, November.

3.10   LOGICAL OR BOOLEAN OPERATORS IN gawk

Another approach to the "missing data" problem is to modify the gawk program selcols4.awk so that records containing the missing data value are skipped, in much the same way that the first record is already skipped because it contains the names of the data fields rather than the data themselves. This approach is exemplified in Program 3.5, selcols5.awk, which is run from the command line as follows:

```
gawk -f selcols5.awk rain98le.dat > le98rain.dat
```
8

Program 3.4 (selcols4.awk) and Program 3.5 (selcols5.awk) differ solely in terms of the pattern specified on line 1 of the code. In selcols5.awk the pattern has been extended so that it matches those records for which the record number is greater than one (NR>1; i.e., the second and subsequent records, but not the first) and

Program 3.5: selcols5.awk

```
(NR>1 && $9!=-999){                                                 1
  printf("%4i%02i%02iT%04i %5.1f\n", $4, $5, $6, $7, $9/10.0);2
}                                                                   3
```

Table 3.7: Logical or Boolean operators in gawk.

Operator	Meaning	Example	Evaluates as ...
&&	and	a && b	True if both a *and* b are true; otherwise, false.
||	or	a || b	True if a *or* b *or* both are true; otherwise, false.
!	not	!a	True if a is false. False if a is true.

for which the value in the ninth field is not equal to -999 ($9!=-999; i.e., the missing data value). The double ampersand symbol (&&) employed in this pattern is known as a logical, or Boolean, operator (Table 3.7) and is interpreted as "and". The symbol != is one of the comparison operators introduced earlier in this chapter (Table 3.4) and means "is not equal to". It is used here to check whether the ninth field of the current record (i.e., the raw precipitation value) contains the missing data value (-999). In short, the revised pattern skips the first record of the input data file, which contains the titles for each of the fields in the remaining records, and any other record in that file for which the accumulated precipitation value is missing.

Exercise 3.3: Using a text editor, write a gawk program to process the file temp98le.dat, which is included on the CD-ROM, so that the output is similar to the contents of the file called le98temp.dat, which was studied in the preceding chapter. Run the program on the command line, redirecting the output to a new file, le98temp.out. Visualize the results in gnuplot, and confirm that the plot is similar to that shown in Figure 2.8.

3.11 SUMMARY

This chapter introduces the gawk programming language and uses it to pre-process an environmental data set. The principal advantage that gawk offers in this context is the comparative ease with which data can be read from a file (or files), and with which specific parts of the data can be selected and processed. As an example, a data set containing information on the accumulated precipitation at Llyn Efyrnwy every 12 hours throughout 1998 is converted into a format suitable for visualization in gnuplot. Further aspects of the gawk programming language are introduced in the chapters that follow, allowing a wider range of environmental data processing and modeling exercises to be performed.

SUPPORTING RESOURCES

The following resources are provided in the `chapter3` sub-directory of the CD-ROM:

`rain98le.dat`	Precipitation data for Llyn Efyrnwy, 1998, in MetOffice format
`selcols.awk`	**gawk** program to print all records and fields in a data file
`selcols2.awk`	**gawk** program to select specific records and fields of data from a file
`selcols3.awk`	**gawk** program to select specific records and fields from a file and print them out in a given format
`selcols4.awk`	**gawk** program to select specific records and fields from a file and print them out in a given format using format modifiers
`selcols5.awk`	**gawk** program to select specific records and fields from a file, skipping records with "missing data" values
`preciptn.bat`	Command line instructions used in this chapter.
`preciptn.plt`	**gnuplot** commands used to generate the figures presented in this chapter
`preciptn.gp`	**gnuplot** commands used to generate figures presented in this chapter as EPS files

Chapter 4

Wind Speed and Wind Power

Topics

- Wind power as a source of natural renewable energy

- Estimating wind energy and wind power

Methods and techniques

- Variables and comments in gawk

- Conditional statements in gawk — the if else construct

- Actions performed *after* reading data from a file — gawk's END block

- Writing, plotting and fitting functions in gnuplot

4.1 INTRODUCTION

This chapter examines the potential for producing electricity from wind power at Llyn Efyrnwy using a combination of a wind turbine and an electricity generator, known as a wind energy conversion system (WECS) (Freris 1990, Manwell *et al.* 2002, Twidell and Weir 2006). Wind speed data recorded by an AWS over the course of a year are analyzed to determine the mean wind speed at Llyn Efyrnwy and the extent to which the wind speed varies around this value. A mathematical model is fitted to these data to estimate the frequency with which wind speeds capable of driving a small WECS are experienced at Llyn Efyrnwy. A separate mathematical model is used to estimate the power that the wind produces and, hence, the amount of electricity that could be generated by a small WECS based at this site. Further aspects of gnuplot and gawk are also introduced.

Figure 4.1: Part of the Taff Ely wind farm, South Wales, UK.

Wind power is an increasingly attractive source of natural renewable energy that may help to reduce the current reliance on fossil fuels, such as coal, oil and gas, and hence limit the emission of carbon dioxide (CO_2) and other "greenhouse" gases into the atmosphere (Golding 1977, Gipe 1995). The emission of greenhouse gases is the subject of an international treaty, commonly referred to as the Kyoto Protocol, which was adopted by the United Nations Framework Convention on Climate Change (UNFCCC) held in Kyoto, Japan, on December 11, 1997. By March 15, 1999, 84 countries had signed up to the Kyoto Protocol, and in so doing agreed to place limits on the emission of various greenhouse gases, such as CO_2, methane (CH_4), nitrous oxide (N_2O), hydrofluorocarbons (HFCs), perfluorocarbons (PFCs) and sulfur hexafluoride (SF_6), into the atmosphere. Signatories to the treaty are also required to develop technologies for carbon sequestration (i.e., the removal of carbon from the atmosphere), to promote energy efficiency and to increase the use of renewable forms of energy, including wind, wave and solar power (http://unfccc.int/resource/convkp. html). The current list of signatories to the Kyoto Protocol (189 as of June 12, 2006) can be found at http://unfccc.int/resource/conv/ratlist.pdf.

Although wind power is a potentially significant source of renewable energy, it is not a panacea. There are many important issues that need to be addressed, not least of which are the intermittent nature of the wind, the general lack of data on the wind "resource" (especially its variability with respect to space and time), the impact that wind "farms" have on landscape quality and visual amenity (a subjective and often highly emotive topic), the large distances that sometimes separate the most suitable sites for wind farms from the major centers of energy demand, and the effects that wind turbines have on both people (e.g., through noise pollution) and wildlife (e.g., via collisions with the turbine rotors) (Twidell and Weir 2006).

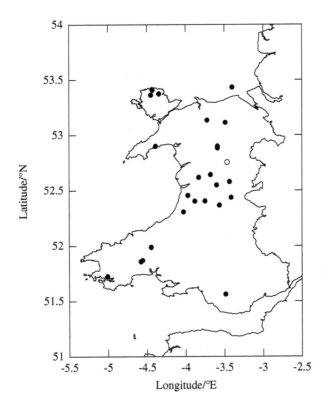

Figure 4.2: Location of onshore and offshore wind farms operating in Wales, UK in 2006 (closed circles). The position of Llyn Efyrnwy is marked by the open circle.

In the UK, most of the wind farms currently in operation are located in the north and west of the country, including Wales (Figure 4.1). The spatial distribution of the 23 wind farms operating in Wales in 2006 is shown in Figure 4.2, while summary statistics about these sites are given in Table 4.1. More wind farms are planned in the near future, including a large off-shore installation in Swansea Bay, South Wales, at a site known as Scarweather Sands. In comparison to these major developments, however, this chapter has a modest objective: to evaluate the energy resource that might be delivered by a small WECS based at Llyn Efyrnwy.

Gipe (1999) identifies three categories of small WECS based on the radius (r) of the turbine rotors: micro ($r \leq 1.25\,\text{m}$), mini ($1.25\,\text{m} < r \leq 2.75\,\text{m}$) and household ($r > 2.75\,\text{m}$). For the purpose of this exercise, it is assumed that the WECS installed at Llyn Efyrnwy is required to meet part of the electricity demand from a small number of residential dwellings and, hence, that the radius of the turbine rotors is 3 m. In this context, it is helpful to note that the average annual consumption of electricity per UK household is estimated to be 4600 kilowatt hours (4600 kW·h; http://www.dti.gov.uk/ energy/statistics/publications/energy-consumption/page17658.html).

The potential for generating electricity from wind power at a particular site, such as Llyn Efyrnwy, is controlled by a number of factors. These factors include the

Table 4.1: Location and characteristics of wind farms operating in Wales, UK in 2006.

Site name	Operational since	Number of turbines	Turbine rating (MW)	Capacity (MW)	Latitude	Latitude(°N)	Longitude	Longitude(°W)
Mynydd Clogau	Apr 2006	17	0.85	14.45	52°34'49"	52.5803°	3°25'59"	3.4331°
Hafoty Ucha 3 extension	Jan 2006	1	0.85	0.85	52°54'01"	52.9003°	4°23'18"	4.3883°
Tir Mostyn and Foel Goch	Sep 2005	25	0.85	21.25	53°06'52"	53.1144°	3°29'17"	3.4881°
Cefn Croes	Apr 2005	39	1.50	58.50	52°24'18"	52.4050°	3°45'03"	3.7508°
Castle Pill Farms	Sep 2004	1	0.50	0.50	51°43'27"	51.7242°	5°00'26"	5.0072°
Llangwryfon	Feb 2004	11	0.85	9.35	52°18'26"	52.3072°	4°01'43"	4.0286°
North Hoyle (Offshore)	Dec 2003	30	2.00	60.00	53°26'00"	53.4333°	3°24'00"	3.4000°
Moel Maelogen	Dec 2002	2	1.30	2.60	53°08'08"	53.1356°	3°43'24"	3.7233°
Haffoty Ucha 2 extension	Dec 2002	2	0.85	1.70	52°54'01"	52.9003°	3°35'18"	3.5883°
Blaen Bowi	Jul 2002	3	1.30	3.90	51°59'17"	51.9881°	4°26'43"	4.4453°
Cemmaes	Mar 2002	18	0.85	15.30	52°38'38"	52.6439°	3°40'45"	3.6792°
Parc Cynog	Feb 2001	5	0.72	3.60	51°51'28"	51.8578°	4°34'40"	4.5778°
Haffoty Ucha 1	Sep 1998	1	0.60	0.60	52°53'01"	52.8836°	3°35'18"	3.5883°
Mynydd Gorddu	Apr 1998	19	0.50	10.20	52°27'18"	52.4550°	3°58'20"	3.9722°
Llyn Alaw	Oct 1997	34	0.60	20.40	53°21'48"	53.3633°	4°26'59"	4.4497°
Centre for Alternative Technology	Apr 1997	1	0.60	0.60	52°37'09"	52.6192°	3°49'54"	3.8317°
Rheidol	Jan 1997	8	0.30	2.40	52°24'12"	52.4033°	3°52'56"	3.8822°
Carno "A" and "B"	Oct 1996	56	0.60	33.60	52°33'02"	52.5506°	3°36'01"	3.6003°
Trysglwyn	Jul 1996	14	0.40	5.60	53°22'28"	53.3744°	4°20'43"	4.3453°
Dyffryn Brodyn	Dec 1994	11	0.50	5.50	51°52'09"	51.8692°	4°33'25"	4.5569°
Bryn Titli	Jul 1994	22	0.45	9.90	52°22'03"	52.3675°	3°33'51"	3.5642°
Taff Ely	Aug 1993	20	0.45	9.00	51°33'46"	51.5628°	3°29'09"	3.4858°
Llandinam "P&L"	Dec 1992	103	0.30	30.90	52°26'11"	52.4364°	3°24'49"	3.4136°
Rhyd-y-Groes	Dec 1992	24	0.30	7.20	53°24'31"	53.4086°	4°25'47"	4.4297°
Mawla (Moel Maelogen)	–	1	1.30	1.30	53°08'08"	53.1350°	3°43'24"	3.723°

Source: http://www.bwea.com/. Latitude and longitude values are given in both degrees, minutes and seconds (DMS) and decimal degrees (DD).

mean wind speed close to ground level ($\overline{u_0}$), typically expressed in meters per second (m·s^{-1}), the variation of wind speed over time (i.e., the frequency with which the wind speed varies and the range of wind speeds encountered), and the mean air density (ρ), expressed in kilograms per cubic meter (kg·m^{-3}), which is in turn dependent on altitude and air temperature. As a general rule, the higher the mean wind speed, the greater the wind power and, hence, the more electricity is generated; however, the actual amount of electricity generated is also controlled by variation of the wind speed about the mean value. Thus, most wind turbines have a "cut in" wind speed at which they start to produce power (Manwell *et al.* 2002, Twidell and Weir 2006). If the wind speed drops below this value, during periods of calm or light breezes for example, no electricity is generated. Similarly, there is often a "cut out" wind speed above which it may be necessary to stop the turbine for safety reasons. In practice, though, many WECS continue to operate at very high wind speeds, albeit at a reduced level of efficiency, with a high power output (Twidell and Weir 2006). On balance, therefore, a site with relatively stable wind conditions is generally preferable to one at which the wind speed is variable, provided that the mean wind speed is sufficiently high (Gipe 1999). Consequently, the two main steps in evaluating the potential power output from a small WECS at Llyn Efyrnwy involve (i) determining the annual mean wind speed at this site and (ii) quantifying the variation of wind speed over time.

4.2 DESCRIPTION OF THE WIND SPEED DATA

The exercises covered in this chapter make use of the file `wind98le.dat`, which can be found on the CD-ROM. This file contains data on the mean wind speed and wind direction recorded at hourly intervals throughout 1998 by an AWS based at Llyn Efyrnwy (Table 4.2 and Figure 4.3). The hourly mean wind speed is reported to the nearest Knot, where one Knot is approximately equal to $0.515\,\text{m·s}^{-1}$. The hourly mean wind direction indicates the direction from which the wind blew, in arc degrees (°) relative to true (geographical) north, to the nearest 10°. The file also contains information on the maximum gust speed, the maximum gust direction and the time of the maximum gust, although these data are not used here and, hence, the fields concerned are omitted from Figure 4.3. The data are supplied by the UK MetOffice in ASCII text format via the BADC web site (http://badc.nerc.ac.uk/).

Hourly mean values of wind speed are not ideal in terms of predicting the likely performance of a WECS because they mask fluctuations in wind speed that take place over shorter timescales: more frequent observations are, however, seldom available (Twidell and Weir 2006). Moreover, the data in `wind98le.dat` describe measurements made at a single height above the ground (10 m; the international standard), so that there is no information on how the wind speed varies with elevation, which would help to determine the optimum configuration of a WECS at Llyn Efyrnwy. Despite these limitations, the data can be used to give a preliminary indication of the potential for electricity generation from wind power at this site, which is often the best that can be achieved in the absence of a dedicated survey (Gipe 1995).

ID	IDTYPE	MET_DOM	YEAR	MON	DAY	END_HOUR	COUNT	MDIR	MSPEED
794801	WIND	ESAWWIND	1998	1	1	0	1	210	9
794801	WIND	ESAWWIND	1998	1	1	100	1	210	8
794801	WIND	ESAWWIND	1998	1	1	200	1	220	9
794801	WIND	ESAWWIND	1998	1	1	300	1	240	2
794801	WIND	ESAWWIND	1998	1	1	400	1	240	1
794801	WIND	ESAWWIND	1998	1	1	500	1	230	6
794801	WIND	ESAWWIND	1998	1	1	600	1	230	3
794801	WIND	ESAWWIND	1998	1	1	700	1	200	3
794801	WIND	ESAWWIND	1998	1	1	800	1	210	11
794801	WIND	ESAWWIND	1998	1	1	900	1	200	11
794801	WIND	ESAWWIND	1998	1	1	1000	1	200	15
794801	WIND	ESAWWIND	1998	1	1	1100	1	190	14
794801	WIND	ESAWWIND	1998	1	1	1200	1	200	21
794801	WIND	ESAWWIND	1998	1	1	1300	1	200	20
794801	WIND	ESAWWIND	1998	1	1	1400	1	200	22
794801	WIND	ESAWWIND	1998	1	1	1500	1	190	22
794801	WIND	ESAWWIND	1998	1	1	1600	1	180	22
794801	WIND	ESAWWIND	1998	1	1	1700	1	190	22
794801	WIND	ESAWWIND	1998	1	1	1800	1	200	20
794801	WIND	ESAWWIND	1998	1	1	1900	1	230	23
794801	WIND	ESAWWIND	1998	1	1	2000	1	230	20
794801	WIND	ESAWWIND	1998	1	1	2100	1	240	22
794801	WIND	ESAWWIND	1998	1	1	2200	1	240	21
794801	WIND	ESAWWIND	1998	1	1	2300	1	240	20
794801	WIND	ESAWWIND	1998	1	1		1	240	21

Figure 4.3: Extract from the file wind981e.dat (first 25 records and first 10 fields only).

Table 4.2: Explanation of the data fields in wind98le.dat.

Field	Name	Contents
1	ID	Meteorological station ID number
2	IDTYPE	Meteorological station type
3	MET_DOM	Meteorological domain
4	YEAR	Year
5	MON	Month (1–12)
6	DAY	Day of the month (1–31)
7	END_HOUR	Hour of observation (00:00–23:00 GMT)
8	COUNT	Number of hours over which measurements are made
9	MDIR	Mean direction from which the wind blows (°)
10	MSPEED	Mean wind speed (Knots; 1 Knot $\approx 0.515\,\mathrm{m \cdot s^{-1}}$)
11	GUST_DIR	Maximum gust direction (°)
12	GUST_SPEED	Maximum gust speed (Knot)
13	GUST_TIME	Time of maximum gust (00:00–23:00 GMT)

4.3 CALCULATING THE ANNUAL MEAN WIND SPEED

The mean wind speed, $\bar{u}$, over a particular period of time can be expressed as

$$\bar{u} = \frac{1}{N} \sum_{i=1}^{N} u_i \tag{4.1}$$

where u_i is the ith observation in a series of N wind speed measurements made at regular intervals of time. Equation 4.1 is translated into gawk code in Program 4.1, which also converts the mean wind speed value from Knots to meters per second (the units conventionally employed in wind energy studies). The structure and operation of this program are explained below.

The pattern on line 1 of Program 4.1 matches those records of the input data file for which the record number is greater than one (NR>1) and (&&) for which the value in the tenth field is not equal to -999 ($10!=-999). The first part of the pattern selects the second and subsequent records of the input data file, but not the first; the second part of the pattern selects only those records that contain a valid hourly mean wind speed measurement, discarding the records in which this value is missing (note that the value -999 is used to indicate missing data in this file). Thus, the two parts of the pattern combine to skip the first record of the file because this contains the field names and any other records in which the wind speed measurement is missing. The corresponding actions, which are grouped within a pair of curly braces on lines 1 and 5, are performed if, and only if, the current record matches both parts of this pattern.

Program 4.1: meanwspd.awk

```
(NR>1 && $10!=-999){                                              1
  wind_speed=$10;                                                 2
  sum_speed=sum_speed+wind_speed;                                 3
  num_obs=num_obs+1;                                              4
}                                                                 5
                                                                  6
END{                                                              7
  mean_wind_speed=0.515*sum_speed/num_obs;                        8
  printf("Mean wind speed=%3.1f m/s\n", mean_wind_speed);         9
}                                                                 10
```

Line 2 of Program 4.1 introduces a new entity, known as a user-defined variable (cf. a built-in variable, such as NR; see Chapter 3). As the name suggests, this type of variable is created by the programmer to store a specific value for subsequent use within a program. The value stored by a variable can be a number, a single character or a text string. Each variable has a name, chosen by the programmer, by which it is identified. As far as possible, the name of a variable should indicate the nature of the data that the variable holds. The name can comprise any sequence of letters (a–z, A–Z), digits (0–9) and underscores (_), but it may not begin with a digit or contain any blank spaces (Robbins 2001). Underscores are useful in this context because they allow several words to be linked together into a single, hopefully more meaningful, name. Thus, the name of the variable on line 2 is wind_speed. It is also important to note that variable names are case sensitive, so that gawk will treat wind_speed and Wind_Speed as entirely separate variables, which may store very different values. Moreover, user-defined variables are initially assigned the value zero (0) or the empty string ("") in gawk. This process, which is performed automatically by the gawk utility, is sometimes known as initialization.

Line 2 of Program 4.1 takes the value from the tenth field ($10) of the current record of the input data file and stores it in the variable wind_speed, replacing whatever value this variable previously held. This process is known as assignment, and the = symbol is referred to as the assignment operator. Here, the variable wind_speed is assigned the value of the tenth field in the current record. Note that a common mistake is to confuse the assignment operator (a single "equals" sign, =) with one of the comparison operators (a double "equals" sign, ==). The former means "make the variable named on the left-hand side of this operator equal to the value given on the right-hand side"; the latter asks the question "is the value on the left-hand side of this operator equal to that on the right hand side". So, for example, the statement wind_speed=4.5 assigns the variable wind_speed the value 4.5 (i.e., it changes the value of the variable), whereas wind_speed==4.5 checks to see whether the value stored in wind_speed is equal to 4.5 (but the value of the variable remains unchanged).

A second variable, sum_speed, is introduced on line 3 of Program 4.1. This is used to store the sum of the wind speed values read from the input data file. Thus, line 3 takes the wind speed value for the current record, which it stored in wind_speed

Table 4.3: Operation of lines 1–5 of Program 4.1 applied to the data in wind98le.dat (Figure 4.3), showing how the values of the variables wind_speed, sum_speed and num_obs change as each record of data is processed.

NR	$10	Pattern match?	wind_speed	sum_speed	num_obs
1	MSPEED	No	0	0	0
2	9	Yes	9	9	1
3	8	Yes	8	17	2
4	9	Yes	9	26	3
5	2	Yes	2	28	4
⋮	⋮	⋮	⋮	⋮	⋮
1599	8	Yes	8	17581	1598
1600	-999	No	8	17581	1598
1601	8	Yes	8	17589	1599
⋮	⋮	⋮	⋮	⋮	⋮
8207	10	Yes	10	72866	8201
8208	10	Yes	10	72876	8202
8209	8	Yes	8	72884	8203

on line 2, and adds this to the sum of the wind speed values examined thus far (sum_speed+wind_speed). The result of this operation is stored in sum_speed, replacing the value previously held by this variable. The syntax of line 3 may seem a little strange at first, but its effect is to add up (sum) the hourly mean wind speed measurements as the data are read in record-by-record from the file, storing the running total in sum_speed.

Line 4 of Program 4.1 introduces a third variable called num_obs. As the name may suggest, this variable is used to count the number of records that contain valid observations of hourly mean wind speed (i.e., that match the pattern specified on line 1 of the program). The value of num_obs is increased ("incremented") by 1 for each such record (num_obs=num_obs+1). Note that Table 4.3 illustrates the way in which the values of the three variables wind_speed, sum_speed and num_obs are updated as each record of the input data file wind98le.dat is read and processed.

The mean wind speed over the entire period for which observations are available is calculated by dividing sum_speed by num_obs once all of the records from the input data file have been processed. gawk provides a special feature, known as the END block (or END rule), that can be used for this purpose. Statements contained in the END block are performed after the main pattern-action block is completed (Figure 4.4). In Program 4.1, the END block is given on lines 7 through 10; it is identified by the keyword END (line 7), which must be typed entirely in uppercase, and a matching pair of opening and closing curly braces, {} (lines 7 and 10, respectively). The curly

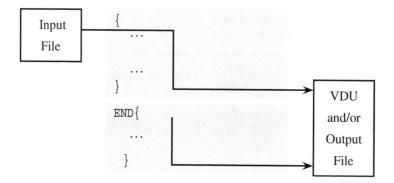

Figure 4.4: gawk's main pattern-action and END blocks.

braces serve to group together all of the statements (i.e., actions) in the END block. In this example, the END block contains two statements (lines 8 and 9 of Program 4.1). Line 8 calculates the mean wind speed in Knots (sum_speed/num_obs), converts this value into units of meters per second by multiplying it by the constant factor 0.515, and stores the result in another new variable, mean_wind_speed. Line 9 employs a printf statement to print some explanatory text followed by the mean wind speed in meters per second, expressed as a three-digit floating-point number reported to one decimal place (%3.1f).

Program 4.1 can be applied to the data in wind98le.dat by typing the following instruction on the command line:

```
gawk -f meanwspd.awk wind98le.dat
```
1

This produces the following output:

```
Mean wind speed=4.6 m/s
```

Thus, the mean wind speed at Llyn Efyrnwy throughout 1998 was approximately $4.6 \, \text{m·s}^{-1}$. Assuming that this value is representative of the longer-term mean, it is too low to consider siting a commercial wind farm there, but it may be sufficient to power a small WECS to meet at least some of the electricity demand from a limited number of households.

4.4 DETERMINING THE MAXIMUM WIND SPEED

The long-term mean wind speed at a particular site is one factor that determines the potential for electricity generation from wind power; variation of the wind speed about the mean value is another. For example, a site that experiences very high winds for relatively short periods of time and calm conditions otherwise is typically less suited to wind energy generation than a location at which the wind speed is more consistent, even if the mean wind speed is slightly lower. The problem of wind speed variability is twofold: first, when the wind speed falls below the cut-in threshold of

Program 4.2: meanmaxw.awk

```
(NR>1 && $10!=-999){                                           1
   wind_speed=$10;                                             2
   sum_speed+=wind_speed;                                      3
   ++num_obs;                                                  4
   if(wind_speed>max_speed){                                   5
     max_speed=wind_speed;                                     6
   }                                                           7
}                                                              8
                                                               9
END{                                                          10
   mean_wind_speed=0.515*sum_speed/num_obs;                   11
   printf("Mean wind speed=%3.1f m/s\n", mean_wind_speed);    12
   max_speed*=0.515;                                          13
   printf("Max. wind speed=%4.1f m/s\n", max_speed);          14
}                                                             15
```

the WECS, no electricity is generated; second, it is not always possible to exploit the full potential of very high winds. Thus, power may have to be "dumped" when the wind speed exceeds the cut-out threshold of the WECS, and electricity generated during high winds that is surplus to immediate requirements has to be stored in some way (e.g., using batteries) to meet the demand in periods of calm or light winds.

The range of wind speeds encountered at a specific site gives an initial indication of the wind speed variability. The minimum wind speed is, of course, likely to be $0\,\text{m·s}^{-1}$ (i.e., calm conditions). The maximum wind speed, on the other hand, varies from site to site and must therefore be determined from meteorological observations. For example, the maximum wind speed at Llyn Efyrnwy can be established from the measurements contained in the file wind98le.dat. One way to find this value is to search through the data set manually, although this method is tedious and error-prone. A more effective approach is to modify Program 4.1 so that it determines the maximum, as well as the mean, wind speed. This task involves comparing the wind speed value for each record of the input data file with the maximum wind speed encountered among the preceding records. If the current wind speed is greater than the previous maximum value, then this becomes the new maximum. This procedure is implemented in Program 4.2 (meanmaxw.awk).

The major difference between Programs 4.1 and 4.2 is that several lines of code have been added to the main pattern-action and END blocks in Program 4.2 (lines 5–7 and 13–14, respectively). These are used to establish and print the maximum wind speed recorded in the input data file. More specifically, lines 5 through 7 introduce another of **gawk**'s programming constructs, the if statement, which is sometimes referred to as a control-flow or branching structure (Figure 4.5); thus, the if statement controls the actions that are performed depending on the outcome of a specified condition. The general syntax of an if statement is if(*condition*){actions}, which is interpreted as "perform the following actions if, and only if, the condition is found

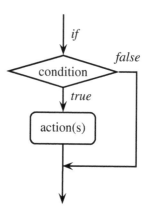

Figure 4.5: gawk's `if` construct.

to be true" (Robbins 2001). In Program 4.2, for example, line 5 checks whether the current wind speed (`wind_speed`) is greater than (`>`) the maximum wind speed encountered thus far (`max_speed`): this is the condition. If this condition is true, `max_speed` is assigned the value of `wind_speed` (i.e., the current wind speed becomes the new maximum), otherwise the value of `max_speed` remains unaltered: this is the action (line 6). Note that a pair of curly braces, `{}`, is used to group the action (or actions) contained within the `if` statement. Also, the action statement is indented by a single tab stop to highlight the fact that it forms part of the `if` statement.

There are several other differences between Programs 4.1 and 4.2, most of which involve small, but important, changes to the syntax of the commands. For instance, line 3 of Program 4.1, which reads `sum_speed=sum_speed+wind_speed`, is replaced by `sum_speed+=wind_speed` on line 3 of Program 4.2. The two statements are functionally equivalent (i.e., they produce the same result), but the latter makes use of the "plus equals" operator (`+=`), which is more concise. Thus, the `+=` operator sums the values given on either side, storing the result in the variable named on the left-hand side (Robbins 2001). So, for example, if the variables x and y initially store the numbers 5 and 3, respectively, the statement `x+=y` modifies x such that, subsequently, its new value is 8; the value of y remains unchanged. Table 4.4 presents the full set of **gawk**'s concise arithmetic assignment operators (Robbins 2001).

A similar modification is made on line 4 of Program 4.2, which keeps track of the number of valid wind speed measurements. The equivalent line of code in Program 4.1 is `num_obs=num_obs+1`, which increases ("increments") the value of `num_obs` by one every time a record matches the pattern specified on line 1. Adding one to the value of a variable is such a common operation in computer programming that **gawk**, in common with many other programming languages, provides a more concise way of expressing it using the "plus plus" symbol, `++`, which is known as the increment operator. Thus, `++num_obs` means "add one to the value of `num_obs` and store the result in `num_obs`". There is a corresponding decrement operator, `--` ("minus minus"), which subtracts one from the value of a variable. So, for example, if x and y initially store the values 5 and 3, respectively, the statements `++x` and `--y` change their values to 6

Table 4.4: gawk's arithmetic assignment operators.

Operator	Syntax	Interpretation
+=	x+=y	Adds the values of x and y, assigns the result to x.
-=	x-=y	Subtracts the value of y from x, assigns the result to x.
*=	x*=y	Multiplies the values of x and y, assigns the result to x.
/=	x/=y	Divides the value of x by y, assigns the result to x.
%=	x%=y	Finds the remainder of x modulo y, assigns the result to x (e.g., if x=10 and y=3 then x%y=1).
**=	x**=y	Raises x to the power of y (i.e., x^y), assigns the result to x.
^=	x^=y	Equivalent to the **= operator.

Table 4.5: gawk's increment and decrement operators.

Operator	Syntax	Example	Interpretation
	x++	y=x++	Assign y the value of x, then increment x by 1
++	++x	y=++x	Increment x by 1, then assign y this value
	x--	y=x--	Assign y the value of x, then decrement x by 1
--	--x	y=--x	Decrement x by 1, then assign y this value

and 2 subsequently. Table 4.5 presents the full set of gawk's increment and decrement operators (Robbins 2001).

Finally, line 13 of Program 4.2 uses the "times equals" operator, *=, to convert the maximum wind speed value from Knots into meters per second. This operator takes the value of the variable given on its left-hand side and multiplies it by the value on the right-hand side; the result is stored in the variable listed on the left-hand side of the operator. So, for example, if x and y initially have values of 5 and 3, respectively, the statement y*=x assigns y the value 15, while the value of x remains unchanged.

Program 4.2 can be applied to the data in wind98le.dat by typing the following instruction on the command line:

```
gawk -f meanmaxw.awk wind98le.dat
```
2

This produces the following output:

```
Mean wind speed=4.6 m/s
Max. wind speed=19.1 m/s
```

Thus, the maximum mean hourly wind speed at Llyn Efyrnwy during 1998 was $19.1\,\mathrm{m\cdot s^{-1}}$.

4.5   EXPLORING WIND SPEED VARIABILITY

A more comprehensive picture of wind speed variability can be gained by calculating the absolute frequency with which winds of different speeds are experienced at Llyn Efyrnwy. To do so, the wind speed data must be grouped into a number of discrete classes with an interval of, for instance, $1\,\text{m·s}^{-1}$ (e.g., $0.0\,\text{m·s}^{-1} \le u_0 < 1.0\,\text{m·s}^{-1}$, $1.0\,\text{m·s}^{-1} \le u_0 < 2.0\,\text{m·s}^{-1}$, and so on). The task then becomes one of counting the number of occasions on which the measured wind speed lies within each of these intervals. Relative frequency values can also be obtained from these data by dividing the raw counts by the total number of observations in the data set. If the number of wind speed measurements in the data set is sufficient, and the observations are representative of the long-term conditions at the site, it is possible to construct a wind speed probability distribution, which can be used to estimate the likelihood with which a given wind speed will occur. This information is important when matching the operational characteristics of a WECS (e.g., its cut-in and cut-out thresholds, and the wind speed at which it is rated for maximum power output) to the local wind regime (Twidell and Weir 2006). Methods for calculating the absolute frequency, relative frequency and probability distributions of wind speed are examined below.

4.5.1   Determining the Absolute Frequency Distribution

There are various ways in which the absolute frequency distribution of different wind speeds can be determined from the data contained in wind98le.dat, one of which is presented in Program 4.3 (windfreq.awk). The code employed in this program sacrifices brevity for simplicity. Thus, the programming constructs are the same as those used in Program 4.2, but the number of lines of code involved is much greater. There are much more concise ways of coding the same problem, but these involve the use of more sophisticated **gawk** programming constructs, notably arrays, which are introduced in subsequent chapters.

The pattern on line 1 of Program 4.3 is identical to that used in Program 4.2, and it serves the same purpose: to skip the first record of the input data file and any other records for which the hourly mean wind speed measurement is missing. Lines 2 and 4, which begin with the hash symbol (#), are known as comments. They are ignored by the **gawk** utility and are included here solely to indicate to the reader the purpose of the subsequent lines of code. The judicious use of comments, such as these, is an example of good programming practice. Used sparingly, comments help the reader to understand what a program is meant to do and how it is intended to function. Line 3 takes the value from the tenth field of the current record of the input data file (i.e., the hourly mean wind speed in Knots), multiplies this value by 0.515 (to convert it into m·s^{-1}) and stores the result in the variable wind_speed.

Lines 5 through 24 employ a series of if statements to count the number of times that different wind speeds occur. Line 5, for example, checks whether the current hourly mean wind speed is less than $1\,\text{m·s}^{-1}$ (wind_speed<1). If this is the case, the value of the variable speed_1 is increased by 1 (i.e., ++speed_1). Thus, speed_1 is used to keep track of the number of times that the wind speed lies in the range

Program 4.3: windfreq.awk (lines 1–46)

```
(NR>1 && $10!=-999){                                              1
 # Wind speed in meters per second                                2
 wind_speed=0.515*$10;                                            3
 # Count frequency of different wind speeds                       4
 if(wind_speed<1){++speed_1}                                      5
 if(wind_speed>=1 && wind_speed<2){++speed_2}                     6
 if(wind_speed>=2 && wind_speed<3){++speed_3}                     7
 if(wind_speed>=3 && wind_speed<4){++speed_4}                     8
 if(wind_speed>=4 && wind_speed<5){++speed_5}                     9
 if(wind_speed>=5 && wind_speed<6){++speed_6}                    10
 if(wind_speed>=6 && wind_speed<7){++speed_7}                    11
 if(wind_speed>=7 && wind_speed<8){++speed_8}                    12
 if(wind_speed>=8 && wind_speed<9){++speed_9}                    13
 if(wind_speed>=9 && wind_speed<10){++speed_10}                  14
 if(wind_speed>=10 && wind_speed<11){++speed_11}                 15
 if(wind_speed>=11 && wind_speed<12){++speed_12}                 16
 if(wind_speed>=12 && wind_speed<13){++speed_13}                 17
 if(wind_speed>=13 && wind_speed<14){++speed_14}                 18
 if(wind_speed>=14 && wind_speed<15){++speed_15}                 19
 if(wind_speed>=15 && wind_speed<16){++speed_16}                 20
 if(wind_speed>=16 && wind_speed<17){++speed_17}                 21
 if(wind_speed>=17 && wind_speed<18){++speed_18}                 22
 if(wind_speed>=18 && wind_speed<19){++speed_19}                 23
 if(wind_speed>=19){++speed_20}                                  24
}                                                                25
                                                                 26
END{                                                             27
 print "# Wind speed distribution, Llyn Efyrnwy (1998)";         28
 print "# Speed (m/s)  Frequency";                               29
 printf("%13.1f %10i\n", 0.5,  speed_1);                         30
 printf("%13.1f %10i\n", 1.5,  speed_2);                         31
 printf("%13.1f %10i\n", 2.5,  speed_3);                         32
 printf("%13.1f %10i\n", 3.5,  speed_4);                         33
 printf("%13.1f %10i\n", 4.5,  speed_5);                         34
 printf("%13.1f %10i\n", 5.5,  speed_6);                         35
 printf("%13.1f %10i\n", 6.5,  speed_7);                         36
 printf("%13.1f %10i\n", 7.5,  speed_8);                         37
 printf("%13.1f %10i\n", 8.5,  speed_9);                         38
 printf("%13.1f %10i\n", 9.5,  speed_10);                        39
 printf("%13.1f %10i\n", 10.5, speed_11);                        40
 printf("%13.1f %10i\n", 11.5, speed_12);                        41
 printf("%13.1f %10i\n", 12.5, speed_13);                        42
 printf("%13.1f %10i\n", 13.5, speed_14);                        43
 printf("%13.1f %10i\n", 14.5, speed_15);                        44
 printf("%13.1f %10i\n", 15.5, speed_16);                        45
 printf("%13.1f %10i\n", 16.5, speed_17);                        46
```

Program 4.3 (continued): windfreq.awk (lines 47–50)

```
    printf("%13.1f %10i\n",  17.5,  speed_18);                        47
    printf("%13.1f %10i\n",  18.5,  speed_19);                        48
    printf("%13.1f %10i\n",  19.5,  speed_20);                        49
}                                                                     50
```

$0\,\text{m·s}^{-1} \leq u_0 < 1\,\text{m·s}^{-1}$. Line 6 performs a similar function for wind speeds in the range $1\,\text{m·s}^{-1} \leq u_0 < 2\,\text{m·s}^{-1}$ and uses the variable speed_2 to store the result. This procedure is replicated for each of the other wind speed classes on lines 7 through 24. In total, 20 if statements and variables are used to record the frequency values for winds ranging in speed from $0\,\text{m·s}^{-1}$ to $20\,\text{m·s}^{-1}$ (recall that the maximum wind speed is $19.1\,\text{m·s}^{-1}$; see Section 4.4) in steps of $1\,\text{m·s}^{-1}$.

The END block (lines 27 to 50) prints out two lines of explanatory text plus the frequency data for each of the wind speed classes. Note that the text output by lines 28 and 29 of Program 4.3 commences with a hash symbol (#). This is intended to make the output data easier to use within **gnuplot**, which treats anything on a line after a hash symbol as a comment (i.e., it ignores lines such as these when plotting data from a file). Lines 30 through 49 use the printf statement to output a pair of values separated by a single blank space. Line 30, for example, prints out the wind speed at the mid-point of the first wind speed class (i.e., $0.5\,\text{m·s}^{-1}$) and the number of records in wind98le.dat which belong to this class (i.e., that exhibit a wind speed in the range of $0\,\text{m·s}^{-1} \leq u_0 < 1\,\text{m·s}^{-1}$). The remaining lines (31 through 49) print out the equivalent values for the other wind speed classes. Note that the format string ("%13.1f %10i\n") used on lines 30 to 49 aligns the columns of data in the output.

Program 4.3 can be applied to the data in wind98le.dat by typing the following instruction on the command line:

```
gawk -f windfreq.awk wind98le.dat > windfreq.out                      3
```

The contents of the output data file, windfreq.out, are presented in Figure 4.6. These data can be visualized in **gnuplot** (Figure 4.7) by typing the following commands:

```
reset                                                                 1
unset key                                                             2
set yrange [0:*]                                                      3
set xrange [0:20]                                                     4
set xlabel "Mean wind speed, u_0/m.s^{-1}"                            5
set ylabel "Absolute frequency"                                      6
plot 'windfreq.out' u 1:2 w boxes lw 2                                7
```

Line 1 resets **gnuplot** to its default behavior (see Chapter 2, page 44), and line 2 indicates that the plot should not contain a key. Line 3 sets the range of values to be displayed on the y-axis, constraining the minimum plotted value to zero, but allowing **gnuplot** to determine the maximum value. Line 4 performs a similar task for the x-axis (i.e., $0\,\text{m·s}^{-1} \leq u_0 < 20\,\text{m·s}^{-1}$). Lines 5 and 6 define the labels to be placed along the x- and y-axes, respectively. Finally, line 7 plots the data in the output file,

```
# Wind speed distribution, Llyn Efyrnwy (1998)
# Speed (m/s)   Frequency
         0.5         753
         1.5         859
         2.5        1216
         3.5        1173
         4.5        1030
         5.5         807
         6.5         679
         7.5         473
         8.5         365
         9.5         268
        10.5         181
        11.5         154
        12.5          95
        13.5          52
        14.5          52
        15.5          30
        16.5          12
        17.5           1
        18.5           2
        19.5           1
```

Figure 4.6: Absolute frequency distribution of the hourly mean wind speed at Llyn Efyrnwy.

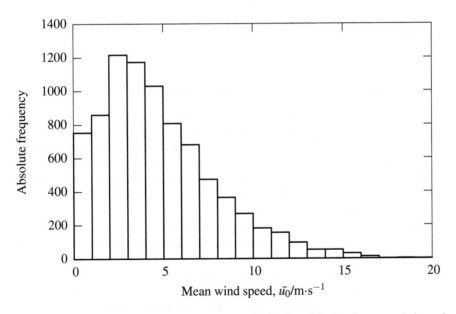

Figure 4.7: Visualization of the absolute frequency distribution of the hourly mean wind speed at Llyn Efyrnwy in 1998.

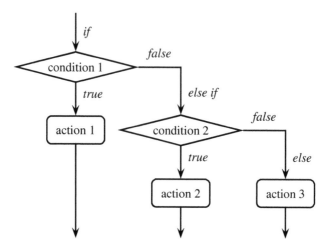

Figure 4.8: gawk's if else construct.

windfreq.out, using the values in field 1 for the x-series and those in field 2 for the
y-series. The data are plotted using the boxes style, which draws a box centered
on the x-axis value, extending from $y = 0$ to the y-axis value. The width of each
box is calculated automatically so that it touches the adjacent boxes (Williams and
Kelly 1998). Here, the line width of the boxes is set to twice the normal value (lw 2)
so that the boxes are more pronounced.

It is clear from Figure 4.6 and Figure 4.7 that the modal wind speed at Llyn
Efyrnwy lies in the range $2\,\mathrm{m\cdot s}^{-1}$ to $3\,\mathrm{m\cdot s}^{-1}$ and that the wind-speed frequency distribution is negatively skewed. The significance of this is that many WECS start to
generate electricity when the wind speed exceeds $3\,\mathrm{m\cdot s}^{-1}$, which suggests that there
will be periods during which no power will be generated by a WECS at Llyn Efyrnwy.

4.5.2 Determining the Relative Frequency Distribution

Relative frequency values are typically more useful than absolute frequency values because they allow comparisons to be made between wind speed observations
recorded at different sites, as well as between measurements made at the same site
on different occasions. Relative frequency values are calculated by dividing the raw
counts by the total number of observations in the data set. This procedure is demonstrated in Program 4.4, which also incorporates a number of minor modifications with
respect to Program 4.3 intended to improve the efficiency of the code.

An extra variable, num_obs, is employed within the main pattern-action block of
Program 4.4 to keep track of the total number of valid wind speed measurements (line
25). The value stored in this variable is used in the END block to convert the absolute
counts for the different wind speed classes into relative frequency values (lines 31 to
50), which are printed as floating-point numbers to four decimal places (%12.4f).

The code on lines 5 through 24 has also been amended to make use of a sequence
of gawk's if else statements (Figure 4.8), instead of the if statements employed in
Program 4.3. To understand why this change has been made, consider the case of a

Program 4.4: windfrq2.awk (lines 1–47)

```
(NR>1 && $10!=-999){                                                  1
  # Wind speed in meters per second                                   2
  wind_speed=0.515*$10;                                               3
  # Count frequency of different wind speeds                          4
  if(wind_speed<1){++speed_1}                                         5
  else if(wind_speed>=1 && wind_speed<2){++speed_2}                   6
  else if(wind_speed>=2 && wind_speed<3){++speed_3}                   7
  else if(wind_speed>=3 && wind_speed<4){++speed_4}                   8
  else if(wind_speed>=4 && wind_speed<5){++speed_5}                   9
  else if(wind_speed>=5 && wind_speed<6){++speed_6}                   10
  else if(wind_speed>=6 && wind_speed<7){++speed_7}                   11
  else if(wind_speed>=7 && wind_speed<8){++speed_8}                   12
  else if(wind_speed>=8 && wind_speed<9){++speed_9}                   13
  else if(wind_speed>=9 && wind_speed<10){++speed_10}                 14
  else if(wind_speed>=10 && wind_speed<11){++speed_11}                15
  else if(wind_speed>=11 && wind_speed<12){++speed_12}                16
  else if(wind_speed>=12 && wind_speed<13){++speed_13}                17
  else if(wind_speed>=13 && wind_speed<14){++speed_14}                18
  else if(wind_speed>=14 && wind_speed<15){++speed_15}                19
  else if(wind_speed>=15 && wind_speed<16){++speed_16}                20
  else if(wind_speed>=16 && wind_speed<17){++speed_17}                21
  else if(wind_speed>=17 && wind_speed<18){++speed_18}                22
  else if(wind_speed>=18 && wind_speed<19){++speed_19}                23
  else {++speed_20}                                                   24
  ++num_obs;                                                          25
}                                                                     26
                                                                      27
END{                                                                  28
  print "# Wind speed distribution, Llyn Efyrnwy (1998)";            29
  print "# Speed (m/s)   Rel. Freq.";                               30
  printf("%13.1f %12.4f\n", 0.5,  speed_1/num_obs);                  31
  printf("%13.1f %12.4f\n", 1.5,  speed_2/num_obs);                  32
  printf("%13.1f %12.4f\n", 2.5,  speed_3/num_obs);                  33
  printf("%13.1f %12.4f\n", 3.5,  speed_4/num_obs);                  34
  printf("%13.1f %12.4f\n", 4.5,  speed_5/num_obs);                  35
  printf("%13.1f %12.4f\n", 5.5,  speed_6/num_obs);                  36
  printf("%13.1f %12.4f\n", 6.5,  speed_7/num_obs);                  37
  printf("%13.1f %12.4f\n", 7.5,  speed_8/num_obs);                  38
  printf("%13.1f %12.4f\n", 8.5,  speed_9/num_obs);                  39
  printf("%13.1f %12.4f\n", 9.5,  speed_10/num_obs);                 40
  printf("%13.1f %12.4f\n", 10.5, speed_11/num_obs);                 41
  printf("%13.1f %12.4f\n", 11.5, speed_12/num_obs);                 42
  printf("%13.1f %12.4f\n", 12.5, speed_13/num_obs);                 43
  printf("%13.1f %12.4f\n", 13.5, speed_14/num_obs);                 44
  printf("%13.1f %12.4f\n", 14.5, speed_15/num_obs);                 45
  printf("%13.1f %12.4f\n", 15.5, speed_16/num_obs);                 46
  printf("%13.1f %12.4f\n", 16.5, speed_17/num_obs);                 47
```

Program 4.4 (continued): windfrq2.awk (lines 48–51)

```
   printf("%13.1f %12.4f\n", 17.5, speed_18/num_obs);        48
   printf("%13.1f %12.4f\n", 18.5, speed_19/num_obs);        49
   printf("%13.1f %12.4f\n", 19.5, speed_20/num_obs);        50
}                                                             51
```

record for which the measured wind speed is $0.5 \, \text{m·s}^{-1}$. In Program 4.3, the record is initially checked against the condition for the if statement on line 5. Since the wind speed is less than $1 \, \text{m·s}^{-1}$, the condition evaluates as true, and the corresponding action is performed (i.e., the value of speed_1 is incremented by one). The record is then checked against each of the other if statements even though, by definition, it cannot fall within any of the other wind speed classes. Thus, the wind speed associated with each record is checked 20 times (i.e., against each of the if statements).

Program 4.4 is a little more intelligent in this respect. The if else statements operate in a nested manner (Figure 4.8). For example, if the condition on line 5 (wind_speed<1) is true for the current record, the corresponding action is performed (++speed_1), after which the program skips lines 6 through 24 and, instead, proceeds to line 25. If the condition on line 5 is not met, then (else) the condition on line 6 (wind_speed>=1 && wind_speed<2) is evaluated. If this condition evaluates as true for the current record, the corresponding action (++speed_2) is performed, after which the program proceeds to line 25, and so on. In this way, the program continues to drop down through the if else statements until one of the conditions evaluates as true, at which point the corresponding action is performed. If none of the conditions is found to be true for the current record, the final else (line 24) acts as a "catch all". Since most of the wind speed values reported in wind98le.dat are quite low, this procedure should increase the efficiency, and hence the speed of execution, of the code.

Program 4.4 can be applied to the data in wind98le.dat by typing the following instruction on the command line:

```
 gawk -f windfrq2.awk wind98le.dat > windfrq2.out      4
```

The contents of the output data file, windfrq2.out, are presented in Figure 4.9. These data can be visualized by typing the following commands in gnuplot:

```
set ylabel "Relative frequency"                       8
plot 'windfrq2.out' u 1:2 w boxes lw 2                9
```

The resulting plot is presented in Figure 4.10.

Exercise 4.1: Write a gawk program to calculate the relative frequency with which the wind blows from the following directions at Llyn Efyrnwy: 315°–15°, 15°–45°, 45°–75°, ... , 275°–315°. Run this program on the data contained in the file wind98le.dat and plot the results in gnuplot.

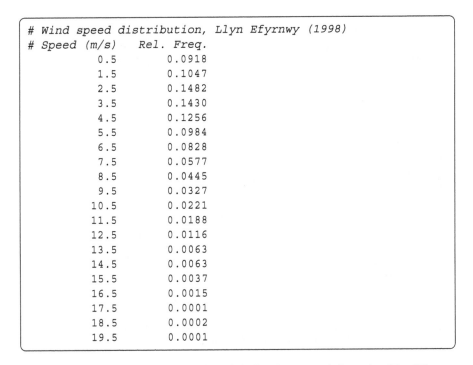

```
# Wind speed distribution, Llyn Efyrnwy (1998)
# Speed (m/s)    Rel. Freq.
        0.5        0.0918
        1.5        0.1047
        2.5        0.1482
        3.5        0.1430
        4.5        0.1256
        5.5        0.0984
        6.5        0.0828
        7.5        0.0577
        8.5        0.0445
        9.5        0.0327
       10.5        0.0221
       11.5        0.0188
       12.5        0.0116
       13.5        0.0063
       14.5        0.0063
       15.5        0.0037
       16.5        0.0015
       17.5        0.0001
       18.5        0.0002
       19.5        0.0001
```

Figure 4.9: Relative frequency distribution of the hourly mean wind speed at Llyn Efyrnwy throughout 1998.

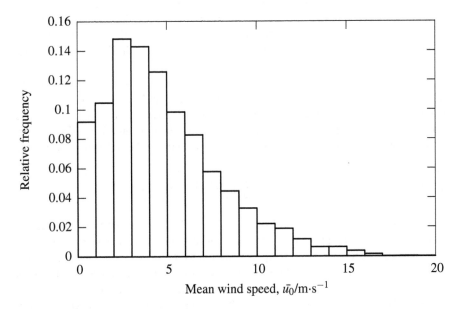

Figure 4.10: Visualization of the relative frequency distribution of hourly mean wind speed at Llyn Efyrnwy throughout 1998.

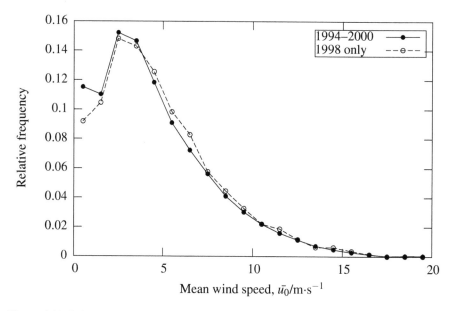

Figure 4.11: Relative frequency distributions of hourly mean wind speed at Llyn Efyrnwy for
1998 and the period 1994 to 2000, inclusive.

4.5.3 Probability Distributions and Probability Density Functions

The file `wind98le.dat` contains a sample of over 8000 wind speed measurements.
Even so, it is possible that the data come from a year in which the observed wind
speeds differ significantly from the long-term norm; that is, the 1998 data may be
unrepresentative of the typical conditions at this site. This possibility can be evaluated
by analyzing a longer time-series of data.

Comprehensive measurements of the hourly mean wind speed at Llyn Efyrnwy
are available for the period from 1994 to 2000, inclusive, from the BADC web site.
The relative frequency distribution of these values is given in the file `wind7yr.dat`,
which is included on the CD-ROM. The data from 1998 only and from 1994 through
to 2000 can be compared by plotting them in gnuplot using the following commands:

```
set key top right Left box                                               10
plot 'wind7yr.dat' u 1:2 t "1994-2000" w lp, \                          11
     'windfrq2.out' u 1:2 t "1998 only" w lp                           12
```

The resulting output is shown in Figure 4.11, which suggests that the 1998 data are
broadly typical of the prevailing conditions at this site. It is reasonable, therefore,
to treat Figure 4.10 as a discrete approximation to the underlying (i.e., continuous)
wind speed probability distribution; that is to say, the values on the y-axis indicate the
probability with which a set of discrete wind speed classes (i.e., $0 \leq u_0 < 1\,\mathrm{m \cdot s^{-1}}$,
$1 \leq u_0 < 2\,\mathrm{m \cdot s^{-1}}, \dots, 19\,\mathrm{m \cdot s^{-1}} \leq u_0 < 20\,\mathrm{m \cdot s^{-1}}$) are experienced at Llyn Efyrnwy.

It is possible to represent the wind speed probability distribution at Llyn Efyrnwy
more accurately using a continuous mathematical function, known as a probability

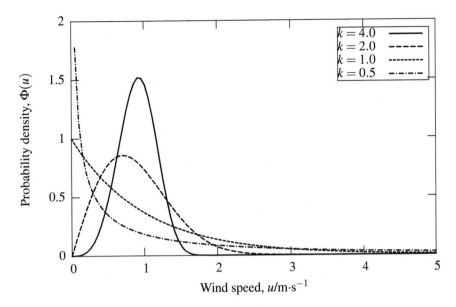

Figure 4.12: Weibull PDFs for various values of the shape parameter (k) and a fixed value of the scale parameter ($c = 1$).

density function (PDF). In studies of wind speed it is common to use either the Weibull distribution or the Rayleigh distribution for this purpose (Bowden *et al.* 1983, Manwell *et al.* 2002, Twidell and Weir 2006). The Weibull distribution is defined in terms of two parameters, c and k:

$$\Phi(u) = \frac{k}{c}\left(\frac{u}{c}\right)^{k-1} \exp\left[-\left(\frac{u}{c}\right)^{k}\right] \tag{4.2}$$

where u is the wind speed, $\Phi(u)$ is the probability density of the wind speed u, c is a scale parameter and k is a shape parameter. The Rayleigh distribution is specified in terms of a single parameter, c, and is a special case of the Weibull PDF for $k = 2$:

$$\Phi(u) = \frac{2u}{c^2} \exp\left[-\left(\frac{u}{c}\right)^{2}\right] \tag{4.3}$$

Figure 4.12 shows a number of Weibull PDFs produced using a fixed value of the scale parameter, c (1.0), but different values of the shape parameter, k (0.5, 1.0, 2.0 and 4.0). This figure was generated using the following gnuplot commands:

```
set xrange [0:5]                                              13
set yrange [0:2]                                              14
set xlabel "Wind speed, u_0/m.s^{-1}"                        15
set ylabel "Probability density, Phi(u)"                     16
set dummy u                                                   17
phi(u,c,k)=(k/c)*((u/c)**(k-1))*exp((-1*(u/c)**k))           18
plot phi(u,1.0,4.0) t "k=4.0" lw 2, \                        19
```

```
phi(u,1.0,2.0)  t  "k=2.0"  lw  2,  \                          20
phi(u,1.0,1.0)  t  "k=1.0"  lw  2,  \                          21
phi(u,1.0,0.5)  t  "k=0.5"  lw  2                              22
```

Lines 13 through 16 specify the ranges of values and the labels to be plotted on the *x*- and *y*-axes. Line 17 instructs gnuplot to employ a "dummy" variable, u, to denote the *x*-axis. This facility makes it possible to refer to the variable plotted on the *x*-axis (i.e., wind speed) by its conventional name or symbol (i.e., *u*), which is often more convenient and memorable than the generic symbol *x* (Williams and Kelly 1998).

Line 18 defines the function that is to be plotted, which is an implementation of Equation 4.2 in gnuplot code. This is known as a user-defined function because it is defined by the user, as distinct from a built-in function, such as sin(), which is provided by default in gnuplot. Thus, the mathematical representation given in Equation 4.2 is translated into gnuplot code. For the most part this is fairly self-explanatory, although it is worth noting that the symbol ** means "raise to the power of" in gnuplot; for instance x**3 means raise x to the power of 3 (x^3). Moreover, parentheses are used more extensively in the gnuplot code than in Equation 4.2 to make absolutely clear the way in which the function is evaluated. For the sake of convenience, gnuplot allows each user-defined function to be given a unique name (phi in this example), so that it can be referred to subsequently without having to type the equation in full each time it is used (e.g., lines 19 through 22).

To the right of the function name on line 18, enclosed in parentheses, is a comma-separated list of values, which are known as the arguments to the function. In this example, the function has three arguments, which means that it expects to receive three input values: the wind speed, *u*, the scale parameter, *c*, and the shape parameter, *k*. The values of these arguments are passed in that order to the function, where they are stored locally in the gnuplot variables u, c and k. This process is analogous to the assignment of variables in gawk (see Chapter 3).

The phi function is employed (or "called") on lines 19 through 22 to plot four separate Weibull PDFs, all of which share the same value of the scale parameter ($c = 1.0$), but each of which has a different value of the shape parameter ($k = 4.0$, 2.0, 1.0 and 0.5, respectively). Note that the value of *u* is not specified here; instead it is allowed to vary freely and is plotted on the *x*-axis. Finally, recall that the backslash (\) character is the line continuation symbol, such that lines 19 through 22 represent a single gnuplot command.

4.5.4   Function Fitting in gnuplot

In practice, rather than specifying the parameter values explicitly in the manner outlined above, the values of *c* and *k* are usually established by fitting the Weibull PDF to an observed wind speed distribution. This offers two important benefits. First, it allows the wind speed distribution at a site to be summarized and reported in terms of just two numbers: the estimated values of *c* and *k*. Second, the estimated values of *c* and *k* can be used to predict the probability with which different wind speeds are likely to occur: for instance calm conditions, when no electricity is generated, or very high winds, which might cause damage to the wind turbine.

```
initial set of free parameter values

c                   = 10
k                   = 1

After 12 iterations the fit converged.
final sum of squares of residuals : 0.000761955
rel. change during last iteration : -1.04466e-08

degrees of freedom (ndf) : 3
rms of residuals       (stdfit) = sqrt(WSSR/ndf)      : 0.0159369
variance of residuals (reduced chisquare) = WSSR/ndf : 0.000253985

Final set of parameters              Asymptotic Standard Error
=======================              =========================


c                   = 5.32883        +/- 0.3177       (5.961%)
k                   = 1.55037        +/- 0.09683      (6.246%)
```

Figure 4.13: Part of the output from the gnuplot function-fitting procedure.

It is, of course, possible to write a short gawk program to fit a mathematical function, such as the Weibull PDF, to a set of observations. Here, however, the curve-fitting capabilities of gnuplot are employed. The precise details of the function-fitting algorithm used by gnuplot need not detain us here, except to note that it makes use of least-squares adjustment (Williams and Kelly 1998). The gnuplot commands needed to achieve this are as follows:

```
set fit logfile "weibull.fit"                                       23
c=10                                                                24
k=1                                                                 25
fit phi(u,c,k) 'windfrq2.out' u 1:2 via c,k                        26
```

Line 23 specifies the name of the file into which the output from the function-fitting procedure is placed; in this example, it is weibull.fit. If a file with this name exists in the working directory, the results of the function-fitting procedure are appended to that file; otherwise, a new file is created with this name. Lines 24 and 25 assign initial (estimated) values to the two parameters, c and k; this process is directly analogous to the assignment of values to variables in gawk (see Chapter 3). Providing initial estimates of c and k in this way helps gnuplot's function-fitting algorithm to find the correction solution. Note that it is sometimes necessary to experiment with these initial values on a trial-and-error basis until a good fit is achieved. Finally, line 26 fits the function phi(u,c,k) to the data contained in fields 1 and 2 of the file windfrq2.out by adjusting the values c and k (i.e., via c,k).

The result of the function-fitting procedure is shown in Figure 4.13. This gives the estimated value of the two parameters as $c = 5.32883$ and $k = 1.55037$. Note that the exact values may differ very slightly between computers (e.g., between 32-bit and 64-bit processors). The output also provides a measure of the uncertainty associated with the estimated values of c (± 0.3177) and k (± 0.09683), which can be interpreted as confidence limits about the estimated values (Williams and Kelly 1998).

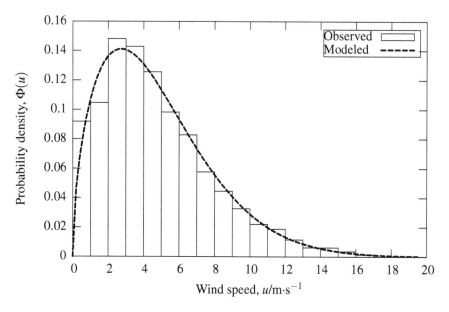

Figure 4.14: Relative frequency distribution of hourly mean wind speed at Llyn Efyrnwy in 1998 and the Weibull PDF fitted to these data ($k \approx 1.536$; $c \approx 5.295$).

The result of the function-fitting procedure can be visualized as follows:

```
set xrange [0:20]                                                          27
set yrange [0:*]                                                           28
plot 'windfrq2.out' u 1:2 t "Observed" w boxes, \                          29
     phi(u,c,k) t "Modeled" w l lt 2 lw 3                                   30
```

Lines 27 and 28 set the range of values to be displayed along the *x*- and *y*-axes, respectively. Lines 29 and 30 plot the relative frequency distribution of the observed wind speeds using the boxes style (to indicate the discrete nature of the data) and the Weibull PDF, which has been fitted to these data, using the line style (because it is a continuous function). Note that the values of c and k for Llyn Efyrnwy have already been established by the function-fitting procedure, and are passed by reference to the function phi: in other words, they do not have to be entered explicitly here. The value of u varies freely and is plotted along the *x*-axis. The result is shown in Figure 4.14.

4.5.5   Probability of the Wind Speed Exceeding a Given Value

Equation 4.2 can be modified to express the probability of the wind speed, *u*, exceeding a given value, u' (Manwell *et al.* 2002, Twidell and Weir 2006):

$$\Phi(u > u') = \exp\left[-\left(\frac{u'}{c}\right)^k\right] \tag{4.4}$$

This equation can be represented in gnuplot by another function, phi_prime:

```
phi_prime(u_prime,c,k)=exp((-1.0*(u_prime/c)**k))
```
31

where `u_prime` equates to u' in Equation 4.4. Thus, the probability of the wind speed exceeding 1 m·s^{-1} is evaluated by typing the following command in **gnuplot**:

```
print phi_prime(1.0,c,k)
```
32

Note that the values of `c` and `k` have already been established by the function-fitting procedure (line 26), and are passed by reference to the user-defined function `phi_prime`. The `print` command prints the output from this function to the screen as follows:

```
0.928000829058524
```

So, the probability of the hourly mean wind speed at Llyn Efyrnwy exceeding 1 m·s^{-1} is approximately 0.928. Put another way, this suggests that hourly mean wind speeds greater than 1 m·s^{-1} are experienced, on average, about 92.8% of the time.

In the example given above, the directive `phi_prime(1.0,c,k)` instructs **gnuplot** to evaluate the user-defined function `phi_prime(u_prime,c,k)` for `u_prime=1.0` and for the values of `c` and `k` previously estimated by the function-fitting procedure. The same result could also have been obtained by giving all three parameter values explicitly, that is:

```
print phi_prime(1.0,5.32883,1.55037)
```
33

Any value of `u_prime` can be used in this context. For instance, the probability that the wind speed exceeds 2.5 m·s^{-1} at Llyn Efyrnwy is evaluated as follows in **gnuplot**:

```
print phi_prime(2.5,c,k)
```
34

This command produces the following result:

```
0.733946698348097
```

Exercise 4.2: Assuming that the cut-in and cut-out wind speeds for the WECS at Llyn Efyrnwy are 3 m·s^{-1} and 20 m·s^{-1}, respectively, calculate the probability with which wind speeds capable of driving the WECS are experienced at this site.

4.6 WIND ENERGY AND POWER

4.6.1 Theoretical Basis

Once the wind speed distribution at Llyn Efyrnwy has been established, the next step is to estimate the power that this produces. It transpires that the energy content of the wind is proportional to the cube of the wind speed, u_0^3. This is converted into power by a wind turbine according to the following equation:

$$P = A \frac{\rho u_0^3}{2}$$
(4.5)

Table 4.6: Factors affecting the power output of a WECS.

Factor	Symbol	Units	Typical value
Power output	P	W	–
Cross-sectional area of turbine rotors	A	m^2	–
Air density	ρ	$kg \cdot m^{-3}$	1.225
Wind speed	u	$m \cdot s^{-1}$	–
Power coefficient	C_p	–	0.4
Generator efficiency	N_g	–	0.75
Mechanical efficiency	N_b	–	0.9

where P is the power of the wind measured in Watts (W), ρ is the density of dry air ($\rho = 1.225\,kg \cdot m^{-3}$ at sea level under average atmospheric pressure conditions and at 15 °C) and A is the cross-sectional area swept by the turbine rotors (Twidell and Weir 2006). The latter, of course, is given by $A = \pi r^2$ where r is the radius in meters of the rotor blades. Careful inspection of Equation 4.5 reveals that if the cross-sectional area of the turbine rotors is doubled, the wind power produced increases by a factor of two, whereas if the wind speed doubles, the power produced by the turbine increases by a factor of eight (Twidell and Weir 2006).

It is impossible to extract all of the power from a free-flowing air mass using a wind turbine, partly because the wind loses momentum as it encounters the rotor blades and partly because it must have some residual kinetic energy to move beyond the turbine (Twidell and Weir 2006). The fraction of wind power extracted varies from one turbine to another and is expressed in terms of the power coefficient, C_P, also known as the coefficient of performance (Table 4.6). The theoretical maximum value of C_P, known as the Betz limit, is 0.5926 (Betz 1926, Manwell *et al.* 2002). The value of C_P for a given turbine is generally much smaller than the Betz limit and varies with wind speed; for a well-designed turbine, C_P might peak around 0.4. There is also some loss of power via the mechanical components of the wind turbine (e.g., the gearbox and bearings) and the electricity generator. These losses are often expressed in terms of efficiency values where, for example, 1.0 indicates no loss and 0.5 indicates 50 % loss of the potentially available power. The mechanical efficiency, N_b, of turbines can differ considerably, and may be as high as 0.95. The efficiency of electricity generators, N_g, also varies, with 0.8 being a common maximum value (Manwell *et al.* 2002). Taking all of these factors together, the power output of a WECS can be expressed as follows:

$$P = A \frac{\rho u_0^3}{2} C_P N_b N_g \qquad (4.6)$$

The relationship between wind speed ($m \cdot s^{-1}$) and wind power (kW) represented by Equation 4.6 can be visualized in gnuplot using the following commands, the output from which is presented in Figure 4.15:

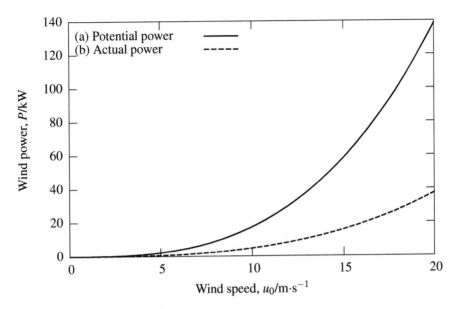

Figure 4.15: Theoretical relationship between wind power (P; kW) and wind speed (u_0; m·s^{-1}), based on assumed values of the cross-sectional area swept by the turbine rotors ($r = 3$ m, $A \approx 28.27$ m^2), air density ($\rho = 1.225$ kg·m^{-3}), power coefficient ($C_P = 0.4$), generator efficiency ($N_g = 0.75$) and mechanical efficiency ($N_b = 0.9$). (a) Potential power (Equation 4.5) and (b) actual power (Equation 4.6).

```
set xtics auto                                                              35
set xrange [0:20]                                                           36
rho=1.225               # Air density kg/m^3 at sea level                   37
radius=3.0              # Rotor radius in meters                            38
area=pi*(radius**2)     # Area swept by turbine rotors                      39
Cp=0.4                  # Power coefficient                                 40
Ng=0.75                 # Generator efficiency                             41
Nb=0.9                  # Mechanical efficiency                            42
Power(u)=area*(rho*(u**3)/2)/1000                                           43
set xlabel "Wind speed, u_0/m.s^{-1}"                                       44
set ylabel "Wind power, P/kW"                                               45
set key top left Left nobox                                                 46
plot Power(u) t "(a) Potential power" lw 2, \                               47
     Power(u)*Cp*Ng*Nb t "(b) Actual power" lw 2                            48
```

A major assumption underlying Figure 4.15 is that the value of C_P is constant at all wind speeds, whereas this is not normally the case. Nevertheless, it serves to highlight the power generation capacity of a small WECS.

A new user-defined function, Power, is declared on line 41. This function has a single argument, u (i.e., the wind speed). The other values required by Power are provided by a number of user-defined variables, which are initialized on lines 35 to 40. The value of π is given by the built-in variable pi (line 37) provided by gnuplot.

Program 4.5: wpower.awk

```
(NR >1  && $10 != -999) {                                           1
   radius = 3.0;              # Length of rotor blades              2
   pi = 3.1415927;            # Pi                                  3
   area=pi*(radius**2);       # Cross-sectional area of rotors      4
   rho = 1.225;               # Air density at sea level            5
   Cp = 0.4;                  # Power coefficient                   6
   Nb = 0.9;                  # Mechanical efficiency               7
   Ng = 0.75;                 # Generator efficiency                8
   knots2ms = 0.515;          # Convert from Kts to m/s             9
   cut_in = 3.0;              # Cut-in wind speed (m/s)            10
   cut_out = 20.0;            # Cut-out wind speed (m/s)           11
   wind_speed=$10*knots2ms;                                        12
   if (wind_speed >= cut_in && wind_speed<cut_out) {               13
      power=area*(rho*(wind_speed**3)/2)*Cp*Nb*Ng;                 14
      printf("%4i%02i%02iT%04i %8.4f\n", \                         15
        $4, $5, $6, $7, power/1000.0);                             16
   }                                                               17
}                                                                  18
```

Note that the multiplication of the various factors on lines 41 and 46, which is implicit in Equations 4.5 and 4.6, must be stated explicitly in gnuplot code using the * operator. Also note that Power outputs values in kW, rather than W, by dividing by 1000 (line 41). The resulting plot compares the potential wind power given by Equation 4.5 with the actual power output of the WECS given by Equation 4.6, taking into account the losses due to C_P, N_b and N_g. Figure 4.15 clearly shows that a relatively small fraction of the potential power of the wind is captured by the WECS.

It is also common to express the power output from a WECS in terms of power per unit area, or power density ($W \cdot m^{-2}$), by dividing the raw power values by the cross-sectional area swept by the rotor blades (A). This allows a direct comparison to be made between the efficiency of wind turbines of different size (Freris 1990, Manwell *et al.* 2002).

Exercise 4.3: Use gnuplot to create a plot of wind power density versus wind speed based on the values of A, C_P, N_b and N_g given in Table 4.6.

4.6.2   Application to Llyn Efyrnwy Data

Equation 4.6 can also be applied to the data in wind98lv.dat, using Program 4.5 (wpower.awk), by typing the following instruction on the command line:

```
gawk -f wpower.awk wind98le.dat > wind98le.pwr                      5
```

Line 1 of Program 4.5 should be familiar from previous programs: it is a pattern that is used to skip the first record of the named data file, which contains textual information as opposed to numerical data, and any record for which the hourly mean wind speed measurement is missing. Lines 2 to 9 initialize a number of the variables used in Equation 4.6. This is not the most efficient way of achieving this, because these actions are repeated for each record that is read from the named data file, but it will suffice for now. Lines 10 and 11 specify the values of the two variables (cut_in and cut_out) that are used to define the wind speeds at which electricity generation using the WECS at Llyn Efyrnwy cuts in (3 m·s^{-1}) and out (20 m·s^{-1}), respectively. Line 12 converts the hourly mean wind speed measurement for the current data record from Knots to meters per second, and stores the result temporarily in the variable wind_speed. Line 13 checks to see whether this value is greater than or equal to the cut-in wind speed and less than the cut-out wind speed. If this condition holds true for the current record, the actions on lines 14 through 16 are performed.

Line 14 calculates the power output arising from the hourly mean wind speed for the current record, taking into account the values assigned to the other variables. This is, in effect, a direct translation of Equation 4.6 into gawk code, and is very similar to the implementation of this equation in gnuplot demonstrated in the preceding section. Note that, as in gnuplot, the operator ** means "raise to the power of" in gawk. Thus, wind_speed**3 means raise the hourly mean wind speed measurement for the current record to the power of 3 (u^3).

Lines 15 and 16 use a printf statement to print out the date and time of the wind speed measurement and the mean power that this produces given the assumed properties of the WECS specified on lines 2 through 11. Recall that the printf format string indicates that the first four arguments — fields 4 ($4, year), 5 ($5, month), 6 ($6, day) and 7 ($7, time) of the current record — should be treated as integer values (%i) and the fifth — the value of the variable power divided by 1000 (i.e., power in kW·h) — as a floating-point number (%f). In addition, these lines of code instruct the gawk utility to allow a minimum width of four characters in the printout for the first and fourth arguments (%4d), two for the second and third (%2d) and eight for the fifth (%8f). The second and third arguments are also padded with leading zeroes if the values that they represent are less than two characters wide (%02d). Thus, the value 1 is printed out as 01. The fifth argument is printed out with four digits after the decimal place (%8.4f). Finally, the printout proceeds to the next line on the screen or output file (\n) after printing all of the arguments for the current record of data. The resulting output, wind98lv.pwr, is shown in Figure 4.16. Note that no electricity is generated at 3 am, 4 am, 6 am or 7 am on January 1, 1998 because the wind speed is too low.

4.6.3 Visualizing the Output

The data file, wind98lv.pwr, output by Program 4.5 can be visualized in gnuplot by entering the following instructions, all of which have been encountered previously:

```
unset key
set xdata time
set timefmt "%Y%m%dT%H%M"
```

```
19980101T0000    0.4656
19980101T0100    0.3270
19980101T0200    0.4656
19980101T0500    0.1380
19980101T0800    0.8501
19980101T0900    0.8501
19980101T1000    2.1555
19980101T1100    1.7525
19980101T1200    5.9148
19980101T1300    5.1094
19980101T1400    6.8007
19980101T1500    6.8007
```

Figure 4.16: Partial contents of the file wind981v.pwr (first 12 records).

```
set xrange ["19980101T0000":"19981231T2359"]          52
set yrange [0:*]                                      53
set format x "%b"                                     54
set xlabel "Month (1998)"                             55
set ylabel "Power, P/kW.h"                            56
plot 'wind981e.pwr' u 1:2 w i                         57
```

The result is presented in Figure 4.17. It is clear from this figure that the power output from a WECS based at Llyn Efyrnwy is likely to be variable, with output exceeding 10 kW·h on occasion, but with prolonged periods when it is much less than this, and sometimes when no power is produced at all. Despite this limitation, electricity generation is possible throughout the year and storage batteries might be used to smooth out some, if not all, of the variations in power output to meet the needs of the intended users.

It is possible to show that the average hourly power output produced over the whole year in this case study is approximately 1.2 kW. The average annual consumption of energy per household in the UK is currently about 4600 kW·h, which roughly equates to 0.5 kW per hour. Thus, all other things being equal, a small WECS based at Llyn Efyrnwy, having the characteristics outlined previously in this chapter, might be expected to meet or supplement the electricity demand of up to two households. In reality the situation is more complex, but this analysis serves as an initial guide.

4.7 SUMMARY

This chapter examines the potential for electricity generation from wind power at Llyn Efyrnwy, based on a WECS with a turbine 3 m in radius. A number of simplifying assumptions are made about energy demand and storage, as well as the performance of the hypothetical WECS. Based on these assumptions, gawk and gnuplot are used to manipulate a set of wind speed data recorded by an AWS to evaluate the magnitude and temporal variability of the wind "resource" at this site and to simulate the likely power output of the WECS. Although it was not expressed as such, these two

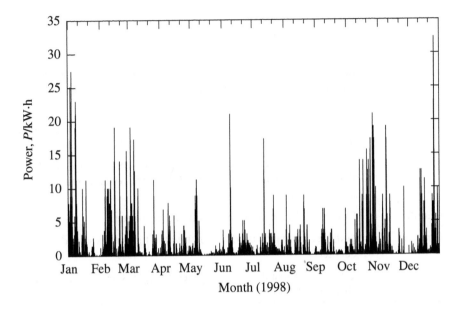

Figure 4.17: Time series of the simulated power output from a small WECS at Llyn Efyrnwy.

tasks involve the specification of a simple model of wind power generation (Equation 4.6), its implementation in computer code, and its application to environmental data recorded at Llyn Efyrnwy. This general approach is explored more extensively in the following chapter.

Exercise 4.4: Write a gawk program to calculate the average hourly power output produced over the entire year by the WECS at Llyn Efyrnwy using the data contained in the file wind98le.pwr.

SUPPORTING RESOURCES

The following resources are provided in the chapter4 sub-directory of the CD-ROM:

wind98le.dat	Hourly mean wind speed and wind direction data set for Llyn Efyrnwy, 1998, in MetOffice format
meanwspd.awk	gawk program to calculate the annual mean wind speed from the data in wind98le.dat
meanmaxw.awk	gawk program to calculate the annual mean and maximum wind speeds from the data in wind98le.dat

windfreq.awk   **gawk** program to determine the absolute frequency with which winds of different speeds occur at Llyn Efyrnwy based on the data in wind98le.dat

windfrq2.awk   **gawk** program to determine the relative frequency with which winds of different speeds occur at Llyn Efyrnwy based on the data in wind98le.dat

wind7yr.dat   Relative frequency distribution of hourly mean wind speeds measured at Llyn Efyrnwy between 1994 and 2000, inclusive

wpower.awk   **gawk** program to calculate the wind power (kW·h) at Llyn Efyrnwy throughout 1998, based on the data contained in wind98le.dat

wind.bat   Command line instructions used in this chapter

wind.plt   **gnuplot** commands used to generate the figures presented in this chapter

wind.gp   **gnuplot** commands used to generate figures presented in this chapter as EPS files

Chapter 5

Solar Radiation at Earth's Surface

Topics

- Environmental significance of solar radiation incident on Earth's surface

- Global, direct and diffuse solar irradiance

- Radiation laws and the electromagnetic spectrum

- Measuring and modeling solar irradiance

Methods and techniques

- Actions performed *before* reading data from a file — the BEGIN block

- Passing parameter values to gawk programs via the command line

- Trigonometric functions in gawk

- More control-flow constructs in gawk — the for loop

5.1 INTRODUCTION

Energy is emitted from the sun in the form of electromagnetic radiation. This energy drives Earth's climate and its environmental systems (Beer *et al.* 2000, IPCC 2001). Thus, for example, Earth's atmosphere, oceans and land surface are heated by the absorption of a fraction of the shortwave solar radiation that is incident upon them. This radiation is subsequently emitted at longer wavelengths causing the atmosphere to be heated further, depending on the balance between the rates at which it absorbs and emits longwave radiation (McGuffie and Henderson-Sellers 1997). The resulting spatial differences in atmospheric heating generate momentum in the air mass, which among other things produces winds at Earth's surface (Twidell and Weir 2006).

Incident solar radiation, in conjunction with air temperature, also plays a key role in the growth and development of vegetation through the process of photosynthesis (Monteith and Unsworth 1995). The latter involves the assimilation of carbon into a range of complex organic molecules, including carbohydrates such as sugars, starch and cellulose (Jones 1983, Campbell and Norman 1998); these organic molecules are subsequently converted into plant tissue, commonly referred to as biomass. The assimilated carbon is drawn down (or "sequestered") from the atmosphere in the form of CO_2. The production of vegetation biomass therefore represents an important "sink" for this major greenhouse gas (IPCC 2001). Consequently, some countries plan to establish forests to sequester atmospheric CO_2 as part of their commitment to the Kyoto Protocol. This represents one element of so-called emission reduction credit (ERC) trading, which allows major new sources of greenhouse gases to be offset by emission reductions elsewhere.

Incident solar radiation is also an important source of renewable energy in its own right. In this context, it can be "harvested" to heat water for domestic or industrial use, to warm air for space heating or for drying crops, and to generate electricity using photovoltaic cells (Twidell and Weir 2006).

The amount of solar radiation incident on Earth's surface varies both spatially and with the time of day and year. The spatial variation is a function of differences in latitude and various properties of the local terrain, such as its altitude, gradient and aspect. The temporal variation is governed by the angle of the sun above the horizon, the turbidity of the atmosphere and the degree of cloud cover.

This chapter explores the temporal variation of incident solar radiation at Llyn Efyrnwy using both *in situ* measurements and a computational model. The former provides a further example of how gawk and gnuplot can be employed to analyze environmental data sets. The latter illustrates how a computational model can be constructed from published mathematical formulae and used to simulate the temporal trends in incident solar radiation observed at Llyn Efyrnwy. Importantly, the model provides a general tool that can, in principle, be applied to other sites, including those for which *in situ* measurements are unavailable.

5.2 DESCRIPTION OF THE SOLAR IRRADIANCE DATA

The exercises covered in this chapter make use of the data file radt981e.dat, which is included on the CD-ROM. This file contains measurements of the total broadband solar radiation incident at Llyn Efyrnwy, also known as the global solar irradiance. These data were recorded at hourly intervals throughout 1998 by an AWS operated by the MetOffice (see Figure 5.1 and Table 5.1). The solar irradiance values are given in Watt hours per square meter ($W{\cdot}h{\cdot}m^{-2}$). This is a non-standard unit of energy, but one that is nevertheless widely used; it describes the solar power in Watts (W) expended over a period of one hour (h) per square meter of Earth's surface (m^{-2}). The data were obtained from the BADC web site in ASCII text format.

ID	IDTYPE	MET_DOM	YEAR	MON	DAY	END_HOUR	COUNT	GLOBAL	DIFFUSE	DIRECT
7948	DCNN	ESAWRADT	1998	1	1	0	1	0	-999	-999
7948	DCNN	ESAWRADT	1998	1	1	100	1	0	-999	-999
7948	DCNN	ESAWRADT	1998	1	1	200	1	0	-999	-999
7948	DCNN	ESAWRADT	1998	1	1	300	1	0	-999	-999
7948	DCNN	ESAWRADT	1998	1	1	400	1	0	-999	-999
7948	DCNN	ESAWRADT	1998	1	1	500	1	0	-999	-999
7948	DCNN	ESAWRADT	1998	1	1	600	1	0	-999	-999
7948	DCNN	ESAWRADT	1998	1	1	700	1	0	-999	-999
7948	DCNN	ESAWRADT	1998	1	1	800	1	-1	-999	-999
7948	DCNN	ESAWRADT	1998	1	1	900	1	13	-999	-999
7948	DCNN	ESAWRADT	1998	1	1	1000	1	44	-999	-999
7948	DCNN	ESAWRADT	1998	1	1	1100	1	24	-999	-999
7948	DCNN	ESAWRADT	1998	1	1	1200	1	27	-999	-999
7948	DCNN	ESAWRADT	1998	1	1	1300	1	33	-999	-999
7948	DCNN	ESAWRADT	1998	1	1	1400	1	13	-999	-999
7948	DCNN	ESAWRADT	1998	1	1	1500	1	8	-999	-999
7948	DCNN	ESAWRADT	1998	1	1	1600	1	5	-999	-999
7948	DCNN	ESAWRADT	1998	1	1	1700	1	2	-999	-999
7948	DCNN	ESAWRADT	1998	1	1	1800	1	2	-999	-999
7948	DCNN	ESAWRADT	1998	1	1	1900	1	0	-999	-999
7948	DCNN	ESAWRADT	1998	1	1	2000	1	0	-999	-999
7948	DCNN	ESAWRADT	1998	1	1	2100	1	0	-999	-999
7948	DCNN	ESAWRADT	1998	1	1	2200	1	0	-999	-999
7948	DCNN	ESAWRADT	1998	1	1	2300	1	0	-999	-999

Figure 5.1: Partial contents of the file radt981e.dat (first 25 records).

Table 5.1: Explanation of the data fields in `radt98lv.dat`.

Field	Name	Contents
1	ID	Meteorological station ID number
2	IDTYPE	Meteorological station type
3	MET_DOM	Meteorological domain
4	YEAR	Year
5	MON	Month (1–12)
6	DAY	Day of the month (1–31)
7	END_HOUR	Hour of observation (00:00–23:00 GMT)
8	COUNT	Number of hours over which measurements are made
9	GLOBAL	Global solar irradiance ($W \cdot h \cdot m^{-2}$)
10	DIFFUSE	Diffuse solar irradiance ($W \cdot h \cdot m^{-2}$)
11	DIRECT	Direct solar irradiance ($W \cdot h \cdot m^{-2}$)

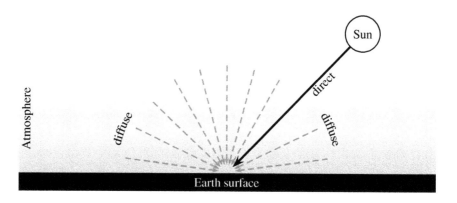

Figure 5.2: Direct and diffuse solar irradiance.

Table 5.1 describes the contents of the 11 fields of data in `radt98le.dat`. Note that fields 9 through 11 contain three separate components of the solar irradiance: global, diffuse and direct (Figure 5.2). Direct irradiance is that which arrives at a point on Earth's surface from the direction of the sun; it is dominated by the direct solar beam (Monteith and Unsworth 1995). Diffuse irradiance is that which arrives from all other directions in the hemisphere above the surface, having been scattered out of the direct solar beam by clouds, gases and particulate matter within the atmosphere, or by reflection from adjacent areas of terrain (Monteith and Unsworth 1995). Global irradiance is the sum of the direct and diffuse components. In this context, therefore, global irradiance refers to the total amount of solar radiation incident at a particular location on Earth's surface, and not to the amount received by the planet as a whole.

Program 5.1: selradt.awk

```
(NR >1 && $9 >=0){                                          1
    printf ("%4i%02i%02iT%04i  %4i\n", $4, $5, $6, $7, $9);  2
}                                                           3
```

```
19980101T0000    0
19980101T0100    0
19980101T0200    0
19980101T0300    0
19980101T0400    0
```

Figure 5.3: First five lines of the output file, radt981e.out, produced using Program 5.1.

Approximately 70 meteorological stations in the UK report solar irradiance data. Most record hourly measurements of global and diffuse irradiance, from which the direct component can be computed by subtraction. Just three stations in the UK report data on direct irradiance. The AWS at Llyn Efyrnwy provides only global irradiance values (field 9 in radt981e.dat), so that fields 10 (diffuse irradiance) and 11 (direct irradiance) of radt981e.dat are filled with the "missing data" value (-999).

5.3 ANALYZING THE OBSERVATIONS

5.3.1 Data Extraction and Pre-Processing

Program 5.1 is designed to extract data on the global solar irradiance at Llyn Efyrnwy, as well as the dates and times at which these observations were made, from the file radt981e.dat. The code is very similar to that employed in the preceding chapters. Line 1 consists of a pattern that matches each record in the data file with the exception of the first (NR>1), provided that the value in the ninth field of the record is greater than or equal to zero ($9>=0). Thus, the pattern skips the first record of the data file, which contains the field names, and discards any other record in which the global solar irradiance value is less than zero. Negative values of global solar irradiance are indicative of missing data (-999) and of unwanted artifacts that are produced by the irradiance sensor in the AWS under very low light conditions (e.g., at dawn and dusk). Provided that the current record matches both parts of the pattern, the corresponding action (line 2) is performed. The action statement prints the values in the fourth (year), fifth (month), sixth (day), seventh (time) and ninth (global solar irradiance) fields of the relevant records in the specified format.

Program 5.1 can be run from the command line as follows:

```
gawk -f selradt.awk radt981e.dat > radt981e.out          1
```

The first few lines of the output file, radt981e.out, are shown in Figure 5.3.

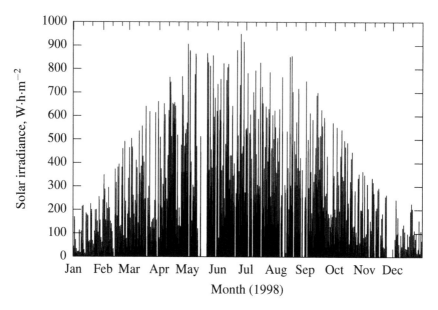

Figure 5.4: Hourly variation in total solar irradiance at Llyn Efyrnwy throughout 1998.

5.3.2 Visualizing the Output

The data contained in the file radt98le.out can be visualized by typing the following commands in gnuplot:

```
reset                                                           1
unset key                                                       2
set xdata time                                                  3
set timefmt "%Y%m%dT%H%M"                                       4
set format x "%b"                                               5
set xlabel "Month (1998)"                                       6
set ylabel "Solar irradiance, W.h.m^-2"                         7
set xrange ["19980101T0000":"19990101T0000"]                    8
plot 'radt98le.out' u 1:2 w i                                   9
```

These commands should be familiar from the preceding chapters. They produce the plot shown in Figure 5.4, which shows the variation in global solar irradiance at Llyn Efyrnwy throughout 1998. There is a clear seasonal component to this variation, as one might expect, with the largest irradiance values recorded during the northern hemisphere summer months, especially between May and August, when the sun is highest in the sky. By comparison, the daily maximum values are much smaller in the northern hemisphere winter months, most notably between November and February, when the sun is lower in the sky. It should also be apparent that there are two extended periods of missing data: one in May, the other at the end of November and the beginning of December. Superimposed on the seasonal trend, there is also significant variation over the diurnal cycle, as well as from day to day, although these effects are difficult to discern in Figure 5.4.

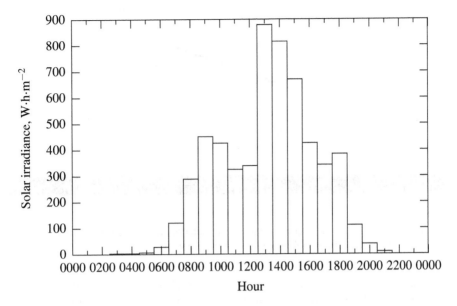

Figure 5.5: Diurnal variation in total solar irradiance at Llyn Efyrnwy on June 21, 1998.

The diurnal variation can be seen more clearly by focusing on measurements made over the course of a single day. For example, gnuplot can be instructed to plot only those data recorded on June 21, 1998, the date of the northern hemisphere summer solstice, by issuing the following commands:

```
set format x "%H%M"                                    10
set xlabel "Hour"                                      11
set xrange ["19980621T0000":"19980621T2359"]           12
plot 'radt98le.out' u 1:2 w boxes                      13
```

These commands produce the output shown in Figure 5.5. Line 10, for example, sets the labels used for each of the tic-marks along the x-axis so that these indicate the time, in hours (%H) and minutes (%M), at which the measurements were made. Line 11 updates the x-axis label accordingly. Line 12 constrains the range of values to be plotted on the x-axis, displaying only those measurements recorded between 00:00 and 23:59 Greenwich Mean Time (GMT) on June 21, 1998. Note that the time and date information specified on this line is placed inside a pair of double quotation marks and is given in the format used in the data file radt98le.out. Finally, line 13 plots the data using the boxes style, taking account of the preceding instructions.

The pattern of variation evident in Figure 5.5 broadly follows the trend that one might expect over the course of a diurnal cycle, with the maximum value of solar irradiance recorded around solar noon, when the sun is at its highest point in the sky, decreasing systematically toward dawn (around 03:00 GMT) and dusk (around 21:00 GMT). The simple diurnal trend is modified, however, by the effect of cloud cover. Clouds affect the solar irradiance signal in two main ways (Figure 5.6). First, they attenuate (i.e., reduce) the direct solar beam by obscuring the solar disk; in general,

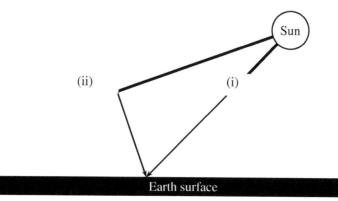

Figure 5.6: Impact of cloud cover on incident solar radiation. (i) Attenuation of the direct solar beam; (ii) reflection from clouds over adjacent terrain.

the greater the cloud cover, the more the direct irradiance is likely to be attenuated. Second, clouds located over adjacent areas of terrain reflect solar radiation onto the target area, increasing the diffuse irradiance that it receives. The trend presented in Figure 5.5 suggests that cloud cover may have reduced the direct solar irradiance, particularly between 10:00 and 12:00 GMT.

Exercise 5.1: Make appropriate modifications to Program 5.1 so that it prints out the global solar irradiance values recorded by the AWS at Llyn Efyrnwy each day at solar noon only (i.e., 12:00 GMT). Run the revised program and redirect the output to a new file. Plot the data in the output file using gnuplot. Comment on the observed trend.

5.4   MODELING SOLAR IRRADIANCE

Globally, the number of sites at which solar irradiance data are routinely collected is very limited. Even in the UK, a comparatively well-instrumented nation, only 70 or so meteorological stations regularly record data of this type on behalf of the MetOffice; this compares to roughly 1200 sites that contribute standard hourly weather observations. The coverage elsewhere in the world is typically much less comprehensive (Gueymard 2003). As a consequence, it is often the case that *in situ* measurements of solar irradiance are unavailable for the site of interest. There are two possible courses of action in these circumstances. The first is to establish a dedicated meteorological station, with the equipment necessary to measure solar irradiance, at the chosen location; this is not always viable, though, for reasons of accessibility and cost. The second is to construct a simulation model that can be used to estimate the solar irradiance at the site. This approach is explored in the remainder of this section.

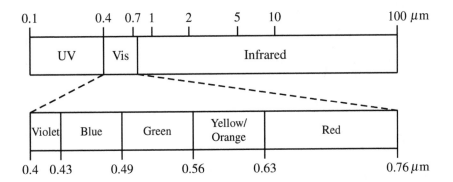

Figure 5.7: Part of the electromagnetic spectrum, showing the major wavelength regions, including an expanded representation of the visible spectrum.

5.4.1 Formulating the Mathematical Model

One of the first steps involved in building an environmental model is to identify the mathematical formulae on which it will be based. Suitable formulae will often have been developed in previous studies reported in the scientific literature and these can be used to construct the mathematical model. Sometimes alternative formulations exist for parts of the model, in which case the task becomes one of selecting the most appropriate formulae and of assembling these into a single coherent model. The selection criteria one might employ in this context include keeping the complexity of the mathematics concerned to a minimum, limiting the number of parameters in the model for which input data are required, and constraining the computational load required to run the model. Many of these issues are considered in the subsections that follow, after a brief introduction to the fundamental principles of solar radiation.

Fundamental Principles of Solar Radiation

The sun emits energy in the form of electromagnetic radiation over a continuous range of wavelengths known as the electromagnetic spectrum (Figure 5.7). The sun's behavior in this respect is similar to that of a perfect (or "blackbody") radiator at a temperature of approximately 5800 K (Campbell and Norman 1998), where K refers to kelvin, a temperature scale closely related to degrees Celsius (K = °C + 273.15). Blackbody radiators absorb and subsequently emit all of the radiation incident upon them. The amount of energy emitted by a blackbody radiator varies with wavelength according to Planck's equation,

$$M(\lambda) = \frac{2\pi hc^2}{\lambda^5 (e^{\frac{hc}{\lambda kT}} - 1)} \qquad (5.1)$$

where λ is the wavelength of the radiation, measured in meters, h is Planck's constant ($h = 6.626 \times 10^{-34}$ J·s where J stands for Joule, a unit of energy), c is the speed of light ($c \approx 3 \times 10^8$ m·s^{-1}), k is Boltzmann's constant ($k = 1.3807 \times 10^{-23}$ J·K^{-1}), T is the

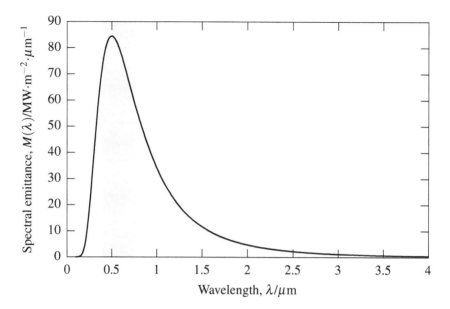

Figure 5.8: Spectral distribution of the radiant energy emitted by a blackbody radiator at 5800 K (e.g., the sun) in Mega-Watts per square meter per micrometer ($MW \cdot m^{-2} \cdot \mu m^{-1}$). The region shaded in gray indicates the visible spectrum.

temperature of the blackbody in kelvin and $M(\lambda)$ is known as the spectral emittance of the blackbody.

Equation 5.1 produces the emittance curve shown in Figure 5.8 for a blackbody radiator at a temperature of 5800 K and for wavelengths up to $4\,\mu m$, where μm denotes a micrometer (i.e., one millionth of a meter). Note that the part of the electromagnetic spectrum that is visible to the human eye lies roughly between $0.4\,\mu m$ and $0.76\,\mu m$ (Figure 5.7), which corresponds closely to the region of maximum emission in the solar spectrum (Figure 5.8).

The exact wavelength of maximum emission by a blackbody radiator is given by Wien's displacement law,

$$\lambda_{max} = \frac{2897}{T} \tag{5.2}$$

where λ_{max} is the wavelength of peak emittance in micrometers. Assuming that the sun has an average surface temperature of 5800 K, $\lambda_{max} = 2897/5800 \approx 0.5\,\mu m$, which corresponds to the part of the electromagnetic spectrum that the human eye perceives as green light. The average temperature of Earth's surface, on the other hand, is roughly 287 K ($\sim 13\,°C$). Consequently, its emission spectrum peaks around $10\,\mu m$, in a part of the spectrum known as the infrared.

The total amount of radiation emitted by a blackbody radiator across all wavelengths, M, is given by the Stefan-Boltzmann equation,

$$M = \sigma T^4 \tag{5.3}$$

where σ is the Stefan-Boltzmann constant ($\sigma = 5.67 \times 10^{-8}\,\text{W} \cdot \text{m}^{-2} \cdot \text{K}^{-4}$). Hence, assuming that the sun behaves like a blackbody radiator at a temperature of 5800 K,

$$
\begin{aligned}
M_{\text{sun}} &\approx (5.67 \times 10^{-8}\,\text{W} \cdot \text{m}^{-2} \cdot \text{K}^{-4}) \times (5800\,\text{K})^4 \\
&\approx 6.4165 \times 10^{7}\,\text{W} \cdot \text{m}^{-2} \qquad\qquad\qquad (5.4)\\
&\approx 64\,\text{MW} \cdot \text{m}^{-2}
\end{aligned}
$$

Note that this value ($64\,\text{MW} \cdot \text{m}^{-2}$) is equivalent to the area under the curve in Figure 5.8. By the same token, assuming that Earth behaves like a blackbody radiator at a temperature of 287 K,

$$
\begin{aligned}
M_{\text{earth}} &\approx (5.67 \times 10^{-8}\,\text{W} \cdot \text{m}^{-2} \cdot \text{K}^{-4}) \times (287\,\text{K})^4 \qquad (5.5)\\
&\approx 385\,\text{W} \cdot \text{m}^{-2}
\end{aligned}
$$

This value is many orders of magnitude smaller than M_{sun}, but it is highly significant in terms of its contribution to warming Earth's atmosphere.

Effect of the Earth–Sun Distance

Because of its finite size and its large distance from the sun, Earth receives only a small fraction of the total radiation emitted by the sun; this fraction is governed by the inverse square law (Schott 1997, Schowengerdt 1997), which states that the intensity of the solar radiation incident on a surface perpendicular to the sun's rays is inversely proportional to the square of its distance from the sun. This can be expressed mathematically as follows:

$$
E_0(\lambda) = M(\lambda) \left(\frac{r_{\text{sun}}}{d} \right)^2 \qquad\qquad\qquad (5.6)
$$

where $E_0(\lambda)$ is the solar radiation incident at the top of Earth's atmosphere, known as the exo-atmospheric solar spectral irradiance, $M(\lambda)$ is the spectral emittance of the sun, r_{sun} is the radius of the sun ($r \approx 6.96 \times 10^8\,\text{m}$) and d is the mean distance between Earth and the sun ($d \approx 1.4 \times 10^{11}\,\text{m}$). Applying Equation 5.6 to the data contained in Figure 5.8 yields the curve shown in Figure 5.9.

It is possible to calculate the total exo-atmospheric solar irradiance across all wavelengths, E_0, by replacing $M(\lambda)$ with M_{sun} in Equation 5.6 as follows:

$$
\begin{aligned}
E_0 &= M_{\text{sun}} \left(\frac{r_{\text{sun}}}{d} \right)^2 \\
&\approx 6.4165 \times 10^7\,\text{W} \cdot \text{m}^{-2} \left(\frac{6.96 \times 10^8\,\text{m}}{1.5 \times 10^{11}\,\text{m}} \right)^2 \qquad (5.7)\\
&\approx 1380\,\text{W} \cdot \text{m}^{-2}
\end{aligned}
$$

This value describes the area beneath the curve in Figure 5.9 and is commonly known as the solar constant, although this name is somewhat misleading because E_0 varies slightly during the course of the solar cycle and more significantly over longer periods of time (Lean 1991, Lee *et al.* 1995, Crommelynck and Dewitte 1997). The term total solar irradiance (TSI) is rather more apposite. Regardless of the name used to refer to

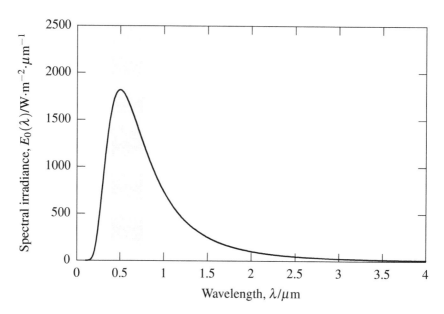

Figure 5.9: Solar spectral irradiance at the top of the atmosphere, assuming that the sun behaves as a blackbody radiator at 5800 K. The region shaded in gray indicates the extent of the visible spectrum.

this quantity, its value is fundamental to understanding global environmental change because it defines the magnitude of the external "forcing" on Earth's climate system.

For various reasons, the actual solar spectrum at the top of Earth's atmosphere is more complex than that shown in Figure 5.9. Numerous reference spectra have been produced to represent this complexity (see Figure 5.10). Coverage of the derivation of these spectra is beyond the scope of this book, except to note that they are based on data obtained from a range of sources, including sensors carried aboard Earth-orbiting satellites, space shuttle missions and high-altitude aircraft (Fligge *et al.* 2001, Gueymard *et al.* 2002).

Attenuation of Incident Solar Radiation by Earth's Atmosphere

Only a fraction of the total exo-atmospheric solar irradiance reaches Earth's surface (Figure 5.10). Some is absorbed by gases and particulate matter, the latter commonly known as aerosols, on its downward path through the atmosphere; some is scattered back out to space by aerosols and clouds (Figure 5.11). The combined effect of these two processes is to attenuate the incident solar radiation as it passes down through the atmosphere. The degree of attenuation depends on the composition of the atmosphere, in terms of its gaseous constituents and the concentrations of different types of aerosol, and on the path length that the solar radiation has to travel through the atmosphere to the ground. The path length is, in turn, dependent on the altitude of the terrain and the angle of the sun above the horizon, which is a function of the latitude of the site and the time of day and year.

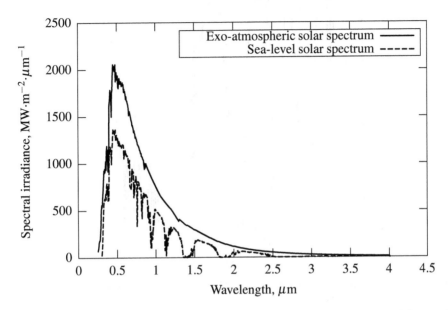

Figure 5.10: Reference solar spectral irradiance at the top of the atmosphere (solid line) and at sea level (dashed line) for a 1.5 air mass (AM1.5) atmospheric path length. (Source of data: http://rredc.nrel.gov/solar/standards/am0/.)

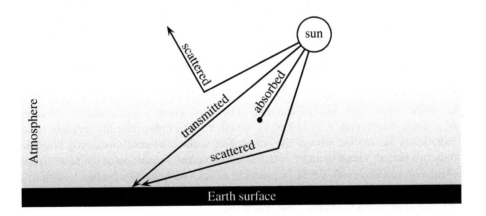

Figure 5.11: Scattering, absorption and transmission of solar radiation on its passage down through Earth's atmosphere.

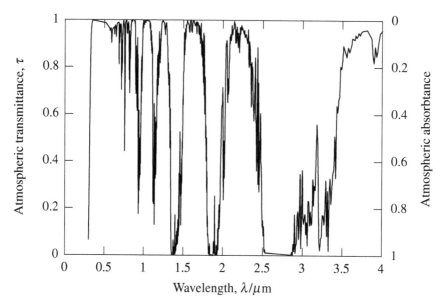

Figure 5.12: Gaseous transmission through Earth's atmosphere as a function of wavelength. (Source of data: Vermote *et al.* 1997.)

The fraction of incident solar radiation that is absorbed by atmospheric gases varies as a function of wavelength (Vermote *et al.* 1997, Figure 5.12), but is relatively stable over space and time. By contrast, the types and amounts of aerosols within the atmosphere, sometimes referred to as the aerosol mass loading, vary considerably from one location to another and over relatively short periods of time. The main natural sources of atmospheric aerosols are dust from volcanic eruptions and desert storms, as well as smoke from forest fires. The principal anthropogenic sources are sulfates produced by the combustion of fossil fuels, such as oil and coal, and smoke resulting from the intentional burning of vegetation biomass.

Accurate representation of the passage of incident solar radiation through Earth's atmosphere requires the use of comprehensive radiative transfer models, such as MODTRAN (Kneizys *et al.* 1996) and 6S (Vermote *et al.* 1997), which are beyond the scope of this book. Instead, the state of the atmosphere is characterized by means of a single coefficient, τ $(0 \leq \tau \leq 1)$, where $\tau = 1$ implies that the incident solar radiation is transmitted through the atmosphere without attenuation (strictly possible only in a perfect vacuum) and $\tau = 0$ denotes complete attenuation of the solar signal (Monteith and Unsworth 1995). In practice, τ varies with wavelength but, to simplify matters further, a single value of τ is used here to describe the average atmospheric transmission across the entire solar spectrum.

Estimating the Direct and Diffuse Components of the Solar Irradiance

Several methods for calculating the direct and diffuse solar radiation incident on a horizontal element of Earth's surface are reported in the scientific literature. The

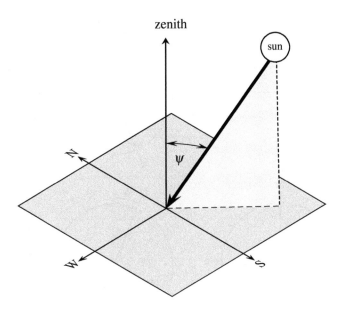

Figure 5.13: Solar zenith angle, ψ, with respect to a flat horizontal surface.

mathematical formulations given by Campbell and Norman (1998), adapted from Liu and Jordan (1960), are used here because of their relative simplicity. Specifically,

$$I_{direct} = E_0 \tau^m \cos \psi \tag{5.8}$$

$$I_{diffuse} = 0.3 \left(1 - \tau^m\right) E_0 \cos \psi \tag{5.9}$$

$$I_{global} = I_{direct} + I_{diffuse} \tag{5.10}$$

where I_{direct}, $I_{diffuse}$ and I_{global} are, respectively, the direct, diffuse and global broad-band solar irradiance at ground level, E_0 is the total exo-atmospheric solar irradiance, τ is the atmospheric transmittance, m is the air mass number and ψ is the solar zenith angle (i.e., the angle of the sun relative to a point directly above the observer; Figure 5.13).

For the purpose of this chapter, E_0 is assumed to be 1380 W·m^{-2} (Equation 5.7) and τ is assumed to be 0.7. The latter value is representative of clear-sky conditions. According to Campbell and Norman (1998), the air mass number, m, is given by

$$m = \frac{p_{alt}}{p_{sea}} \cdot \frac{1}{\cos \psi} \tag{5.11}$$

for $\psi \leq 80°$; where p_{alt} and p_{sea} are the atmospheric pressure at the altitude of the study site and at sea level, respectively, and ψ is the solar zenith angle.

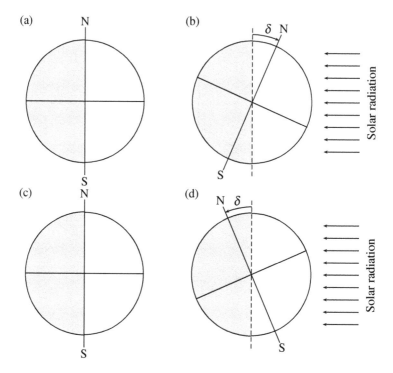

Figure 5.14: Variation in the solar declination angle at four times of the year: (a) March 21 ($\delta = 0°$), (b) June 21 ($\delta = +23.4°$), (c) September 21 ($\delta = 0°$) and (d) December 21 ($\delta = -23.4°$). Light areas indicate sunlit parts of Earth's surface. (After Twidell and Weir 2006.)

Calculating the Solar Zenith Angle and the Solar Declination Angle

To evaluate Equations 5.8 and 5.9, the solar zenith angle, ψ, must either be known or be calculated. The solar zenith angle varies with latitude, time of year and time of day as follows:

$$\cos \psi = \sin \phi_L \sin \delta + \cos \phi_L \cos \delta \cos \theta \qquad (5.12)$$

where ϕ_L is the latitude of the study site ($\phi_L = 52.756°$N at Llyn Efyrnwy), δ is the solar declination angle and θ is the hour angle of the sun (Twidell and Weir 2006).

   The solar declination angle describes the angle of the plane of the sun with respect to Earth's equator (Figure 5.14). This angle varies as a function of the time of year between approximately $+23.4°$ on June 21 (i.e., the northern hemisphere summer solstice) and $-23.4°$ on December 21 (i.e., the northern hemisphere winter solstice; Figure 5.15). Once again, numerous methods for calculating δ are reported in the literature but approximate values, suitable for use here, may be obtained from the following equation:

$$\delta \approx -23.4 \cos \left(\frac{360(\text{DoY} + 10)}{365} \right) \qquad (5.13)$$

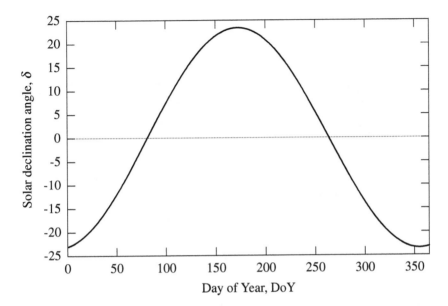

Figure 5.15: Variation in the solar declination angle, δ, as a function of the day of year, DoY.

where DoY is the day of year (Jones 1983), and where DoY $= 1$ on January 1, DoY $=$ 2 on January 2, and so on. More accurate values of the solar declination angle can be obtained from the standard meteorological tables (List 2000).

The hour angle of the sun, θ, describes the angle through which Earth has rotated with respect to local solar noon (Twidell and Weir 2006). Since Earth rotates $360°$ every 24 hours, or $15°$ per hour, approximate values of θ are given by the following equation:

$$\theta \approx 15(12 - h) \tag{5.14}$$

where h is the local solar time in hours, ranging from 0 to 24 (Figure 5.16), such that $\theta = 0°$ at local solar noon.

5.4.2 Implementing the Computational Model

Once an appropriate formulation for the mathematical model has been selected, the next step is to convert the equations into gawk code; that is, to implement the computational model. This process involves a translation between the vocabulary, grammar and syntax of one language (i.e., the symbols and equations of mathematics) and those of another (i.e., the programming language gawk). Just as in the translation between natural languages, such as English and French, there is rarely a single "correct" solution, although some solutions might be regarded as being intrinsically better than others. The flexibility of gawk means that the implementation can often be realized in several ways. In some instances, the differences may amount to little more than a matter of programming style, broadly analogous to the variations in the style of

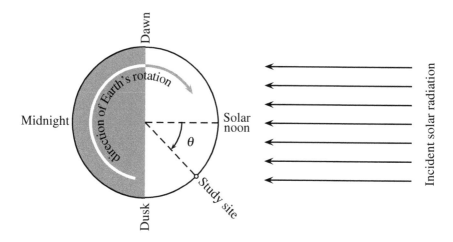

Figure 5.16: Solar hour angle, θ, as though viewing Earth vertically downward from a point
directly above the North Pole.

written prose among literary authors. On other occasions, however, the differences
may have a significant impact in terms of the efficiency of the resulting code, such
as its computational speed or the demands that it places on various aspects of the
computer's resources, including memory usage and hard-disk access. Thus, some
implementations may be considered to be more elegant or otherwise preferable to
others.

Program 5.2, `solarrad.awk`, represents a fairly straightforward implementation
of the mathematical equations outlined above, which tries to balance brevity versus
clarity of the code.

The BEGIN *Block*

The whole of Program 5.2 is set within a BEGIN block, or BEGIN rule, which is a special
feature of the **gawk** programming language. The BEGIN block is used to group together
a set of statements that are processed *before* those contained in the main pattern-
action block (i.e., before reading data from a file or files; Figure 5.17); it is analogous
to the END block, introduced in Chapter 4, which is used to group statements that are
processed *after* the main pattern-action block has been completed. In this particular
example, however, the program does not contain a main pattern-action block, only
a BEGIN block; this is because the aim is to generate estimated values of the solar
irradiance at Llyn Efyrnwy, rather than to process data input from a file. This type of
computational model is sometimes referred to as a simulation model.

The BEGIN block commences on line 1 of Program 5.2 with the keyword BEGIN,
which must be typed entirely in upper case and must be followed immediately, on
the same line, by an opening curly brace ({). The BEGIN block ends on line 35 with
a closing curly brace (}). The **gawk** statements between the curly braces (lines 2
through 34) constitute the body of the BEGIN block.

Program 5.2: solarrad.awk

```
BEGIN{                                                                   1
   latitude=52.756;           # Latitude (degrees)                      2
   E_0=1380.0;                # Exo-atmos. solar irradiance             3
   tau=0.7;                   # Atmos. transmittance                    4
   hour_angle=0;              # Solar hour angle (degrees)              5
   DOY=1;                     # Day of year                            6
   p_alt=1000;                # Atmos. pressure (altitude)             7
   p_sea=1013;                # Atmos. pressure (sea level)            8
                                                                        9
   deg2rad=(2*3.1415927)/360; # Degrees to radians                     10
                                                                       11
   latitude*=deg2rad;         # Latitude in radians                    12
   hour_angle*=deg2rad;       # Hour angle in radians                   13
                                                                       14
   # Solar declination angle (Eq. 5.13)                               15
   declination = (-23.4*deg2rad)* \                                    16
      cos(deg2rad*(360*(DOY+10)/365));                                 17
                                                                       18
   # Cosine of the solar zenith angle (Eq. 5.12)                      19
   cos_zenith=sin(latitude)*sin(declination)+ \                       20
      cos(latitude)*cos(declination)*cos(hour_angle);                 21
                                                                       22
   # Atmospheric air mass (Eq. 5.11)                                  23
   air_mass=(p_alt/p_sea)/cos_zenith;                                 24
                                                                       25
   # Direct, diffuse and global (total) solar                         26
   # irradiance (Eq. 5.8, 5.9 and 5.10)                               27
   I_direct=(E_0)*(tau**air_mass)*cos_zenith;                         28
   I_diffuse=0.3*(1-(tau**air_mass))*(E_0)*cos_zenith;                29
   I_global=I_direct+I_diffuse;                                       30
                                                                       31
   # Output results                                                   32
   printf("%3i %7.3f %7.3f %7.3f\n", \                                33
      DOY, I_global, I_direct, I_diffuse);                            34
}                                                                      35
```

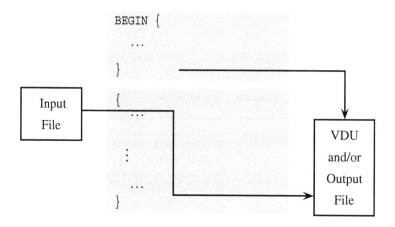

Figure 5.17: gawk's BEGIN and main pattern-action blocks.

Naming Conventions for Variables

Lines 2 through 8 of Program 5.2 initialize a number of variables that are used within the program. The names given to these variables are intended, as far as possible, to be indicative of the corresponding terms in the mathematical equations (see Table 5.2). Several of the variable names have been constructed from two words joined together by an underscore symbol (_). In the case of p_{alt}, for example, p and alt are conjoined to form p_alt. This convention is quite commonly used, although other formats for variable names are also employed. As a general rule, the use of very short variable names, such as x, y and z, is not recommended because they are not self-explanatory. By the same token, very long variable names are best avoided because they tend to inhibit, rather than improve, the readability of the code. Above all, it is important to be consistent when choosing names for variables, whichever convention is adopted. Also remember, as far as gawk is concerned, p_alt, p_Alt and P_alt are three entirely separate variables that may store completely different values.

Note that Program 5.2 uses fixed values for several of the key parameters (lines 2 through 8). More specifically, it is assumes that $\phi_L = 52.756°$, $E_0 = 1380\,\text{W·m}^{-2}$, $\tau = 0.7$, $\theta = 0°$ and DoY $= 1$. Thus, Program 5.2 simulates the direct and diffuse broadband solar irradiance received at Llyn Efyrnwy ($\phi_L = 52.756°$) under clear-sky conditions ($\tau = 0.7$) at solar noon ($\theta = 0°$) on the first day of January (DoY $= 1$). Similarly, the atmospheric pressure at the site (p_{alt}) and at sea level (p_{sea}) is assumed to be 1000 mb and 1013 mb, respectively, where mb denotes pressure in millibars (lines 7 and 8). This is a reasonable assumption in the circumstances, although the code could be modified to use data on atmospheric pressure recorded by the AWS.

Trigonometric Functions in gawk

One characteristic of **gawk**, which is common to other programming languages used for scientific computing, is that the arguments to trigonometric functions (e.g., sin,

Table 5.2: Symbols used in the mathematical model (Equations 5.7 through 5.14) and variables used in the computational model (Program 5.2) of solar irradiance at Llyn Efyrnwy.

Term	Symbol	Variable	Equations
Latitude	ϕ_L	`latitude`	5.12
Total exo-atmospheric solar irradiance	E_0	`E_0`	5.7
Atmospheric transmittance	τ	`tau`	5.8, 5.9
Atmospheric pressure at altitude of site	p_{alt}	`p_alt`	5.11
Atmospheric pressure at sea level	p_{sea}	`p_sea`	5.11
Solar hour angle	θ	`hour_angle`	5.12, 5.14
Day of year	DoY	`DOY`	5.13
Solar declination angle	δ	`declination`	5.12, 5.13
Cosine of solar zenith angle	$\cos(\psi)$	`cos_zenith`	5.8, 5.9, 5.12
Air mass number	m	`air_mass`	5.8, 5.9, 5.11
Direct solar irradiance	I_{direct}	`I_direct`	5.8
Diffuse solar irradiance	$I_{diffuse}$	`I_diffuse`	5.9
Global solar irradiance	I_{global}	`I_global`	5.10

cos and tan) must be expressed in radians. Several of the values that are given as input to the model, however, are expressed in arc degrees (e.g., the latitude of Llyn Efyrnwy and the solar hour angle). Consequently, a multiplication factor is established on line 10 of Program 5.2, which is used to convert between arc degrees and radians. Since there are 2π radians and 360° in a circle, the conversion factor is simply $2\pi/360$; where $\pi \approx 3.1415927$. This value is stored in the variable deg2rad. Thus, lines 12 and 13 multiply the original values of the variables latitude and hour_angle by deg2rad to convert them into radians. Recall that latitude*=deg2rad is merely a more concise form of the statement latitude=latitude*deg2rad.

Converting Equations into Code

Equation 5.13, which calculates the solar declination angle on any day of the year, is implemented on lines 16 and 17 of Program 5.2. This single **gawk** statement is split over two lines of code for presentation purposes using the line continuation symbol (\), which is the last character on line 16. The conversion factor deg2rad is used twice in this statement to transform the angles concerned from arc degrees to radians. Also note that the multiplication operator (×), which is implicit in Equation 5.13, must be entered explicitly using the * symbol in the corresponding **gawk** code; for example, $360(\text{DoY} + 10)$ in the equation becomes 360*(DOY+10) in the code.

Lines 20 and 21 represent an implementation of Equation 5.12, which is used to determine the cosine of the solar zenith angle. One of the terms in this equation is the solar declination angle (δ or declination), which explains why that value must be calculated first (i.e., on lines 16 and 17). Once again, this single long gawk statement is split over two lines of the program for presentation purposes; in all other ways lines 20 and 21 represent a direct translation from Equation 5.12 into gawk code.

Line 24 implements Equation 5.11. This line of code therefore calculates the air mass number given the cosine of the solar zenith angle (lines 20 and 21) and the atmospheric pressure at the site and at sea level (lines 7 and 8).

Lines 28 through 30 calculate the direct, diffuse and global broadband solar irradiance at Llyn Efyrnwy based on Equations 5.8 through 5.10. Recall that in gawk the operator ** means "raise to the power of", so that the term τ^m in Equations 5.8 and 5.9 can be expressed as tau**air_mass (lines 28 and 29). Finally, lines 33 and 34 print out the resulting values for the global, direct and diffuse irradiance and the day of year in the chosen format.

Running the Computational Model

Program 5.2 can be run from the command line as follows:

```
gawk -f solarrad.awk > solarrad.out
```
2

This command line produces the following output:

```
1 158.991   81.425   77.566
```

Thus, the model indicates that approximately 159 W·m^{-2} of global solar irradiance is received at Llyn Efyrnwy under clear-sky conditions at solar noon on January 1. This total comprises roughly 81.5 W·m^{-2} of direct solar flux and 77.5 W·m^{-2} of diffuse solar flux.

Strictly speaking, the results output by the model are instantaneous values; that is, the model predicts the solar irradiance at a given instant in time (i.e., solar noon). Consequently, the model output is not directly comparable with the measurements made by the AWS, which are integrated over the preceding hour. For the purpose of this exercise, however, it is assumed that the instantaneous values generated by the model represent the mean irradiance conditions during the preceding hour. This is a reasonable first approximation in most instances. Given this assumption, it should be noted that the predicted value of total solar irradiance (159 W·h·m^{-2}) is much higher than the measured value recorded by the AWS at solar noon on January 1, 1998 (27 W·h·m^{-2}), possibly because of the prevailing cloud and atmospheric conditions on that day, but it is quite similar to the value recorded by the AWS at the same time on the following day (169 W·m^{-2}).

5.4.3   Enhancing the Implementation

Entering Parameter Values via the Command Line

One of the limitations of Program 5.2 is that the file containing the code must be opened, edited and saved using a text editor to alter the value of any of the variables. To simulate the global solar irradiance at solar noon for every day of the year, for example, the code would have to be edited 365 times, changing the value of DOY (line 6) on each occasion. This is clearly unsatisfactory. A better approach would be to assign values to selected variables when the model is run, via the command line for instance.

gawk allows the value of one or more variables to be specified on the command line using the -v switch (or flag), which has the general syntax -v `variable=value`. For example, the following command line assigns the variable DOY the value 2 at the start of the program:

```
gawk  -f  solarrad.awk  -v  DOY=2
```
3

To take advantage of this facility, however, line 6 must be removed from Program 5.2; otherwise, it would override the value of DOY entered on the command line.

Control-Flow Constructs in gawk

The ability to initialize a variable on the command line, in the manner outlined above, is a very useful feature of **gawk**, one that will be used extensively throughout this book. Nevertheless, it does relatively little to help where the objective is to simulate the variation in solar irradiance throughout the year. Even using the -v flag, the program must still be run 365 times in its current configuration, with a different value of DOY entered via the command line on each occasion. Clearly, this is a very time-consuming process and one that is likely to be error-prone. A more efficient approach would be to get the computer to do the hard work by building this procedure into the program itself, so that the value of DOY is varied automatically between 1 and 365. gawk provides a number of simple mechanisms to perform this type of repetitive task. These mechanisms are known as looping or control-flow constructs.

gawk has three control-flow constructs: the for, while and do while statements. Only the first of these is examined here; the other two are introduced in subsequent chapters. The for statement performs a particular sequence of operations a given number of times. It has the following general syntax (Robbins 2001):

```
for(initialization; condition; increment){
     body
}
```

The following fragment of **gawk** code, which is represented graphically in Figure 5.18, presents a practical example of a for loop:

```
for(DOY=1;  DOY<=365;  ++DOY){
     body
}
```

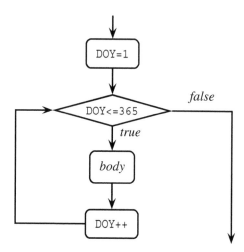

Figure 5.18: A `for` loop designed to perform a set of instructions contained in the body 365 times. The value of the variable DOY increases by 1 each time, from 1 to 365.

This code is intended to perform the set of instructions contained in the body of the loop (which is not specified here) 365 times, using a different value for the variable DOY on each occasion, commencing with 1, culminating with 365, and increasing in steps of 1. Thus, the code fragment starts by initializing the value DOY to 1 (DOY=1). The code fragment then assesses whether the value of DOY is less than or equal to 365 (DOY<=365); this is known as the condition. If the condition DOY<=365 is true, the code in the body is executed. The value of DOY is then increased by 1 (++DOY), so that DOY=2 subsequently. The condition is then re-evaluated using the new value of DOY. If the condition remains true (i.e., the value of DOY is still less than or equal to 365), the code in the body of the loop is executed again taking into account the new value of DOY, and so on. The variable DOY is, in effect, used to count the number of times the loop has been traversed thus far, and a variable used in this way is sometimes known as a counter. The program continues to loop around these statements until the condition is false (i.e., the value of DOY exceeds 365). At this point the program either proceeds to the next line of code after the `for` loop or, if there are no further lines of code, stops.

Program 5.3 (solarrd2.awk) presents a modified version of solarrad.awk that makes use of the example `for` loop outlined above. The loop starts on line 15 with the keyword `for`, the control statements and an opening curly brace, and ends on line 35 with a closing curly brace. Lines 16 through 34 of this program therefore constitute the body of the loop. These have been indented by an extra tab space to make it clear to the reader where the loop begins and ends. The other difference with respect to Program 5.2 is that the value of DOY is no longer initialized in the first few lines of the program, since this is now performed by the `for` loop (line 15 in Program 5.3). Program 5.3 can be run from the command line as follows:

```
gawk -f solarrd2.awk > solarrd2.out
```

Program 5.3: solarrd2.awk

```
BEGIN{                                                              1
    latitude=52.756;            # Latitude (degrees)               2
    E_0=1380.0;                 # Exo-atmos. solar irradiance      3
    tau=0.7;                    # Atmos. transmittance             4
    hour_angle=0;               # Solar hour angle (degrees)       5
    p_alt=1000;                 # Atmos. pressure (altitude)       6
    p_sea=1013;                 # Atmos. pressure (sea level)      7
                                                                   8
    deg2rad=(2*3.1415927)/360;  # Degrees to radians               9
                                                                   10
    latitude*=deg2rad;          # Latitude in radians              11
    hour_angle*=deg2rad;        # hour angle in radians            12
                                                                   13
    # for-loop to cycle through each day of year                   14
    for(DOY=1;DOY<=365;++DOY){                                     15
        # Solar declination angle (Eq. 5.13)                       16
        declination = (-23.4*deg2rad)* \                           17
            cos(deg2rad*(360*(DOY+10)/365));                       18
                                                                   19
        # Cosine of the solar zenith angle (Eq. 5.12)              20
        cos_zenith=sin(latitude)*sin(declination)+ \               21
            cos(latitude)*cos(declination)*cos(hour_angle);        22
                                                                   23
        # Atmospheric air mass (Eq. 5.11)                          24
        air_mass=(p_alt/p_sea)/cos_zenith;                         25
                                                                   26
        # Direct, diffuse and global (total) solar                 27
        # irradiance (Eq. 5.8, 5.9 and 5.10)                       28
        I_direct=(E_0)*(tau**air_mass)*cos_zenith;                 29
        I_diffuse=0.3*(1-(tau**air_mass))*(E_0)*cos_zenith;        30
        I_global=I_direct+I_diffuse;                               31
                                                                   32
        # Output results                                           33
        printf("%3i %7.3f %7.3f %7.3f\n", \                        34
            DOY, I_global, I_direct, I_diffuse);                   35
    }                                                              36
}                                                                  37
```

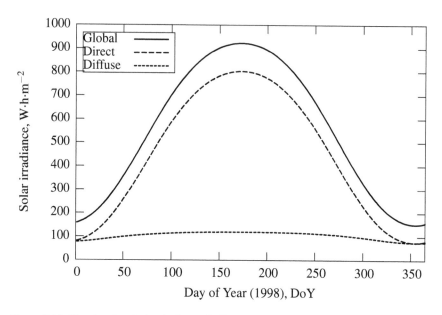

Figure 5.19: Simulated variation in the total, direct and diffuse solar irradiance as a function of the day of year (DoY).

5.4.4   Visualizing the Simulated Variation in Solar Irradiance

The data contained in the output file, solarrd2.out, can be visualized in gnuplot (Figure 5.19) by entering the following commands:

```
reset                                                              14
set xrange [0:365]                                                 15
set style data lines                                               16
set key top left Left box                                          17
set xlabel "Day of Year (1998), DoY"                               18
set ylabel "Solar irradiance, W.h.m^-2"                            19
plot 'solarrd2.out' u 1:2 t "Global" lw 2, \                       20
     'solarrd2.out' u 1:3 t "Direct" lw 2, \                       21
     'solarrd2.out' u 1:4 t "Diffuse" lw 2                         22
```

Line 14 instructs gnuplot to reset all of the plotting parameters (e.g., the axis labels, plotting ranges, data types and so on) to their default values. Line 15 sets the range of values plotted on the *x*-axis, commencing at 0 (i.e., the start of the year) and ending at 365 (i.e., the final day of the year). Line 16 sets the default data style. Line 17 places a key in the top left-hand corner of the plot, enclosed in a box in which the text is left-justified. Lines 18 and 19 define the labels to be placed along the *x*- and *y*-axes, respectively. Lines 20 through 22 plot the data for the day of year (*x*-axis) against the global, direct and diffuse irradiance values (*y*-axis), respectively. These data are contained in the first, second, third and fourth fields of the file solarrd2.out. Note that the final gnuplot command is split over three lines (20 through 22) for presentation purposes using the line continuation symbol ($\backslash$).

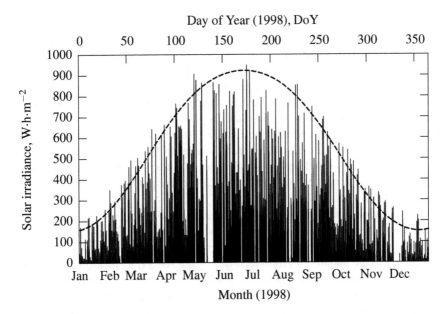

Figure 5.20: Comparison between observed (vertical impulses) and simulated (solid line) total
solar irradiance as a function of the time of year.

The seasonal pattern evident in Figure 5.19 is similar to that seen in Figure 5.5.
Moreover, it appears that the majority of the variation in global solar irradiance is
accounted for by changes in the level of direct solar irradiance, which is a function
of the solar zenith angle (Equation 5.8). The diffuse solar irradiance displays the
same basic trend, but the range of values is much smaller. Indeed, a reasonable first
approximation would be to assume that the diffuse component is constant (roughly
$100\,\mathrm{W\cdot h\cdot m^{-2}}$) throughout the year on clear days (Campbell and Norman 1998).

It is possible to make a direct comparison between the measured and simulated
variation in total solar irradiance by plotting the two data sets together (Figure 5.20).
This can be achieved by entering the following commands in **gnuplot**:

```
unset key                                                          23
set xdata time                                                     24
set timefmt "%Y%m%dT%H%M"                                          25
set format x "%b"                                                  26
set xlabel "Month (1998)"                                          27
set x2label "Day of Year (1998), DoY"                              28
set xrange ["19980101T0000":"19990101T0000"]                       29
set x2range [0:365]                                                30
set xtics nomirror                                                 31
set x2tics nomirror auto                                           32
plot 'radt98le.out' u 1:2 w i, \                                   33
     'solarrd2.out' u 1:2 w l axes x2y1 lw 2                       34
```

Line 23 removes the key from the figure. Line 24 indicates that the data plotted on the
first *x*-axis (i.e., the measured irradiance values) should be treated as a time series, and

line 25 indicates the format in which the time and date information are given. Line 26 indicates that the labels for the tic-marks on the first x-axis should be given by the shortened name of the month (i.e., Jan, Feb, ... , Dec). Lines 27 and 28 specify the labels that should be placed along the first (lower) and second (upper) x-axes, while lines 29 and 30 control the range of values to be plotted on each of these axes. Lines 31 and 32 ensure that the tic-marks used on each of the x-axes are not mirrored on the other one (i.e., only one set of tic-marks is plotted on each x-axis). Finally, lines 33 and 34 plot the measured and simulated total solar irradiance values: the former are read from the file `radt981e.out`, created earlier in this chapter, and are plotted using the `impluses` data style on the standard (`x1y1`) axes; the latter are read from the file `solarrd2.out`, generated using Program 5.3, and are plotted using the `lines` data style on the `x2y1` axes.

The observations presented in Figure 5.20 are recorded at hourly intervals throughout the day, whereas the simulation model estimates the irradiance values at solar noon only. Taking this into account, it appears that the estimated values correspond quite closely to the maximum values recorded by the AWS (i.e., measurements likely to be made at, or close to, solar noon) on some days. It is likely that the atmosphere was particularly clear and cloud-free on these occasions, consistent with the assumptions made in the computational model (i.e., $\tau = 0.7$). The lower solar irradiance values measured by the AWS on the other days may result from either a more turbid atmosphere (i.e., $\tau < 0.7$) or the influence of cloud cover. These possibilities could be tested by examining data on atmospheric visibility and cloud cover recorded by the AWS, but this is left as an exercise for the reader.

Exercise 5.2: Make appropriate modifications to Program 5.3 so that it simulates the total solar irradiance at Llyn Efyrnwy as a function of the time of day (hour) over a single day of the year. Assume that $E_0 = 1380 \text{W·m}^{-2}$, $\tau = 0.7$, $\phi_L = 52.756°$, $p_{alt} = 1000 \text{mb}$ and $p_{sea} = 1013 \text{mb}$. Run the revised computational model for June 21, 1998 (DoY $= 172$). Use the data that this generates to produce a plot similar to the one shown in Figure 5.20, which presents a comparison between the observed and the simulated hourly variation in total solar irradiance. Comment on the likely reasons for the differences between the observed and simulated responses.

5.5   SUMMARY

This chapter explores the temporal variation of incident solar radiation received at Llyn Efyrnwy, using *in situ* measurements and a simulation model. The simulation model is constructed from mathematical formulae that are published in the scientific literature. Alternative mathematical formulations exist for various parts of the model considered here; in each case, the simplest of these is selected. Sometimes, however, the appropriate mathematical formulae will not have been developed previously, or else the existing formulations may be considered too complex, may require too many input values, or too much computing power, for the task at hand. In these circumstances one may be required to formulate a new mathematical model *ab initio*. This is exemplified in the following chapter, which illustrates how one might begin with an extremely simple mathematical model, gradually enhancing this as necessary so that its output conforms ever more closely to observations of the environmental process or system that it is intended to represent.

SUPPORTING RESOURCES

The following resources are provided in the `chapter5` sub-directory of the CD-ROM:

`radt981e.dat`	Hourly mean wind speed and wind direction data set for Llyn Efyrnwy, 1998, in MetOffice format
`selradt.awk`	**gawk** program to extract the measurements of global solar irradiance from `radt981e.dat` and output them in a format suitable to be plotted using **gnuplot**
`solarrad.awk`	**gawk** program to estimate the global, direct and diffuse solar irradiance received at Llyn Efyrnwy at solar noon on a specific day of the year
`solarrd2.awk`	**gawk** program to estimate the global, direct and diffuse solar irradiance received at Llyn Efyrnwy at solar noon on each day of the year
`solar.bat`	Command line instructions used in this chapter
`solar.plt`	**gnuplot** commands used to generate the figures presented in this chapter
`solar.gp`	**gnuplot** commands used to generate figures presented in this chapter as EPS files

Chapter 6

Light Interaction with a Plant Canopy

Topics

- Light interaction with plant canopies

- Significance of light interaction in energy budget and plant growth studies, and for Earth observation

- Developing a model of light interaction with a plant canopy

- Stages in the model development cycle: specification, formulation, implementation and evaluation

- Representing the multiple scattering of light within a plant canopy

6.1 INTRODUCTION

In the preceding chapter, attention was focused on the amount of solar radiation that reaches Earth's surface, using both measurements and models. This chapter builds on those foundations, examining how incident solar radiation interacts with different materials on Earth's surface. Emphasis is placed on plant canopies, in part because of the important dynamic role that they play in Earth's climate system and, in particular, the global carbon cycle.

Knowledge of the principles and processes underlying radiation interaction with Earth's surface is central to the study of many environmental issues, including those concerned with Earth's energy budget (McGuffie and Henderson-Sellers 1997). In that context, solar radiation incident on Earth's surface is ultimately either absorbed

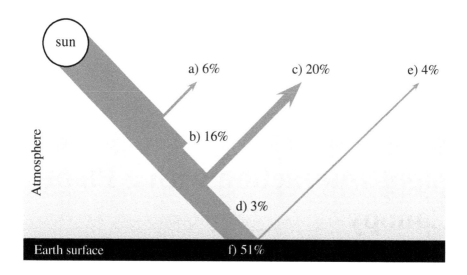

Figure 6.1: Earth's shortwave radiation budget, showing the approximate percentages of the incident solar radiation that are (a) reflected by the atmosphere, (b) absorbed by the atmosphere, (c) reflected by clouds, (d) absorbed by clouds, (e) reflected by the land and ocean surface and (f) absorbed by the land and ocean surface.

by the surface or reflected from it, the latter passing back up through the atmosphere and out to space (Figure 6.1). The balance between absorption and reflection is an important factor in Earth's climate system (McGuffie and Henderson-Sellers 1997), in part because the absorbed fraction is subsequently re-radiated at longer wavelengths, contributing to the warming of Earth's atmosphere (Figure 6.2).

Earth's surface is composed of various types of material, including vegetation, soil, water, snow and ice, which differ in terms of physical and chemical composition. As a result, each reflects and absorbs different fractions of the incident solar radiation. The fractions reflected and absorbed also vary as a function of wavelength. Healthy green vegetation, for example, reflects very little of the incident solar radiation at blue ($\sim 0.46 \mu$m) and red ($\sim 0.69 \mu$m) wavelengths because plant pigments, particularly chlorophyll a and b, absorb strongly at these wavelengths (Figure 6.3). By contrast, vegetation generally reflects a slightly larger fraction of the incident solar radiation at green wavelengths ($\sim 0.55 \mu$m) and almost half of the incoming solar radiation at near infra-red (NIR) wavelengths ($\sim 0.85 \mu$m) (Gausman 1977, Tucker and Garratt 1977, Grant 1987, Jacquemoud and Baret 1990). The latter is due to the greatly reduced absorption by, and hence the intense scattering of radiation within and between, plant leaves in the NIR. Thus, the spatial disposition of surface materials, together with their spectral reflectance and absorptance characteristics, plays an important role in determining Earth's energy balance.

A fraction of the incident solar radiation at wavelengths between 0.4μm and 0.7μm is absorbed by vegetation for photosynthesis (Campbell and Norman 1998). This quantity is often referred to as the fraction of absorbed photosynthetically ac-

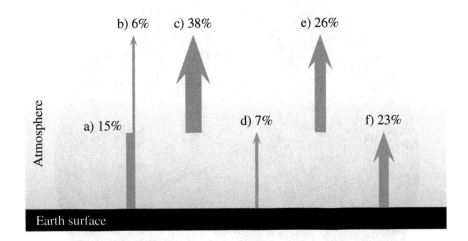

Figure 6.2: Earth's longwave radiation budget, showing the approximate percentages of long-wave radiation emission from the surface that are (a) absorbed by atmospheric gases or (b) escape to space, (c) the net emission by atmospheric gases, (d) the sensible heat flux from the surface, (e) the net emission by clouds and (f) the latent heat flux from the surface.

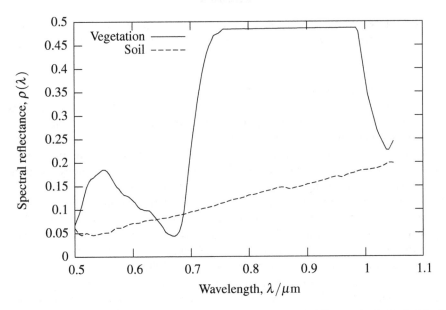

Figure 6.3: Typical spectral reflectance curves for vegetation (solid line) and soil (dashed line) in the range $0.5 \, \mu$m (blue/green) to $1.05 \, \mu$m (near-infrared).

Figure 6.4: Upward-looking hemispherical photograph taken from within a deciduous tree canopy, giving an indication of the horizontal and vertical distribution of light within the canopy.

tive radiation (fAPAR). In conjunction with temperature and the availability of water and nutrients, the fAPAR governs the growth and development of vegetation canopies (Jones 1983, Monteith and Unsworth 1995). The fAPAR is a function of the amount of above-ground plant material (biomass) and its 3D structure (i.e., the size and shape of the leaves and the other plant parts, their inclination and orientation in terms of zenith and azimuth angles, and their spatial distribution within the canopy). In turn, the fAPAR controls the uptake of CO_2 by plants through photosynthesis, as well as the release of water by evapotranspiration. Understanding the interception of solar radiation by vegetation canopies is therefore central to studies of the global carbon cycle and carbon sequestration (Lal *et al.* 1999, Cao *et al.* 2003), of plant growth and crop yield (Russell *et al.* 1989, Bouman 1992, Goudriaan and van Laar 1994) and of hydrometeorology (Ward and Robinson 1999, Beven 2001).

The vertical and horizontal distribution of light within a vegetation canopy is also an important ecological factor, which has significant implications for the strategies adopted by plants in terms of their competition for the available light resource (Figure 6.4). Thus, some species seek competitive advantage by growing tall quickly to gain access to the maximum amount of solar radiation, which is found at the top of the canopy (Whittaker 1975). These species, however, must use much of the energy that they derive from photosynthesis in the production of woody biomass for the stem (or trunk) and branches, which are required to support their foliage. In contrast, other species may become adapted to life at the base of the canopy, where the level of available light is generally greatly reduced, but much less effort has to be put into the production of woody biomass (Whittaker 1975). They may also be able to take advantage of light that penetrates through gaps in the crown of the canopy, which produces "sun flecks" at its base.

Lastly, there is considerable interest in the interaction of solar radiation with plant canopies in the context of satellite remote sensing (Goel 1987, Gao and Lesht 1997, Pinty and Verstraete 1998). This arises from the need to develop inventories of the amount, condition and productivity of vegetation, in both managed and unmanaged landscapes, at regional and global scales (Barnsley 1999). Among other things, this information is required by the dynamic global vegetation models (DGVMs) and land-surface parameterization schemes incorporated within the present generation of global climate models (Woodward *et al.* 1995, Sellers *et al.* 1996, Friend *et al.* 1997, Cox *et al.* 1998, Cramer *et al.* 2001). Data from sensors mounted aboard Earth-orbiting satellites can assist in initializing these models and in constraining their output through the use of data assimilation techniques. These data describe the amount of solar radiation reflected from Earth's surface, however, rather than the properties of interest. Turning them into useful information therefore demands a model that relates specific properties of the land surface to measurements of reflected radiation, and *vice versa*. Once again, the development of such models requires an understanding of how solar radiation interacts with Earth's surface.

6.2 DEVELOPING A MODEL OF LIGHT INTERACTION WITH PLANT CANOPIES

A model was constructed in the preceding chapter to simulate the amount of solar radiation incident on Earth's surface. The model is based on mathematical equations that are widely reported in the relevant scientific literature. A similar approach could easily be adopted in the present chapter, where the aim is to represent the interaction of solar radiation with a plant canopy. There are various candidate formulae and models from which to choose in this context; see, for example, Verhoef (1984), Camillo (1987), Goel (1987), Verstraete *et al.* (1990), Gao (1993), Gastellu-Etchegorry *et al.* (1996), Kuusk (1996) and North (1996). There are many other situations, however, in which appropriate mathematical formulae and models have yet to be developed, or where the existing formulations are considered to be too complex, or else demand too much input data or too many computational resources, for the task at hand. In these circumstances, one must formulate the mathematical model from scratch (*ab initio*). To illustrate what this involves, existing models of radiation interaction with plant canopies are eschewed here. Instead, a new model is developed from first principles. The primary objective in doing so is to highlight the main stages involved in model development and, more specifically, to demonstrate that the model-building process often requires several iterations around the cycle of model specification, formulation, implementation and evaluation. Thus, a very simple model is constructed at first. The sophistication of the model is subsequently increased so that its output conforms ever more closely to reality. The reader might like to note that the final version of the model reported in this chapter is similar to the "Adding method" model developed by Cooper *et al.* (1982).

6.2.1 Specifying the Conceptual Model

The first step in building a model is to incorporate knowledge of the environmental processes and parameters involved, together with any assumptions, approximations and simplifications that need to be made, in a conceptual model. The reasons for doing so are twofold. First, it clarifies the nature and purpose of the model, not least in the mind of the modeler. Second, it provides information to help potential users of the model understand its scope and limitations, to challenge its underlying assumptions, approximations and simplifications, and to develop improved versions of the model in the future.

Here, the conceptual model is summarized by the following three statements:

1. Different materials on Earth's surface reflect different fractions of the incident solar radiation (i.e., reflectance varies according to the type of surface material).

2. The fraction of incident solar radiation reflected by a given surface material is dependent on the wavelength of the radiation (i.e., reflectance also varies spectrally).

3. A given area of Earth's surface may be covered by a mixture of materials.

In short, the fraction of incident solar radiation that is reflected from a given area of Earth's surface is a function of both the wavelength of the radiation and the relative proportions of ground covered by different surface materials.

Four simplifying assumptions are also made, namely that (i) the sun is located directly overhead, (ii) the surface area is flat and level, (iii) radiation travels either vertically downward from the sun to Earth's surface or vertically upward from Earth's surface into space, and (iv) the atmosphere has no appreciable effect on either the downwelling or the upwelling radiation. It should be evident that these are quite drastic simplifications. Nevertheless, they help to focus attention on the core elements of the problem, ignoring for now the more complex effects produced by variations in the angle of the sun, the 3D structure of the surface materials, and the physical and chemical composition of the atmosphere. These can, of course, be added into the model at a later stage, if so required.

6.2.2 Formulating the Mathematical Model

The second stage is to represent the conceptual model in mathematical terms, using functions and equations. This is sometimes known as formulating the mathematical model because it involves translating the conceptual model into mathematical formulae (Edwards and Hamson 1989). It is also frequently the most challenging stage in

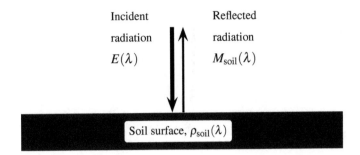

Figure 6.5: Reflection from a soil surface. See text for details.

the development of a model. Sometimes this is because the solution to the problem demands the use of advanced mathematical techniques (although each of the models developed here requires only basic skills in algebra). More often it is because there are several different ways in which the system can be represented mathematically and it is not immediately apparent which approach is best. Consequently, deriving a suitable mathematical formulation is often a trial-and-error process, but it is also a skill that improves with practice. Finally, it is worth noting that some models cannot be expressed mathematically, at least not in concise analytical terms. These must therefore be solved numerically. This issue will not be considered further here, but it is addressed extensively in Chapter 7.

The ability to "see" a model in mathematical terms is often aided by the use of diagrams to represent the various component elements of the model system and the relationships between them. This is frequently achieved using Forrester diagrams (Forrester 1961, 1969, 1973), which represent the structure of a system in terms of sources, sinks, reservoirs (also known as pools or stocks), flows (or processes), converters and inter-relationships (also known as links or connectors). These are widely used to study system dynamics in ecological modeling (Ford 1999, Deaton and Winebrake 2000). Here, though, a less formal approach is adopted, starting with a simple schematic representation of the reflection of solar radiation from a bare soil surface (Figure 6.5).

Reflection from a Soil Surface

Figure 6.5 shows two fluxes (or streams) of radiation. These are (i) the incident flux, which travels vertically downward from the sun to the soil surface and (ii) the reflected flux, which travels vertically upward from the soil surface to space. Note that the arrows are intended to represent aggregate fluxes across the whole of the soil surface, not just the locations to and from which they point. Strictly speaking, therefore, they should be referred to as flux densities because they describe radiant fluxes over a given surface area (Monteith and Unsworth 1995). The incident flux density is usually known as irradiance (E; see also Chapter 5), while the reflected flux density is referred to as radiant exitance (M). Both are measured in units of Watts per square meter (W·m^{-2}).

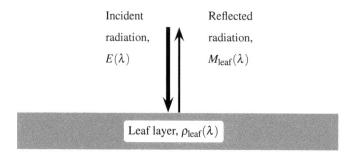

Figure 6.6: Reflection from a layer of leaves.

Irradiance and radiant exitance typically vary as a function of the wavelength of the radiation, which is denoted by λ. Thus, the terms spectral irradiance, $E(\lambda)$, and spectral radiant exitance, $M(\lambda)$, are used to refer to values of these quantities at a specific wavelength or measured over a narrow range of wavelengths, known as a waveband. Both are measured in units of Watts per square meter per micrometer $(\text{W·m}^{-2}\text{·}\mu\text{m}^{-1})$. Recall from the preceding chapter that a micrometer, μm, is a unit of wavelength denoting one millionth of a meter.

The fraction of incident solar radiation that is reflected from a soil surface is, by definition, governed by the reflectance of the soil. This is, in turn, a function of its physical and chemical composition. The reflectance of soil also varies according to wavelength, which can be expressed as follows:

$$\rho_{\text{soil}}(\lambda) = \frac{M_{\text{soil}}(\lambda)}{E(\lambda)} \tag{6.1}$$

where ρ denotes reflectance. It should be evident from this that the spectral reflectance of soil, $\rho_{\text{soil}}(\lambda)$, must lie in the range 0 to 1 $(0 \leq \rho_{\text{soil}}(\lambda) \leq 1)$, because it is physically impossible for the amount of radiation reflected from the soil surface at a given wavelength to be negative (which would imply that the soil absorbs more radiation than is incident upon it) or to exceed unity (which would imply that it reflects more radiation than it receives). Equation 6.1 can be rearranged to yield the amount of radiation reflected from the soil given values of $E(\lambda)$ and $\rho_{\text{soil}}(\lambda)$, as follows:

$$M_{\text{soil}}(\lambda) = E(\lambda) \times \rho_{\text{soil}}(\lambda) \tag{6.2}$$

Reflection from a Leaf Layer

Now consider another area of ground that is covered entirely by a layer of healthy green leaves through which no soil is visible from above (Figure 6.6). Assume for now that the leaves are sufficiently thick and closely spaced, without overlapping, so that none of the incident solar radiation penetrates down to the soil substrate in which the vegetation grows. In these circumstances, the fraction of incident solar radiation that is reflected from the leaf layer is governed by the spectral reflectance

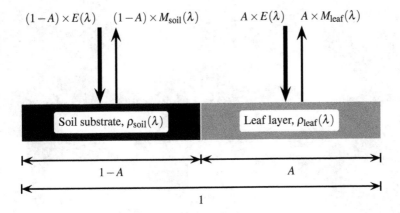

Figure 6.7: Reflection from a mixed soil and leaf surface.

of the leaves, $\rho_{\text{leaf}}(\lambda)$. This is, in turn, a function of their physical and chemical composition. The spectral reflectance of the leaves can therefore be expressed as

$$\rho_{\text{leaf}}(\lambda) = \frac{M_{\text{leaf}}(\lambda)}{E(\lambda)} \tag{6.3}$$

and hence

$$M_{\text{leaf}}(\lambda) = E(\lambda) \times \rho_{\text{leaf}}(\lambda) \tag{6.4}$$

where the value of $\rho_{\text{leaf}}(\lambda)$ lies in the range 0 to 1 ($0 \leq \rho_{\text{leaf}}(\lambda) \leq 1$).

Reflection from a Mixed Soil and Leaf Surface

Now consider the case where part of the surface area is covered by a dense layer of leaves, while the remainder is exposed soil (Figure 6.7). Assume that the fractional area of ground covered by leaves is A, where $0 \leq A \leq 1$, so that the fractional area of exposed soil is $1 - A$. This is a reasonable first approximation to a simple row crop at a relatively early stage of growth (Figure 6.8).

If solar radiation at a given wavelength, $E(\lambda)$, is incident uniformly across the entire area, the fraction of the radiation that is reflected from the surface as a whole can be written as follows:

$$\rho_{\text{surface}}(\lambda) = \frac{M_{\text{surface}}(\lambda)}{E(\lambda)} \tag{6.5}$$

where

$$M_{\text{surface}}(\lambda) = ((1 - A) \times M_{\text{soil}}(\lambda)) + (A \times M_{\text{leaf}}(\lambda)) \tag{6.6}$$

Figure 6.8: Photograph of a sugar beet crop, showing the mixture of leaves (dark areas) and soil (light areas) visible when the canopy is viewed vertically downward (image size approximately $1.4\,\text{m} \times 0.8\,\text{m}$).

since the soil covers $1 - A$, and the leaves cover A, of the total surface area. Moreover, substituting Equations 6.2 and 6.4 into Equation 6.6 gives

$$M_{\text{surface}}(\lambda) = (A \times E(\lambda) \times \rho_{\text{leaf}}(\lambda)) + ((1 - A) \times E(\lambda) \times \rho_{\text{soil}}(\lambda)) \qquad (6.7)$$

Equation 6.7 states that the amount of solar radiation incident on the leaf layer is proportional to the area of ground that its covers, A. It then reflects a fraction, $\rho_{\text{leaf}}(\lambda)$, of this radiation back out to space. The total amount of radiation reflected from the leaf layer is therefore $A \times E(\lambda) \times \rho_{\text{leaf}}(\lambda)$. Likewise, the soil receives some fraction of the total incident solar radiation, $E(\lambda)$, proportional to the area that it covers, $(1 - A)$, and reflects a fraction of this, $\rho_{\text{soil}}(\lambda)$, back out to space, which equates to $(1 - A) \times E(\lambda) \times \rho_{\text{soil}}(\lambda)$. The total amount of radiation reflected from the surface as a whole is therefore the sum of these two components.

Equation 6.7 can be rearranged to determine the average spectral reflectance for the surface as a whole by dividing by $E(\lambda)$.

$$\rho_{\text{surface}}(\lambda) = \frac{M_{\text{surface}}(\lambda)}{E(\lambda)} = A\rho_{\text{leaf}}(\lambda) + (1 - A)\rho_{\text{soil}}(\lambda) \qquad (6.8)$$

So, for example, if $A = 0.5$ (i.e., one half of the surface area is covered by leaves), $\rho_{\text{leaf}}(\lambda) = 0.475$ (i.e., the leaves reflect just under one half of the solar radiation that is incident on them) and $\rho_{\text{soil}}(\lambda) = 0.125$ (i.e., the soil reflects one eighth of the solar radiation incident on it), the spectral reflectance for the surface as a whole is given by

$$
\begin{aligned}
\rho_{surface}(\lambda) &= (0.5 \times 0.475) + ((1-0.5) \times 0.125) \\
&= 0.2375 + 0.0625 \\
&= 0.3
\end{aligned}
\tag{6.9}
$$

The values of $\rho_{leaf}(\lambda)$ and $\rho_{soil}(\lambda)$ used in this example are typical of those at NIR wavelengths ($\sim 0.85 \mu m$); that is, wavelengths slightly longer than the human eye can see (Figure 6.3). By comparison, at red wavelengths ($\sim 0.65 \mu m$), which are visible to the human eye, typical values of $\rho_{leaf}(\lambda)$ and $\rho_{soil}(\lambda)$ are 0.04 and 0.08, respectively. These produce a value of $\rho_{surface}(\lambda) = 0.06$ for the same fractional areas of leaves and soil.

Checking the Mathematical Model

It is good practice to get into the habit of checking a model, in the manner outlined above, at a number of different stages during its development, to confirm that it is performing as expected. The mathematical model given in Equation 6.8, for example, can be tested by presenting it with a range of input values. These should, at the very least, consist of extreme or "special" case values. For instance, when $A = 0$ (i.e., there are no leaves present), $\rho_{surface}(\lambda)$ should equal $\rho_{soil}(\lambda)$. Similarly, when $A = 1$ (i.e., no soil is visible), $\rho_{surface}(\lambda)$ should equal $\rho_{leaf}(\lambda)$.

Clearly, few environmental models are as simple as the one considered here, where the formulation can be checked quickly and simply by mental arithmetic. More complex models generally demand much greater rigor and more extensive testing to establish that they operate as intended. Indeed, it may only be possible to do so after they have been implemented in computer code, perhaps by performing a large number of model "runs" (simulations) using various sets of input values.

6.2.3 Implementing the Computational Model

The third step is to implement the mathematical model (Equation 6.8) in computer code; that is, to create the computational model. There are various ways that this can be realized in **gawk**, one of which is given in Program 6.1 (`mixture.awk`). This implementation is intended to simulate the reflectance of a mixed soil and leaf surface for a range of values of A (i.e., the fractional area of ground covered by leaves), varying between 0 and 1 in increments of 0.1.

Program Structure

The first 13 lines of code in Program 6.1 comprise a series of comments, denoted by the hash (#) symbol at the start of each line. These are used to indicate the purpose of the code, to show how the program should be invoked via the command line, and to identify the major variables concerned. Note that the terms A, $\rho_{leaf}(\lambda)$, $\rho_{soil}(\lambda)$ and $\rho_{surface}(\lambda)$ from the mathematical model (Equation 6.8) are represented by the

Program 6.1: mixture.awk

```
# Simple model of solar radiation interaction with a mixed     1
# soil and vegetation surface. The program calculates the      2
# average spectral reflectance of a surface covered by the     3
# specified areal fractions of soil and vegetation (leaves).   4
#                                                               5
# Usage: gawk -f mixture.awk -v rho_leaf=value \               6
#    -v rho_soil=value [ > output_file ]                       7
#                                                               8
# Variables:                                                    9
# area_leaf    Fractional area covered by leaves              10
# rho_leaf     Leaf spectral reflectance                      11
# rho_soil     Soil spectral reflectance                      12
# rho_surface Average spectral reflectance of surface         13
                                                               14
BEGIN{                                                         15
  for(area_leaf=0;area_leaf<=1;area_leaf+=0.1){               16
    rho_surface=(area_leaf*rho_leaf)+ \                       17
      ((1-area_leaf)*rho_soil);                               18
    print area_leaf, rho_surface;                             19
  }                                                            20
}                                                              21
```

variables area_leaf, rho_leaf, rho_soil and rho_surface, respectively, in the code. More generally, a combination of carefully chosen variable names, instructive comments and clear layout tends to produce code that is easier to read, understand and maintain; it is, in effect, self-documenting (British Computer Society 1998).

The rest of the code is contained within the BEGIN block (lines 15 to 21) because the intention is to use the computational model to generate data (i.e., to perform a simulation), rather than to process data stored in a file. A for loop (line 16) is used to vary the fractional area of ground covered by leaves (area_leaf) between 0 (bare soil) and 1 (complete vegetation cover) in steps of 0.1. For each iteration around the loop, the reflectance of a surface with that particular mixture of leaves and soil is calculated (rho_surface; lines 17 and 18). Recall that the backslash symbol (\) at the end of line 17 is the line continuation symbol. This allows a single long **gawk** statement to be split over two or more lines of code. Finally, the program prints out the fractional area covered by leaves and the reflectance of the mixed soil and vegetation surface (line 19).

It is important to note that values for the spectral reflectance of the leaves and the soil are not specified ("hard wired") in this program. These values must therefore be provided via the command line using the -v switch (e.g., -v rho_leaf=0.475). The intention in doing so is to keep the program as flexible as possible, so that it can be used to simulate the reflectance of the mixed leaf and soil surface at any given wavelength. If appropriate values of these variables are not provided on the command line, **gawk** will set them to zero (0), with possibly unintended consequences.

6.2.4 Running the Model

This implementation of the model can be run from the command line by typing

```
gawk -f mixture.awk -v rho_leaf=0.475 -v rho_soil=0.125 > ↻    1
     ↻mixture.nir
```

to simulate the variation in spectral reflectance at NIR wavelengths as a function of the fractional area covered by leaves, and

```
gawk -f mixture.awk -v rho_leaf=0.04 -v rho_soil=0.08 > ↻    2
     ↻mixture.red
```

to perform an equivalent simulation at red wavelengths. Notice that a separate -v flag is used for each variable whose value is initialized on the command line.

6.2.5 Evaluating the Output from the Computational Model

The next step is to examine graphically the results that the model produces. This can be achieved by typing the following commands in gnuplot:

```
set xlabel "Fraction of ground covered by leaves, A"     1
set ylabel "Spectral reflectance, rho_surface(lambda)"   2
set key top left                                         3
set yrange [0:0.7]                                       4
set data style lp                                        5
plot 'mixture.red' t "Red (0.65 micrometers)", \         6
     'mixture.nir' t "NIR (0.85 micrometers)"            7
```

These commands assume that the data files mixture.red and mixture.nir, generated by Program 6.1, are located in the working directory; otherwise, their full or relative pathnames must be given. The results are presented in Figure 6.9.

It is evident from Figure 6.9 that the model formulation presented in Equation 6.8, and implemented in Program 6.1, defines a straight-line relationship between the fraction of ground covered by leaves (A; area_leaf) and the spectral reflectance for the surface as a whole ($\rho_{surface}(\lambda)$; rho_surface). Indeed, it might be described as a linear mixture model (Ichoku and Karnieli 1996). Detailed observations of real plant canopies, however, suggest that the actual trends are more like those shown in Figure 6.10. Specifically, spectral reflectance exhibits a curvilinear relationship with increasing vegetation amount at both red and NIR wavelengths: for the former, the relationship is indirect (negative); for the latter, it is direct (positive). Both exhibit asymptotic trends, with the asymptote being reached at lower levels of vegetation cover in the red than in the NIR. Note that spectral reflectance is plotted as a function of leaf area index (LAI) in Figure 6.10. The difference between this and fractional ground cover is explored more fully in the next section.

The differences between Figures 6.9 and 6.10 suggest that the approach adopted thus far has been too simplistic. Consequently, the next section revisits the conceptual model and considers how it might be made to correspond more closely to reality, updating the mathematical and computational models accordingly.

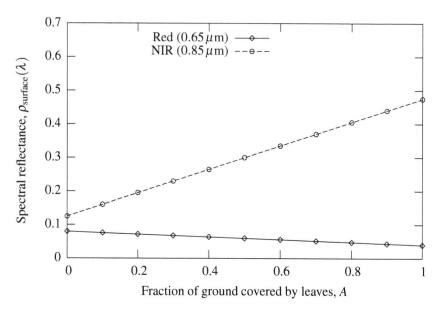

Figure 6.9: Output from a simple light-interaction model for a surface comprising a mixture of soil and leaves (Program 6.1). Red: $\rho_{\text{leaf}}(\lambda) = 0.04$, $\rho_{\text{soil}}(\lambda) = 0.08$. NIR: $\rho_{\text{leaf}}(\lambda) = 0.475$, $\rho_{\text{soil}}(\lambda) = 0.125$.

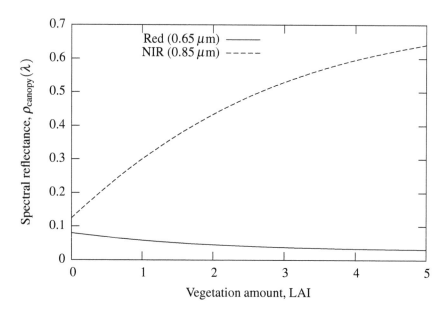

Figure 6.10: Expected variation in the spectral reflectance of a simple plant canopy at red and NIR wavelengths as a function of vegetation amount, LAI.

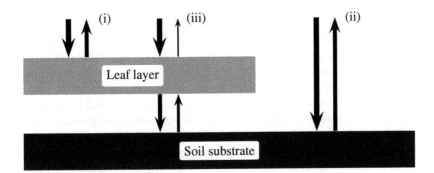

Figure 6.11: Two-layer model of light interaction with a plant canopy.

Exercise 6.1: Develop a revised version of the simple light-interaction model that simulates the spectral reflectance of an area of ground covered by a mixture of three surface materials (e.g., vegetation, soil and snow or ice). Begin by specifying the conceptual model, stating any assumptions that must be made. Next, formulate the mathematical model, basing this on Equation 6.8. Implement the model in gawk, using Program 6.1 as a template. Finally, run the revised model and examine the output.

6.3 A TWO-LAYER LIGHT INTERACTION MODEL

6.3.1 Improving the Conceptual Model

The model that has been developed thus far simulates the spectral reflectance for an area of ground comprising a mixture of leaves and soil. It portrays the surface as a single layer, or slab, in which the leaves and soil lie side by side (Figure 6.7). Clearly, this is not a very realistic representation of the physical structure of most plant canopies, in which the leaves are usually located some distance above the soil substrate. In terms of the conceptual model, this would be better represented by two discrete layers; that is, by a separate leaf-layer suspended above the soil substrate (Figure 6.11).

In the revised conceptual model, the leaf layer covers some fraction (A) of the total surface area when viewed vertically downward from above, while the remainder $(1 - A)$ represents one or more gaps through which the soil substrate can been seen. In Figure 6.11 the leaf layer is represented as a single continuous slab, but it might equally well be a discontinuous fragmented layer covering the same fractional area of ground. Below the leaf layer, the soil substrate extends across the entire surface area. As a result of this modification, incoming solar radiation can now traverse three different pathways through the plant canopy before being reflected back out to space (Figure 6.11). Thus, it may (i) be reflected directly from the leaf layer, (ii) pass down through a gap in the leaf layer, reflect from the soil substrate and escape back out to space through the same gap in the leaf layer, or (iii) pass down through the leaf layer, reflect from the soil substrate and pass back up through the leaf layer.

The revised conceptual model introduces two new phenomena: one explicitly, namely transmission; the other implicitly, namely absorption. Transmission refers to the passage of radiation through an object, such as a leaf, while transmittance is the fraction of solar radiation incident upon an object that is transmitted through it (Glickman 2000). Absorption refers to radiation that is retained by a substance (i.e., radiation that is neither reflected nor transmitted), while absorptance is the fraction of the incident radiation that is absorbed (Glickman 2000). Both transmittance and absorptance vary as a function of wavelength in most instances. It is common, therefore, to refer to spectral transmittance and spectral absorptance when dealing with radiation at a particular wavelength or in a narrow spectral waveband.

6.3.2 Reformulating the Mathematical Model

The symbols A, $\rho_{\text{leaf}}(\lambda)$ and $\rho_{\text{soil}}(\lambda)$ are employed with the same meaning as before, while $\tau_{\text{leaf}}(\lambda)$ is used to refer to the spectral transmittance of the leaf layer, $\alpha_{\text{leaf}}(\lambda)$ the spectral absorptance of the leaf layer, and $\alpha_{\text{soil}}(\lambda)$ the spectral absorptance of the soil substrate, noting that

$$\rho_{\text{leaf}}(\lambda) + \tau_{\text{leaf}}(\lambda) + \alpha_{\text{leaf}}(\lambda) = 1 \tag{6.10}$$

Thus, solar radiation incident upon the leaf layer is partitioned three ways: some is reflected from it, some is transmitted through it, and some is absorbed by it. Following the law of conservation of energy, the reflected, transmitted and absorbed fractions must account for the total radiation striking the leaf layer (i.e., they must sum to 1). Similarly,

$$\rho_{\text{soil}}(\lambda) + \alpha_{\text{soil}}(\lambda) = 1 \tag{6.11}$$

which implies that the soil substrate is assumed to be completely opaque, that is $\tau_{\text{soil}}(\lambda) = 0$.

The two-layer model can therefore be formulated in terms of the three pathways shown in Figure 6.11, namely

(i) $A\rho_{\text{leaf}}(\lambda)$

(ii) $(1-A)\rho_{\text{soil}}(\lambda)$

(iii) $A\tau_{\text{leaf}}(\lambda)\rho_{\text{soil}}(\lambda)\tau_{\text{leaf}}(\lambda)$

and where the spectral reflectance from the vegetation canopy as a whole is the sum of these three terms, i.e.,

$$\begin{aligned} \rho_{\text{canopy}}(\lambda) &= A\rho_{\text{leaf}}(\lambda) + (1-A)\rho_{\text{soil}}(\lambda) \\ &\quad + A\tau_{\text{leaf}}(\lambda)\rho_{\text{soil}}(\lambda)\tau_{\text{leaf}}(\lambda) \end{aligned} \tag{6.12}$$

While the profusion of Greek symbols in this formulation might make it look rather daunting, bear in mind that it involves nothing more demanding than simple addition and multiplication.

It may be helpful to think of Equation 6.12 in terms of fractions and percentages. Thus, the fraction of incident solar radiation intercepted by some part of the leaf layer is A. For example, if the leaf layer completely covers the ground below ($A = 1$), the fraction intercepted is 1 (all of the incident solar radiation) because there are no gaps through which the solar radiation can pass uninterrupted down to the soil substrate. By contrast, if the leaf layer covers half of the area of ground below ($A = 0.5$), the fraction intercepted is 0.5. Put another way, 50% of the solar radiation will be intercepted by the leaf layer, while the remainder will pass down through a gap toward the soil substrate. Similarly, $\rho_{leaf}(\lambda)$ describes the fraction of solar radiation striking the upper surface of the leaf layer that is reflected back out to space. So, for example, if $\rho_{leaf}(\lambda) = 0.475$ at NIR wavelengths, 47.5% of the solar radiation striking the leaf layer will be reflected upward from it; the remainder will be absorbed by it or transmitted through it. Moreover, it is possible to combine these two values to determine the fraction of incident radiation that is reflected upward from the leaf layer. This relates to pathway (i) in Figure 6.11 and is given by $A\rho_{leaf}(\lambda)$. Thus, if $A = 0.5$ and $\rho_{leaf}(\lambda) = 0.475$ at NIR wavelengths, the result is $0.5 \times 0.475 = 0.2375$. In other words, 23.75% of the incident NIR radiation is reflected directly from the leaf layer. Note that the equivalent figure at red wavelengths is 2%, assuming that $\rho_{leaf}(\lambda) = 0.04$.

For incident solar radiation to traverse pathway (ii) in Figure 6.11 it must pass down through a gap in the leaf layer $(1 - A)$, be reflected from the soil substrate $(\rho_{soil}(\lambda))$ and pass back up through the same gap in the leaf layer. Note that in this model the fraction of radiation reflected from the soil substrate that passes back up through the gap in the leaf layer is 1 (i.e., all of it) because the model assumes that radiation travels vertically downward and upward through the canopy. So, if the radiation has passed down through a gap in the leaf layer it must, by definition, pass back up through the same gap on its return path. The fraction of incident solar radiation that traverses this pathway through the canopy is therefore $(1 - A)\rho_{soil}(\lambda)$. Assuming that $A = 0.5$ and $\rho_{soil}(\lambda) = 0.125$ at NIR wavelengths, $(1 - A)\rho_{soil}(\lambda) = 0.5 \times 0.125 = 0.0625$. Thus, 6.25% of the incident NIR radiation takes this route through the model canopy. The equivalent figure at red wavelengths is 4%, assuming $A = 0.5$ and $\rho_{leaf}(\lambda) = 0.08$.

Finally, for incident solar radiation to traverse pathway (iii) in Figure 6.11 it must be intercepted by and transmitted through the leaf layer on its downward path, be reflected upward from the soil substrate, and then be transmitted back up through the leaf layer. Again, the fraction of this radiation that is intercepted by the leaf layer on its upward path from the soil substrate is 1 because it passed down through the leaf layer to reach the soil substrate in the first place. Given the assumptions made in the model, it must retrace the same route on its way back up. Nevertheless, the second interception leads to further attenuation of the upwelling radiation because only a fraction of this is transmitted through the leaf layer. Therefore, the fraction of NIR radiation that traverses this particular pathway is $A\tau_{leaf}(\lambda)\rho_{soil}(\lambda)\tau_{leaf}(\lambda) \approx$

Program 6.2: twolayer.awk

```
# Simple model of light interaction with a two-layer     1
# plant canopy (i.e., a leaf layer suspended above a     2
# soil substrate).                                        3
#                                                         4
# Version 1.                                              5
#                                                         6
# Usage: gawk -f twolayer.awk -v rho_leaf=value \         7
#               -v tau_leaf=value -v rho_soil=value \     8
#               [ > output_file ]                         9
#                                                        10
# Variables:                                             11
# area_leaf   Fractional area covered by leaves          12
# rho_leaf    Leaf spectral reflectance                  13
# tau_leaf    Leaf spectral transmittance               14
# rho_soil    Soil spectral reflectance                  15
# rho_canopy  Canopy spectral reflectance                16
# gap         Fractional area of gaps in leaf layer      17
#                                                        18
BEGIN{                                                   19
   for(area_leaf=0;area_leaf<=1;area_leaf+=0.1){         20
      gap=1-area_leaf;                                   21
      rho_canopy=area_leaf*rho_leaf + (gap*rho_soil) + \ 22
         (area_leaf*tau_leaf*rho_soil*tau_leaf);         23
      print area_leaf, rho_canopy;                       24
   }                                                     25
}                                                        26
```

0.0141, assuming that $A = 0.5$, $\rho_{leaf}(\lambda) = 0.475$, $\tau_{leaf}(\lambda) = 0.475$ and $\rho_{soil}(\lambda) = 0.125$. Note that $\tau_{leaf}(\lambda) = \rho_{leaf}(\lambda)$ is a reasonable approximation in most instances. Thus, approximately 1.41% of the incident NIR radiation takes this route through the model canopy. The equivalent figure at red wavelengths is approximately 0.006%, assuming that $\tau_{leaf}(\lambda) = \rho_{leaf}(\lambda) = 0.04$ and $\rho_{soil}(\lambda) = 0.08$.

6.3.3   Implementing the Two-Layer Model in gawk

Implementing the two-layer model given in Equation 6.12 involves making a few minor modifications to Program 6.1. These are presented in Program 6.2. They consist of (i) the introduction of a new variable, gap, the value of which is set to 1-area_leaf (line 21), which is intended solely to improve the readability of the code, (ii) some changes to the model calculation and the output of results (lines 22 and 23) that are required to represent Equation 6.12, and (iii) a small change to the variables whose values are printed out (line 24). Once again, recall that the backslash (\) at the end of line 22 is the line continuation symbol.

6.3.4 Running the Two-Layer Model

The two-layer model can be run from the command line by typing

```
gawk -f twolayer.awk -v rho_leaf=0.475 -v tau_leaf=0.475 -↩   3
    ↩v rho_soil=0.125 > 2layer.nir
```

assuming that $\rho_{\text{leaf}}(\lambda) = 0.475$ and $\rho_{\text{soil}}(\lambda) = 0.125$, and that $\tau_{\text{leaf}}(\lambda) = \rho_{\text{leaf}}(\lambda)$ at NIR wavelengths. Note, though, that Program 6.2 allows one to evaluate the effect of differences between $\tau_{\text{leaf}}(\lambda)$ and $\rho_{\text{leaf}}(\lambda)$, if so desired. The equivalent simulation can be performed for red wavelengths by typing the following instruction on the command line:

```
gawk -f twolayer.awk -v rho_leaf=0.04 -v tau_leaf=0.04 -v ↩   4
    ↩rho_soil=0.08 > 2layer.red
```

assuming that $\rho_{\text{leaf}}(\lambda) = 0.04$, $\rho_{\text{soil}}(\lambda) = 0.08$ and that $\tau_{\text{leaf}}(\lambda) = \rho_{\text{leaf}}(\lambda)$.

6.3.5 Evaluating the Two-Layer Model

It is possible to generate a plot of the output from the two-layer model by typing the following commands in **gnuplot** (Figure 6.12):

```
set ylabel "Spectral reflectance, rho_canopy(lambda)"        8
plot '2layer.red' t "Red (0.65 micrometers)", \              9
     '2layer.nir' t "NIR (0.85 micrometers)"                10
```

These commands assume that the files 2layer.red and 2layer.nir, created using Program 6.2, are located in the working directory; otherwise, their full or relative pathnames must be provided.

It is clear from Figure 6.12 that the results produced using the two-layer light interaction model do not correspond well to observations of real vegetation canopies (Figure 6.10). Indeed, they do not differ significantly from those of the one-layer model. The reasons for this are examined in the following two sections, in which several further modifications are made to the model so that its output conforms more closely to expectations.

6.4 ACCOUNTING FOR MULTIPLE SCATTERING

6.4.1 Enhancing the Conceptual and Mathematical Models

It is instructive to note that the first and second terms in Equation 6.12, $A\rho_{\text{leaf}}(\lambda)$ and $(1-A)\rho_{\text{soil}}(\lambda)$, relate to solar radiation that has interacted once only with the plant canopy, either with the leaf layer or with the soil substrate, but not both. These terms therefore describe what are known as single scattering events. In contrast, the third term in Equation 6.12, $A\tau_{\text{leaf}}(\lambda)\rho_{\text{soil}}(\lambda)\tau_{\text{leaf}}(\lambda)$, relates to radiation that has interacted three times with elements of the vegetation canopy. More specifically, it describes radiation that is first transmitted down through the leaf layer, then reflected from the soil substrate, and finally transmitted back up through the leaf layer. This is

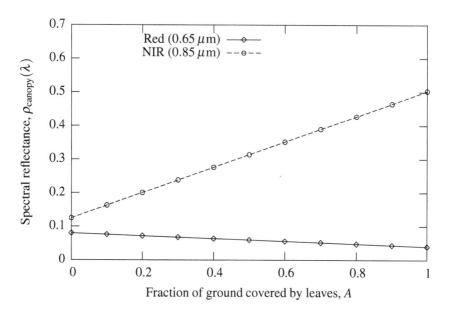

Figure 6.12: Spectral reflectance of a plant canopy as a function of the fractional cover of leaves based on the two-layer light interaction model given in Program 6.2.

commonly known as multiple scattering, since the radiation is scattered (i.e., reflected or transmitted) more than once on its path through the canopy. It is also common to refer to the "order" of scattering. Thus, first-order scattering refers to radiation that has been scattered once, second-order scattering refers to radiation that has been scattered twice, and so on.

Bearing this in mind, it is possible that the performance of the two-layer model could be improved by incorporating higher-order scattering effects. Figure 6.13, for example, illustrates the path that incident solar radiation might take through the two-layer canopy as a result of fifth-order scattering. This can be represented as follows:

$$A \times \tau_{leaf}(\lambda) \times \rho_{soil}(\lambda) \times \rho_{leaf}(\lambda) \times \rho_{soil}(\lambda) \times \tau_{leaf}(\lambda) \qquad (6.13)$$

Note that, here, $\rho_{leaf}(\lambda)$ denotes reflectance downward from the underside (or abaxial surface) of the leaf layer, as well as upward from the upper (or adaxial) surface. This assumes that the reflectance of the abaxial and adaxial surfaces of the leaf layer are identical; while this is not always the case for individual leaves in real vegetation canopies, it is a reasonable approximation in this instance.

Adding the fifth-order scattering term (Equation 6.13) into the two-layer light interaction model gives

$$\begin{aligned} \rho_{canopy}(\lambda) \quad = \quad & A\rho_{leaf}(\lambda) + (1-A)\rho_{soil}(\lambda) \\ & + A\tau_{leaf}(\lambda)\rho_{soil}(\lambda)\tau_{leaf}(\lambda) \\ & \boxed{+A\tau_{leaf}(\lambda)\rho_{soil}(\lambda)\rho_{leaf}(\lambda)\rho_{soil}(\lambda)\tau_{leaf}(\lambda)} \end{aligned} \qquad (6.14)$$

where the fifth-order term is highlighted in the box.

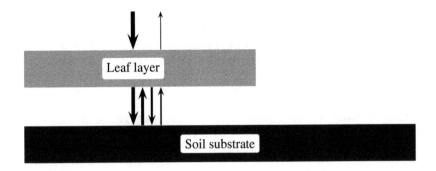

Figure 6.13: Fifth-order multiple scattering in a two-layer model of light interaction. Note that
the arrows, which indicate the radiation pathway through the model canopy, have
been offset slightly with respect to one another for the sake of clarity.

6.4.2 Implementing the Revised Two-Layer Model

An implementation of the revised two-layer model (Equation 6.14) is presented in
Program 6.3. It differs only marginally from Program 6.2. Specifically, the spectral
reflectance of the canopy (rho_canopy) is now calculated over four lines of code (22
through 25), with the fifth-order scattering term given on lines 24 and 25.

6.4.3 Running the Revised Two-Layer Model

The revised model can be run from the command line by typing

```
gawk -f twolayr2.awk -v rho_leaf=0.475 -v tau_leaf=0.475 -↳    5
    ↳v rho_soil=0.125 > 2layer2.nir
```

assuming that $\rho_{\text{leaf}}(\lambda) = 0.475$, $\rho_{\text{soil}}(\lambda) = 0.125$ and that $\tau_{\text{leaf}}(\lambda) = \rho_{\text{leaf}}(\lambda)$ at NIR
wavelengths. It is also possible to perform an equivalent simulation for red wave-
lengths by typing

```
gawk -f twolayr2.awk -v rho_leaf=0.04 -v tau_leaf=0.04 -v ↳    6
    ↳rho_soil=0.08 > 2layer2.red
```

assuming that $\rho_{\text{leaf}}(\lambda) = 0.04$, $\rho_{\text{soil}}(\lambda) = 0.08$ and that $\tau_{\text{leaf}}(\lambda) = \rho_{\text{leaf}}(\lambda)$.

6.4.4 Evaluating the Revised Two-Layer Model

A plot of the output produced by the revised two-layer model (Program 6.3) can be
generated in gnuplot using the following commands:

```
plot '2layer2.red' t "Red (0.65 micrometers)", \     11
     '2layer2.nir' t "NIR (0.85 micrometers)"          12
```

assuming that the data files 2layer2.red and 2layer2.nir are located in the working
directory; otherwise, their full or relative pathnames must be given (Figure 6.14).

Program 6.3: twolayr2.awk

```
# Simple model of light interaction with a two-layer          1
# plant canopy (i.e., a leaf layer suspended above a          2
# soil substrate), incorporating fifth-order scattering.      3
#                                                             4
# Version 2                                                   5
#                                                             6
# Usage: gawk -f twolayr2.awk -v rho_leaf=value \             7
#                 -v tau_leaf=value -v rho_soil=value \       8
#                 [ > output_file ]                           9
#                                                            10
# Variables:                                                 11
# area_leaf   Fractional area covered by leaves              12
# rho_leaf    Leaf spectral reflectance                     13
# tau_leaf    Leaf spectral transmittance                   14
# rho_soil    Soil spectral reflectance                     15
# rho_canopy  Canopy spectral reflectance                   16
# gap         Fractional area of gaps in leaf layer         17
#                                                            18
BEGIN{                                                       19
   for (area_leaf=0; area_leaf<=1; area_leaf+=0.1){          20
      gap=1-area_leaf;                                       21
      rho_canopy=area_leaf*rho_leaf + (gap*rho_soil) + \     22
         (area_leaf*tau_leaf*rho_soil*tau_leaf) + \          23
         (area_leaf*tau_leaf*rho_soil*rho_leaf* \            24
         rho_soil*tau_leaf);                                 25
      print area_leaf, rho_canopy;                           26
   }                                                         27
}                                                            28
```

Disappointingly, the results still do not correspond well to the responses expected for real plant canopies (Figure 6.10); the inclusion of fifth-order scattering in the model has not had a significant impact in this instance. The reason for this is clear in hindsight; multiple scattering is represented in the model as the arithmetic product of a number of fractional values. For example, if $\rho_{\text{leaf}}(\lambda) = 0.475$, $\tau_{\text{leaf}}(\lambda) = 0.475$ and $\rho_{\text{soil}}(\lambda) = 0.125$, as is typical at NIR wavelengths, and $A = 0.5$, the fraction of the incident solar radiation that traverses the pathway through the model canopy shown in Figure 6.13 is $0.5 \times 0.475 \times 0.125 \times 0.475 \times 0.125 \times 0.475 \approx 0.00084$, i.e., less than 0.1% of the total incident flux. By comparison, the fraction that is scattered once only, by the leaf layer or the soil substrate, is $A\rho_{\text{leaf}}(\lambda) + (1-A)\rho_{\text{soil}}(\lambda) = 0.3$. The contribution of single scattering to the total canopy reflectance at NIR wavelengths is therefore approximately 357 times larger than that of fifth-order scattering (i.e., $0.3/0.00084$). The equivalent value at red wavelengths is greater than 300,000. For the two-layer model, therefore, the inclusion of fifth-order multiple scattering does not significantly affect the total spectral reflectance of the model canopy.

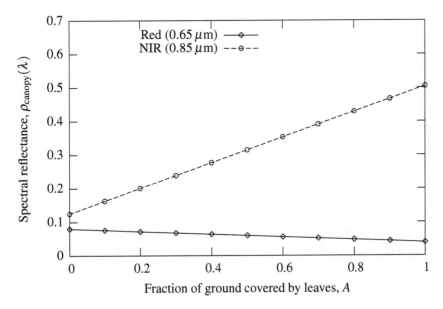

Figure 6.14: Spectral reflectance of a plant canopy as a function of the fractional cover of leaves based on the two-layer model including fifth-order scattering.

6.5 MULTIPLE LEAF-LAYER MODELS

6.5.1 Enhancing the Conceptual Model

Since the inclusion of fifth-order multiple scattering has not, on its own, significantly increased the realism of the two-layer model, other enhancements must be sought. One possibility is to increase the number of leaf layers in the model to generate, for example, a three-layer model that contains two leaf-layers and a soil substrate. This is a slightly more realistic representation of the 3D structure of real vegetation canopies. It also has four important ramifications for the model. First, it increases the number of pathways that radiation can trace through the model canopy (Figure 6.15). Second, it raises the possibility of specifying separately the fractional area of ground covered by leaves in each layer. This allows variations in the density of plant material to be modeled as a function of depth into the canopy, if so desired. Third, it enables variations in the spatial clumping of plant material to be modeled by representing different degrees of overlap between the leaf layers. Fourth, a different measure must be used to represent the total amount of vegetation in the canopy as a whole. The LAI, which is the total single-sided area of leaves per unit area of ground, is widely employed in this context. If, for example, both leaf layers in Figure 6.15 completely cover the ground below ($A_{\text{leaf layer}1} = 1$ and $A_{\text{leaf layer}2} = 1$), then LAI $= 2 \times 1 = 2$. If, instead, each leaf layer covers only half of the ground below ($A_{\text{leaf layer}1} = 0.5$ and $A_{\text{leaf layer}2} = 0.5$), then LAI $= 2 \times 0.5 = 1$. Thus, $0 \leq \text{LAI} \leq N$ in this model, where N is the number of leaf layers.

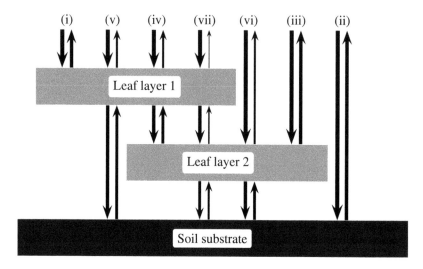

Figure 6.15: Schematic representation of the various pathways (i–vii) that incident solar radiation can take through a three-layer model of a plant canopy.

The remainder of this section examines the three-layer light interaction model shown in Figure 6.15. To simplify things slightly, it is assumed that the fraction of ground covered by each of the leaf layers is identical. Otherwise, the conceptual basis and assumptions of the model remain the same as for the two-layer case.

6.5.2 Formulating the Three-Layer Model

The mathematical formulation of the three-layer model is simply an extension of the two-layer case and is given by Equation 6.15.

$$
\begin{aligned}
\rho_{\text{canopy}}(\lambda) \;=\; & A\rho_{\text{leaf}}(\lambda) \\
& + (1-A)(1-A)\rho_{\text{soil}}(\lambda) \\
& + (1-A)A\rho_{\text{leaf}}(\lambda) \\
& + A\tau_{\text{leaf}}(\lambda)A\rho_{\text{leaf}}(\lambda)\tau_{\text{leaf}}(\lambda) \\
& + A\tau_{\text{leaf}}(\lambda)(1-A)\rho_{\text{soil}}(\lambda)\tau_{\text{leaf}}(\lambda) \\
& + (1-A)A\tau_{\text{leaf}}(\lambda)\rho_{\text{soil}}(\lambda)\tau_{\text{leaf}}(\lambda) \\
& + A\tau_{\text{leaf}}(\lambda)A\tau_{\text{leaf}}(\lambda)\rho_{\text{soil}}(\lambda)\tau_{\text{leaf}}(\lambda)\tau_{\text{leaf}}(\lambda)
\end{aligned}
\tag{6.15}
$$

Note that the seven terms listed in Equation 6.15 correspond, in order, to the pathways numbered (i) to (vii) in Figure 6.15. These terms and pathways are explained more fully in Table 6.1. Note that the fifth and sixth terms, pathways (v) and (vi), are arithmetically equivalent. Also remember that if radiation passes down through a gap in a leaf layer on its way into the canopy it must pass back up through that gap on its return path because it is assumed that radiation travels vertically upward and

Table 6.1: Explanation of the pathways that radiation can traverse through the three-layer model of a plant canopy presented in Figure 6.15 and the corresponding terms in Equation 6.15.

Path	Term	Explanation
(i)	$A\rho_{\text{leaf}}(\lambda)$	Intercept and reflect up from top leaf-layer.
(ii)	$(1-A)(1-A)\rho_{\text{soil}}(\lambda)$	Pass down through gaps in both leaf layers, reflect up from soil substrate and pass back up through the same gaps.
(iii)	$(1-A)A\rho_{\text{leaf}}(\lambda)$	Pass down through gap in top leaf-layer, reflect up from lower leaf-layer, and pass back up through gap in top leaf-layer.
(iv)	$A\tau_{\text{leaf}}(\lambda)A\rho_{\text{leaf}}(\lambda)\tau_{\text{leaf}}(\lambda)$	Transmit down through top leaf-layer, reflect up from lower leaf-layer, transmit back up through top leaf-layer.
(v)	$A\tau_{\text{leaf}}(\lambda)(1-A)\rho_{\text{soil}}(\lambda)\tau_{\text{leaf}}(\lambda)$	Transmit down through top leaf-layer, pass down through gap in lower leaf layer, reflect up from soil substrate, pass back up through gap in lower leaf-layer and transmit up through top leaf-layer.
(vi)	$(1-A)A\tau_{\text{leaf}}(\lambda)\rho_{\text{soil}}(\lambda)\tau_{\text{leaf}}(\lambda)$	Pass down through gap in top leaf-layer, transmit down through lower leaf-layer, reflect up from soil, transmit up through lower leaf-layer and pass back up through gap in top leaf-layer.
(vii)	$A\tau_{\text{leaf}}(\lambda)A\tau_{\text{leaf}}(\lambda)\rho_{\text{soil}}(\lambda)\tau_{\text{leaf}}(\lambda)\tau_{\text{leaf}}(\lambda)$	Transmit down through both leaf layers, reflect up from soil substrate, and transmit up through both leaf layers.

downward through the canopy. A similar logic applies to radiation that is intercepted by, and transmitted through, one or both of the leaf layers. For this reason, the parameters A and $(1 - A)$ only influence the radiation on its downward path into the model canopy.

Spatial overlap between the two leaf-layers is handled implicitly in Equation 6.15. Thus, for any value of $A > 0$, some of the incident solar radiation interacts with both leaf layers (terms (iv) and (vii) in Equation 6.15), which implies that the two leaf-layers overlap to some extent, even at very low levels of vegetation cover. Moreover, the fraction of radiation that intercepts both leaf layers (and by implication the degree of overlap between them) increases as the value of A grows.

One of the limitations of this model is that there is no way in which radiation can be multiply scattered an even number of times, even though this is possible in real vegetation canopies. So, for example, the first, second and third terms in Equation 6.15 describe single scattering events; the fourth, fifth and sixth terms pertain to third-order multiple scattering; while the final term relates to fifth-order multiple scattering. Although Equation 6.15 encompasses all possible first-order and third-order scattering terms, the same cannot be said for the fifth-order terms. The latter can also occur as a result of radiation being reflected upward and downward several times between the leaf layers, or between the bottom leaf-layer and the soil substrate. If all of the pathways that result in fifth-order scattering are included, three extra terms (shown in boxes below) must be added to Equation 6.15 to give Equation 6.16.

$$
\begin{aligned}
\rho_{\text{canopy}}(\lambda) \quad = \quad & A\rho_{\text{leaf}}(\lambda) \\
& + (1-A)(1-A)\rho_{\text{soil}}(\lambda) \\
& + (1-A)A\rho_{\text{leaf}}(\lambda) \\
& + A\tau_{\text{leaf}}(\lambda)A\rho_{\text{leaf}}(\lambda)\tau_{\text{leaf}}(\lambda) \\
& \boxed{+ A\tau_{\text{leaf}}(\lambda)A\rho_{\text{leaf}}(\lambda)\rho_{\text{leaf}}(\lambda)\rho_{\text{leaf}}(\lambda)\tau_{\text{leaf}}(\lambda)} \\
& + A\tau_{\text{leaf}}(\lambda)(1-A)\rho(\lambda)_{soil}\tau_{\text{leaf}}(\lambda) \\
& \boxed{+ A\tau_{\text{leaf}}(\lambda)(1-A)\rho(\lambda)_{soil}\rho_{\text{leaf}}(\lambda)\rho(\lambda)_{soil}\tau_{\text{leaf}}(\lambda)} \\
& + (1-A)A\tau_{\text{leaf}}(\lambda)\rho(\lambda)_{soil}\tau_{\text{leaf}}(\lambda) \\
& \boxed{+ (1-A)A\tau_{\text{leaf}}(\lambda)\rho(\lambda)_{soil}\rho_{\text{leaf}}(\lambda)\rho(\lambda)_{soil}\tau_{\text{leaf}}(\lambda)} \\
& + A\tau_{\text{leaf}}(\lambda)A\tau_{\text{leaf}}(\lambda)\rho_{\text{soil}}(\lambda)\tau_{\text{leaf}}(\lambda)\tau_{\text{leaf}}(\lambda)
\end{aligned}
$$

$$(6.16)$$

Note that the second and third of these new terms are arithmetically equivalent.

6.5.3 Implementing the Three-Layer Model

Program 6.4 is an implementation of the three-layer model represented in Figure 6.15, taking into account up to fifth-order scattering, based on Equation 6.16. The code differs only slightly from that of the two-layer model (Program 6.3). Specifically, a new variable, LAI, is used to store the total area of leaves in the canopy (line 21),

Program 6.4: 3layers.awk

```
# Program to calculate the spectral reflectance of a          1
# vegetation canopy modeled as two plane-parallel layers      2
# of leaves above a soil substrate.                           3
#                                                             4
# Usage: gawk -f 3layers.awk -v rho_leaf=value \             5
#              -v tau_leaf=value -v rho_soil=value \         6
#              [ > output_file ]                             7
#                                                             8
# Variables:                                                  9
# area_leaf     Fractional area of leaves in each leaf layer 10
# LAI           Leaf Area Index                              11
# rho_leaf      Leaf spectral reflectance                    12
# tau_leaf      Leaf spectral transmittance                  13
# rho_soil      Soil spectral reflectance                    14
# rho_canopy    Canopy spectral reflectance                  15
# gap           Fractional area of gaps in each leaf layer   16
#                                                            17
BEGIN {                                                      18
  for(area_leaf=0;area_leaf<=1;area_leaf+=0.1){              19
    # Initialize variables                                  20
    LAI=2*area_leaf;                                        21
    gap=(1-area_leaf);                                      22
    # Calculate and print total canopy reflectance          23
    rho_canopy=area_leaf*rho_leaf + (gap*gap*rho_soil) + \  24
      (gap*area_leaf*rho_leaf) + \                          25
      area_leaf*tau_leaf*area_leaf*rho_leaf*tau_leaf + \    26
      area_leaf*tau_leaf*area_leaf*rho_leaf*rho_leaf*\      27
        rho_leaf*tau_leaf + \                               28
      2*(area_leaf*tau_leaf*gap*rho_soil*tau_leaf) + \      29
      2*(gap*area_leaf*tau_leaf*rho_soil*rho_leaf*\         30
        rho_soil*tau_leaf) + \                              31
      area_leaf*tau_leaf*area_leaf*tau_leaf*rho_soil* \     32
        tau_leaf*tau_leaf;                                  33
    print LAI, rho_canopy;                                  34
  }                                                         35
}                                                           36
```

while the fraction of incident solar radiation traversing each of the pathways through the model canopy is calculated on lines 24 through 33.

6.5.4   Running the Three-Layer Model

Program 6.4 can be run from the command line by typing

```
gawk -f 3layers.awk -v rho_leaf=0.475 -v tau_leaf=0.475 -v↴    7
  ↪ rho_soil=0.125 > 3layers.nir
```

assuming that $\rho_{leaf}(\lambda) = 0.475$, $\rho_{soil}(\lambda) = 0.125$ and $\tau_{leaf}(\lambda) = \rho_{leaf}(\lambda)$ at NIR wavelengths. It is possible to perform an equivalent simulation for red wavelengths by entering the following command line:

```
gawk -f 3layers.awk -v rho_leaf=0.04 -v tau_leaf=0.04 -v ↴    8
  ↪rho_soil=0.08 > 3layers.red
```

assuming that $\rho_{leaf}(\lambda) = 0.04$, $\rho_{soil}(\lambda) = 0.08$ and $\tau_{leaf}(\lambda) = \rho_{leaf}(\lambda)$ in this part of the electromagnetic spectrum.

6.5.5   Evaluating the Multiple-Layer Model

Figure 6.16 presents the output generated by Program 6.4, which has been plotted in gnuplot using the following commands, assuming that the data files 3layers.red and 3layers.nir are located in the working directory; otherwise, their full or relative pathnames must be provided:

```
set xlabel "Leaf Area Index (LAI)"                            13
plot '3layers.red' u 1:2 t "Red (0.65 micrometers)", \        14
     '3layers.nir' u 1:2 t "NIR (0.85 micrometers)"           15
```

The main thing to note is that the simulated responses produced by the three-layer model are very similar to the curvi-linear asymptotic relationships observed over real plant canopies, shown in Figure 6.10, in terms of both form and magnitude. Thus, the extra leaf-layer appears to have improved significantly the realism of the model.

It is possible that adding even more leaf layers would enhance the model still further. The code given in Program 6.4 could certainly be amended to explore this possibility. The way in which the problem has been approached, however, means that the complexity of the code grows every time the number of leaf layers in the model is increased. This is because each of the pathways that radiation traverses through the canopy must be stated explicitly. There are a very large number of such pathways through a four-layer model (i.e., three leaf-layers plus the soil substrate), taking into account up to seventh-order multiple scattering. The potential for making mistakes in the formulation of the mathematical model, and in the implementation of this as computer code, therefore grows concomitantly. Consequently, the approach exemplified in this chapter will not be pursued any further. Instead, two other ways of tackling the same problem are explored in the next chapter, both of which enable simulations to be performed for model canopies containing any number of leaf layers and taking into account very high orders of multiple scattering.

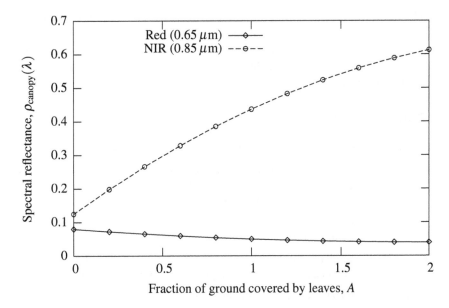

Figure 6.16: Spectral reflectance of a mixed soil and vegetation canopy as a function of LAI, based on the three-layer model given in Program 6.4.

Exercise 6.2: Run the `3layers.awk` implementation of the three-layer canopy reflectance model (Program 6.4) using various values of $\tau(\lambda)_{\text{leaf}} \neq \rho_{\text{leaf}}(\lambda)$ over both a bright soil ($\rho_{\text{soil}}(\lambda) > \rho_{\text{leaf}}(\lambda)$) and dark soil background ($\rho_{\text{soil}}(\lambda) < \rho_{\text{leaf}}(\lambda)$). Plot the results using gnuplot.

6.6 SUMMARY

A relatively sophisticated model is developed from scratch in this chapter. This model represents the interaction of solar radiation with a plant canopy. In its final form, the model reproduces reasonably well the variation in spectral reflectance as a function of LAI observed for real vegetation canopies. This result does not guarantee, of course, that the model adequately represents the processes that generate the responses observed in nature. It does, however, demonstrate the benefits of starting with a simple model and gradually extending this as required. It also shows that a real-world entity as complex in structure as a plant canopy can be represented quite effectively by a series of plane-parallel leaf-layers suspended above a soil substrate, and that the resultant model can be formulated and implemented using remarkably simple mathematical methods and computational techniques.

It is possible to envisage a number of modifications and extensions that might be made to the multiple-layer model, which would allow it to be used over a much wider range of situations. For instance, a further layer might be incorporated to represent reflection, absorption and transmission by clouds above the vegetation canopy, or snow blanketing the soil substrate. Equally, one might allow each of the leaf layers to cover different fractions of the soil substrate, or the reflectance of the adaxial and abaxial surfaces of the leaf layers to differ. These possibilities, however, are left as an exercise for the reader to explore.

SUPPORTING RESOURCES

The following resources are provided in the chapter6 sub-directory of the CD-ROM:

mixture.awk	gawk program to simulate the spectral reflectance of a linearly mixed soil and vegetation surface
2layer.awk	gawk program to simulate the spectral reflectance of a two-layer plant canopy
2layr2.awk	gawk program to simulate the spectral reflectance of a two-layer plant canopy, incorporating fifth-order multiple scattering
3layers.awk	gawk program to simulate the spectral reflectance of a three-layer plant canopy
interact.bat	Command line instructions used in this chapter
interact.plt	gnuplot commands used to generate the figures presented in this chapter
interact.gp	gnuplot commands used to generate figures presented in this chapter as EPS files

Chapter 7

Analytical and Numerical Solutions

Topics

- Analytical and numerical solutions to environmental models

- Bouguer's Law and radiation extinction in plant canopies

Methods and techniques

- Arrays in gawk

- Iteration and iterative methods

- More control-flow structures in gawk: the do-while loop

7.1 INTRODUCTION

A simple model was developed in the preceding chapter to represent the interaction of solar radiation with a plant canopy. The model canopy initially consisted of a single layer, comprising a linear mixture of leaves and soil. The sophistication of the model was subsequently increased, in stages, by (i) placing the leaves in a separate layer suspended above the soil substrate, (ii) taking into account multiple scattering within the canopy, and (iii) adding an extra leaf layer. These modifications, particularly the last one, improved the ability of the model to replicate the asymptotic relationship observed between spectral reflectance and vegetation amount in real plant canopies at both red and NIR wavelengths. At the same time, however, the modifications increased the complexity of the model and, hence, its computational implementation. This is because the model represents explicitly all of the possible pathways that solar radiation can trace through the plant canopy, which makes it difficult to account for very high order multiple scattering or to add further leaf layers into the model.

Two solutions to these problems are examined in this chapter. These solutions illustrate the principal alternative approaches that are commonly employed to tackle a wide range of complex modeling problems. The first returns to the two-layer case (i.e., a single leaf-layer suspended above a soil substrate) to demonstrate how the contribution of multiple scattering to the total canopy reflectance can be solved by analytical means; that is, by reformulating the model so that it is expressed concisely in terms of a set of mathematical functions and constants. This approach is referred to as an analytical or closed-form solution. The second approach involves modifying the computational model so that it can accommodate any number of leaf layers and take into account very high order multiple scattering. This approach does not attempt to find an analytical solution to the model. It relies, instead, on the computer's ability to solve the problem through "brute force", by performing a pre-defined sequence of numerical operations repeatedly. This approach is known as a numerical solution.

7.2   AN EXACT ANALYTICAL SOLUTION TO THE TWO-LAYER MODEL

7.2.1   Reformulating the Two-Layer Model

The two-layer model developed in the preceding chapter can be used to simulate the spectral reflectance of a simple plant canopy, taking into account single scattering from the leaf and soil layers, as well as third-order and fifth-order multiple scattering between the layers. In reality, some of the incident solar radiation is scattered more than five times between the leaf layer and the soil substrate before it eventually escapes the canopy. A decision was taken, however, to exclude these higher-order multiple scattering terms from the model formulation based on the results presented in Section 6.4.4, which suggest that fifth-order scattering makes a relatively small contribution to the total canopy reflectance at NIR wavelengths and a negligible one at red wavelengths. Nevertheless, it cannot be assumed that this is always the case; the model was tested using only a limited range of input values. It is possible that high-order multiple scattering makes a more significant contribution to the spectral reflectance of the canopy in other situations. One might imagine that this is the case, for example, when the substrate is highly reflective (e.g., when the substrate is snow).

Although the formulation of the two-layer model used in the preceding chapter does not prevent one from including higher-order scattering terms, the task is made more difficult because of the need to specify explicitly each of the radiation pathways concerned. An alternative formulation of the same model is explored in this section, one that provides an exact analytical solution to the multiple scattering problem. The solution holds only for the special case in which the leaf layer completely covers the ground below ($A = 1$). To simplify the presentation slightly, the terms R_C, R_L and R_S are used to denote the spectral reflectance of the canopy, the leaf layer and the soil substrate, respectively, and T_L to denote the spectral transmittance of the leaf layer. These replace the terms $\rho_{\text{canopy}}(\lambda)$, $\rho_{\text{leaf}}(\lambda)$, $\rho_{\text{soil}}(\lambda)$ and $\tau_{\text{leaf}}(\lambda)$, respectively, employed in the preceding chapter.

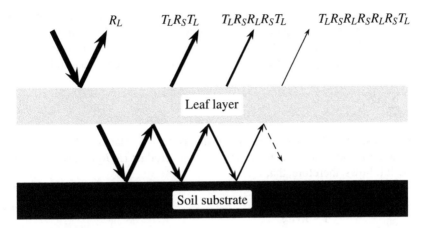

Figure 7.1: Multiple scattering of incident solar radiation in a two-layer plant canopy. R_L and T_L denote the spectral reflectance and transmittance of the leaf layer, respectively, and R_S is the spectral reflectance of the soil substrate.

Figure 7.1 shows a number of different pathways that incident solar radiation can take through the two-layer plant canopy. Note that, while the model assumes radiation travels vertically upward and downward through the canopy, the arrows that represent the different radiation pathways are drawn at oblique angles with respect to the leaf layer and the soil substrate for the sake of diagrammatic clarity. Thus, incident solar radiation can be reflected from the leaf layer (R_L) or be scattered numerous times between the leaf layer and the soil substrate before escaping from the canopy. The terms corresponding to third-order ($T_L R_S T_L$), fifth-order ($T_L R_S R_L R_S T_L$) and seventh-order ($T_L R_S R_L R_S R_L R_S T_L$) multiple scattering are indicated in Figure 7.1, but it should be noted that radiation can also be scattered nine times, eleven times and so on, *ad infinitum*. The spectral reflectance of the canopy (R_C) is therefore given by the sum of the infinite series presented in Equation 7.1.

$$R_C = R_L + T_L R_S T_L + T_L R_S R_L R_S T_L + T_L R_S R_L R_S R_L R_S T_L + \ldots \qquad (7.1)$$

Herein lies the problem: to determine R_C exactly, the sum of this infinite series must be calculated. If the calculation is truncated at a particular order of multiple scattering, as in the preceding chapter, the result is only an approximation to R_C, i.e.,

$$R_C \approx R_L + T_L R_S T_L + T_L R_S R_L R_S T_L \qquad (7.2)$$

Harte (1988) presents an exact analytical solution to this problem, based on the approach of Rasool and Schneider (1971), which is adopted here. It starts by noting that the second and successive terms of the infinite series presented in Equation 7.1 are products (multiples) of $T_L R_S$ and T_L. Thus, Equation 7.1 can be rewritten as

$$R_C = R_L + T_L R_S \left(1 + R_L R_S + R_L R_S R_L R_S + \ldots \right) T_L \qquad (7.3)$$

which can be further simplified to

$$R_C = R_L + T_L R_S \left(1 + R_L R_S + (R_L R_S)^2 + \ldots + (R_L R_S)^\infty \right) T_L \qquad (7.4)$$

Note that the term in parentheses in Equation 7.4 describes a convergent geometric series. In other words, the quotient of successive elements of this term is a constant value (i.e., $R_L R_S$), since $(R_L R_S)^0 = 1$ and $(R_L R_S)^1 = R_L R_S$. This is important because the sum of an infinite geometric series of this form can be expressed as follows:

$$\sum_{i=0}^{\infty} (x)^i = \frac{1}{1-x} \qquad (7.5)$$

when the absolute value of x is less than 1; that is, $|x| < 1$ (Harris and Stocker 1998). Now since R_L and R_S are fractional values, their product ($R_L R_S$) must also be less than 1. It follows, therefore, that

$$\sum_{i=0}^{\infty} (R_L R_S)^i = \frac{1}{1-R_L R_S} \qquad (7.6)$$

Consequently, we can restate Equation 7.4 as

$$R_C = R_L + T_L R_S \left(\frac{1}{1-R_L R_S} \right) T_L \qquad (7.7)$$

which further simplifies to

$$R_C = R_L + \frac{T_L R_S T_L}{1-R_L R_S} \qquad (7.8)$$

This represents an exact analytical solution to the two-layer model for the special case in which $A = 1$. This solution accounts for infinite-order scattering of solar radiation between the leaf layer and the soil substrate.

7.2.2 Implementing and Running the Exact Analytical Solution

A computational implementation of the exact analytical solution to the two-layer model (Equation 7.8) is given in Program 7.1. The terms R_C, R_L, T_L and R_S are represented by the variables R_Canopy, R_Leaf, T_Leaf and R_Soil, respectively. As usual, the values of these variables need to be supplied via the command line, using a separate -v flag for each variable. Program 7.1 represents a direct translation of the mathematical formula given in Equation 7.8 into **gawk** code. Apart from the comments (lines 1 to 14), all of the code is contained within the BEGIN block (lines 16 to 20). This is because the intention is to generate data (i.e., to perform a simulation), rather than to process data contained in a file. Note that a line continuation symbol ($\backslash$) is used to break a long statement over lines 17 and 18.

Program 7.1 can be run from the command line by typing the following instruction to determine the spectral reflectance of the two-layer canopy at NIR wavelengths, assuming that $R_L = 0.475$, $R_S = 0.125$ and $T_L = R_L$:

```
gawk -f analytic.awk -v R_Leaf=0.475 -v T_Leaf=0.475 -v ↩          1
    ↪R_Soil=0.125
```

Program 7.1: analytic.awk

```
# Exact analytical solution to the simple, two-layer (leaf     1
# + soil substrate) model of light interaction with a plant    2
# canopy, for the special case in which the leaf layer          3
# completely covers the soil substrate. Accounts for            4
# infinite-order multiple scattering.                           5
#                                                               6
# Usage: gawk -f analytic.awk -v R_Leaf=value \                 7
#             -v T_Leaf=value -v R_Soil=value [ > output_file ] 8
#                                                               9
# Variables:                                                    10
# R_Leaf     Leaf spectral reflectance                          11
# T_Leaf     Leaf spectral transmittance                        12
# R_Soil     Soil spectral reflectance                          13
# R_Canopy   Total canopy spectral reflectance                  14
                                                                15
BEGIN{                                                          16
    R_Canopy=R_Leaf+(T_Leaf*T_Leaf*R_Soil) \                    17
        /(1-(R_Leaf*R_Soil));                                   18
    print R_Canopy;                                             19
}                                                               20
```

Similarly, the following command line can be used to calculate the equivalent value at red wavelengths, assuming that $R_L = 0.04$, $R_S = 0.08$ and $T_L = R_L$:

```
gawk -f analytic.awk -v R_Leaf=0.04 -v T_Leaf=0.04 -v ↪      2
   ↪R_Soil=0.08
```

7.2.3 Evaluating the Exact Analytical Solution

Table 7.1 presents a comparison of the results obtained using alternative formulations and implementations of the two-layer model: (i) twolayr2.awk (Program 6.3, Chapter 6), the approximate solution, taking into account up to fifth-order multiple scattering and (ii) analytic.awk (Program 7.1), the exact analytical solution. For the purpose of comparability, the results derived from both implementations relate to the $A = 1$ case; that is, where the leaf layer completely covers the ground below. Data are presented for a snow substrate and a soil substrate (i.e., high and low reflectance backgrounds, respectively), and for red and NIR wavelengths (i.e., for low and high leaf reflectance, respectively). The values used for the spectral reflectance of the snow substrate are $R_S = 0.9$ at red wavelengths and $R_S = 0.75$ in the NIR.

Table 7.1 suggests that, in most of the cases considered, the inclusion of very high order multiple scattering has relatively little impact on the canopy spectral reflectance predicted by the two-layer model. The exception is the case of the snow substrate in the NIR, where the approximate solution underestimates the exact solution by about 0.034 (3.4%) reflectance.

Table 7.1: Spectral reflectance of a simple plant canopy with either a soil or a snow substrate derived using alternative formulations and implementations of the two-layer model.

Substrate	λ	Parameter Values			Formulation/Implementation	
		R_L	T_L	R_S	twolayr2.awk	analytic.awk
Soil	Red	0.040	0.040	0.080	0.0401	0.0401
	NIR	0.475	0.475	0.040	0.5049	0.5050
Snow	Red	0.040	0.040	0.900	0.0415	0.0415
	NIR	0.475	0.475	0.750	0.7045	0.7379

Note that R_L and T_L are, respectively, the spectral reflectance and transmittance of the leaf layer, and R_S is the spectral reflectance of the substrate.

More importantly, these results demonstrate that there is frequently more than one solution to an environmental modeling problem. The approximate solution is the more flexible in this instance because it allows canopy reflectance to be modeled as a function of the fraction of ground covered by vegetation. By contrast, the exact analytical solution is limited to the special case in which the leaf layer completely covers the ground below ($A = 1$). Nevertheless, it allows one to gauge the error associated with the approximate solution. It is rare, though, to enjoy the luxury of having both exact and approximate solutions.

Exercise 7.1: Run the exact analytical solution to the two-layer model (Program 7.1) to calculate the canopy spectral reflectance at blue (0.5 μm) and green (0.55 μm) wavelengths. Estimate the values of R_L and R_S at these wavelengths from Figure 6.3 (page 145). Assume that $T_L = R_L$.

7.3   AN ITERATIVE NUMERICAL SOLUTION TO THE MULTIPLE LEAF-LAYER MODEL

The example discussed in Section 7.2 demonstrates that it is sometimes possible to find an exact analytical solution to a model. Not all models are amenable to this type of solution though; some must be solved numerically. The latter involves a sequence of operations that are performed repeatedly and, each time, applied to the result of the previous iteration (British Computer Society 1998). Sometimes an estimate, or guess, is made regarding the correct solution, to which the iterative procedure is initially applied. Provided that the model has been formulated and implemented correctly, the expectation is that the procedure will move ever closer to the correct solution with each successive iteration. Depending on the nature of the model, and in some cases the accuracy of the initial estimate, the iterative procedure will eventually reach the correct solution or a very close approximation to it. Perhaps the best known example of this approach is Newton's method for calculating $\sqrt{x}$ (Harris and Stocker 1998).

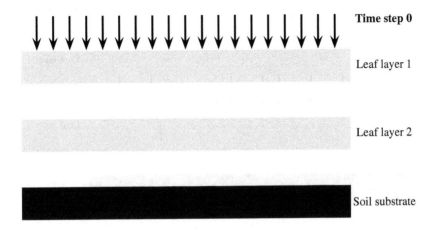

Figure 7.2: Interaction of a photon stream with a multiple leaf-layer canopy (time step 0).

A numerical solution to the light-interaction model, which can accommodate any number of leaf layers and take into account very high order multiple scattering within the canopy, is explored in this section. The computational implementation of this model relies heavily on looping (control-flow) constructs of the type introduced in the preceding chapters. A new control-flow construct, gawk's do while statement, is also introduced here. The circumstances in which this construct or the for loop is most appropriate are examined. Before proceeding to that stage, however, the underlying conceptual model is revisited once more.

7.3.1 Revisiting the Conceptual Model

Imagine that solar radiation arrives at the top of a multiple layer plant canopy as a stream of photons; that is, as individual particles of light (Figure 7.2). Assume that each of the leaf layers in the canopy completely covers the soil substrate (equivalent to the $A = 1$ case in the preceding section and chapter). Imagine, also, that one is able to observe the passage of these photons over infinitesimally small intervals in time, as they travel through the canopy. Now consider what one might see.

At first, the photons strike the top leaf-layer (Figure 7.2). As a result, some of the photons are reflected back out to space, some are transmitted downward into the canopy, while the remainder are absorbed by the top leaf-layer (Figure 7.3, top). The photons that are transmitted down through the first leaf-layer are then intercepted by the second leaf-layer. Some of these photons are reflected back up toward the first leaf-layer, some are transmitted down toward the soil substrate, and some are absorbed (Figure 7.3, middle). Eventually, some of the photons reach the soil substrate and are either absorbed by it or are reflected back up through the canopy (Figure 7.3, bottom). At the same time, other photons are scattered upward and downward between the leaf layers. This process continues until, ultimately, the photons are either absorbed by the leaf layers and the soil substrate or they escape from the canopy via

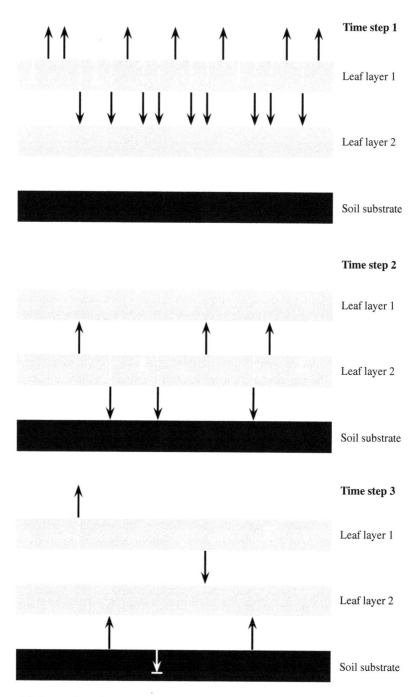

Figure 7.3: Interaction of a photon stream with a multiple leaf-layer canopy at time steps 1 (top), 2 (middle) and 3 (bottom).

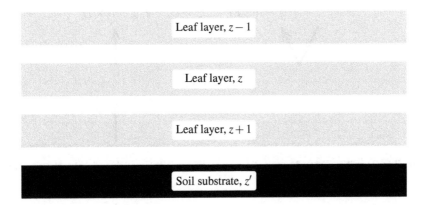

Figure 7.4: Schematic representation of a plant canopy comprising multiple plane-parallel leaf layers ($z-1$, z, and $z+1$) and a soil substrate (z').

the top leaf-layer. The reflectance of the canopy as a whole is therefore given by the fraction of incident photons that eventually exit the canopy via the top leaf-layer.

Re-stating the conceptual model in this way highlights the following points:

1. There are effectively two "streams" of photons, which travel through the canopy in opposite directions — one passes vertically downward into the canopy, the other passes vertically upward and eventually out of the canopy.

2. Photons continue to travel in the same direction after interacting with a leaf layer if they are transmitted through it.

3. The direction in which a photon travels changes (reverses) if it interacts with and is reflected from either a leaf layer or the soil substrate.

4. Photons are lost altogether (the photon stream is attenuated) if they interact with and are absorbed by either a leaf layer or the soil substrate.

Importantly, these points are independent of the number of leaf layers in the model canopy. In other words, our understanding of the nature of the problem has been further generalized. With this in mind, the mathematical model is now reformulated for a final time to account for the interaction of solar radiation with a plant canopy comprising any number of plane-parallel leaf layers.

7.3.2 Formulating the Mathematical Model

Consider an individual leaf-layer, z, located somewhere within a plant canopy that comprises several such plane-parallel layers plus a soil substrate, z' (Figure 7.4). To determine the reflectance of the canopy as a whole, the numbers of photons that are scattered both upward and downward after interacting with this layer (and each of the other canopy layers) must be established. The former comprises photons traveling

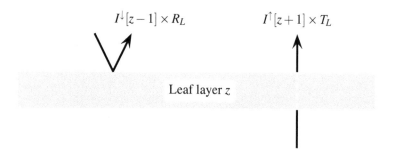

Figure 7.5: Contributions to flux traveling upward from leaf-layer z in a simple, plane-parallel, plant canopy.

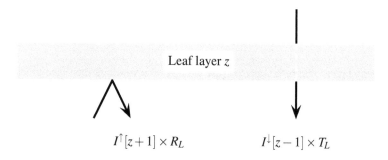

Figure 7.6: Contributions to flux traveling downward from leaf-layer z in a simple, plane-parallel, plant canopy.

upward from the layer below (layer $z+1$) that are transmitted through layer z, as well as photons traveling downward from the layer above (layer $z-1$) that are reflected upward from layer z (Figure 7.5). The latter consists of photons traveling downward from the layer above $(z-1)$ that are transmitted through layer z, plus photons traveling upward from the layer below $(z+1)$ that are reflected downward from layer z (Figure 7.6). The soil substrate (layer z') represents a special case in this context because it contributes only to the upward flux stream (Figure 7.7). Note that the arrows representing the pathways of the reflected flux in Figure 7.5 through Figure 7.7 are drawn at oblique angles with respect to the leaf and soil layers solely for the purpose of diagrammatic clarity; the model assumes that radiation travels either vertically upward or vertically downward only.

The symbols $I^\uparrow[z]$ and $I^\downarrow[z]$ are used here to refer to the photons (flux) traveling upward and downward, respectively, from layer z in the model canopy. Based on the descriptions above, $I^\uparrow[z]$ can therefore be expressed as

$$I^\uparrow[z] = \left(I^\downarrow[z-1] \times R_L\right) + \left(I^\uparrow[z+1] \times T_L\right) \tag{7.9}$$

where R_L and T_L are the spectral reflectance and transmittance, respectively, of leaf layer z (Figure 7.5). For the purpose of this example, it is assumed that each of the

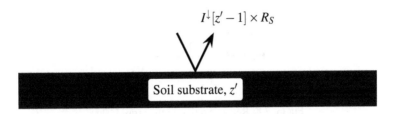

Figure 7.7: Contribution to flux traveling upward from the soil substrate (z') in a simple, plane-parallel, plant canopy.

leaf layers has the same spectral reflectance and transmittance properties, and that the values of these properties are identical for both the upper (adaxial) and lower (abaxial) surfaces of each layer. By the same token, $I^\downarrow[z]$ (Figure 7.6) is given by

$$I^\downarrow[z] = \left(I^\uparrow[z+1] \times R_L\right) + \left(I^\downarrow[z-1] \times T_L\right) \tag{7.10}$$

Moreover, the flux traveling upward from the soil substrate is given by

$$I^\uparrow[z'] = I^\downarrow[z'-1] \times R_S \tag{7.11}$$

where $z'-1$ is the leaf layer immediately above the soil substrate (Figure 7.7).

It should be evident from Equation 7.9 through Equation 7.11 that the number of photons traveling upward or downward from any given layer in the canopy is partly dependent on the number of photons that are incident on that layer from the layers immediately above and below it. Thus, for example, the values of both $I^\downarrow[z-1]$ and $I^\uparrow[z+1]$ are required to calculate $I^\uparrow[z]$. The problem is that $I^\downarrow[z-1]$ is dependent, in turn, on $I^\uparrow[z]$. Consequently, to determine $I^\downarrow[z-1]$, $I^\uparrow[z]$ must already be known, and the converse is also true. At first glance, the circularity involved in this argument appears to pose an insuperable mathematical conundrum, but it transpires that the problem can be solved simply and effectively using iterative computational techniques, which are demonstrated in the following section.

7.3.3 Implementing the Multiple Leaf-Layer Model

There are several important issues that need to be addressed when implementing the multiple layer model given in Equation 7.9 through Equation 7.11. In particular, the program must be able to store and manipulate information on the spectral reflectance and transmittance properties of the leaf layers, as well as the number of photons leaving each layer in both the upward and downward directions. As noted previously, it is assumed that each of the leaf layers has the same spectral reflectance and transmittance properties, and that the adaxial and abaxial surfaces are identical in terms of spectral reflectance and transmittance. Thus, only two gawk variables, R_Leaf and T_Leaf, are required to represent these properties and to store their values. A further variable, R_Soil, is needed to represent the spectral reflectance of the soil substrate.

Representing the number of photons leaving each layer is more problematic. One solution is to use separate variables for each layer; for example, I_Up_Layer1,

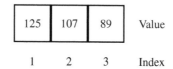

Figure 7.8: Representation of the array I_Up.

I_Down_Layer1, I_Up_Layer2, I_Down_Layer2 and so on, where I_Up_Layer1 denotes the number of photons traveling upward from layer 1, and I_Down_Layer1 refers to the number traveling downward from the same layer. The problem with this approach is that it becomes less efficient and more error-prone as the number of leaf layers increases. Thus, one must keep track of 20 similarly named variables for a model incorporating 10 leaf-layers. More importantly, the objective is to implement a model that can represent a canopy containing any number of leaf layers. This means that one does not know in advance how many leaf layers will be specified and, hence, how many variables are required. A much better solution, therefore, is to employ **gawk**'s array capabilities.

Using Arrays to Store Collections of Related Data Values

An array is a computational structure that allows a collection of related data items to be grouped together and referred to by a single identifier or name (British Computer Society 1998). An array can be thought of as a table of cells, or elements, each of which is capable of storing a single numerical value or character string (i.e., text). Each element of an array is uniquely referenced by its index. For example, rather than storing the number of photons passing upward from each leaf layer using separate variables, I_Up_Layer1 = 125, I_Up_Layer2 = 107, I_Up_Layer3 = 89, and so on, it is possible to group these values together and store them in a single array called I_Up (Figure 7.8). In this context, I_Up[1] refers to element 1 of the array, which contains the value 125, I_Up[2] refers to element 2 of the array, which contains the value 107, and I_Up[3] refers to element 3 of the array, which contains the value 89. In the example studied here, the index is used to distinguish the different leaf layers. Note that the array index is enclosed by square brackets ([]).

Unlike some other computer programming languages, **gawk** is very flexible about how it sets up and handles arrays. For example, arrays need not be declared before use, nor is there any need to specify in advance the number of elements that are contained within an array (Robbins 2001). Moreover, further elements may be added to an array at any time (i.e., dynamically). The technical details of how this is achieved need not detain us here, but it is useful to note that **gawk**'s arrays are associative (Robbins 2001). In practice, this means that the index of an array is not restricted to positive integer values, they can also be character strings. For example, one could store monthly rainfall data in an array called rainfall using the name of the month as the index (e.g., rainfall[january], rainfall[february],...). Of course, one could also have used rainfall[1] to refer to the rainfall in January, rainfall[2] to refer to

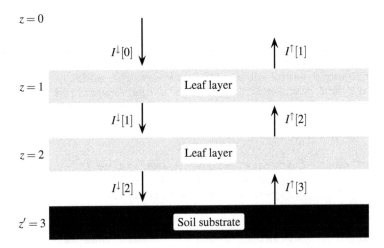

Figure 7.9: Two-stream model of radiation transport through a multiple leaf-layer model of a vegetation canopy (where `layers=2`; see text for explanation).

that in February, and so on, but associative arrays in which the indexes are character strings are arguably more intuitive.

Two other important features of **gawk**'s arrays are (i) that they cannot share the same name as a variable used in the same **gawk** program and (ii) unless one specifies otherwise, each element of the array is automatically assigned a value of `""`, known as the null (empty) string. When used in mathematical calculations, **gawk**'s null string equates to the value zero (0).

Putting Arrays into Practice

An implementation of the multiple leaf-layer model given in Equation 7.9 through Equation 7.11, which makes use of arrays, is presented in Program 7.2. Ignoring comments, this program consists of just six lines of code. These are examined in detail below with the help of Figure 7.9, which illustrates the situation for a three-layer model (two leaf layers and a soil substrate).

Program 7.2 requires the user to enter via the command line the number of leaf layers in the model canopy. This information is stored in the variable `layers`. The user must also supply values for the spectral reflectance and transmittance of the leaf layers (stored in the variables `R_Leaf` and `T_Leaf`, respectively), as well as the spectral reflectance of the soil substrate (`R_Soil`).

The program uses two separate arrays (`I_Up` and `I_Down`; lines 16 and 17) to store data on the upward and downward fluxes of radiation through the model canopy. In this context, `I_Down[0]` denotes downwelling solar radiation incident at the top of the canopy, `I_Down[1]` refers to flux traveling downward from the top leaf-layer, and so on through to `I_Down[layers]`, which indicates flux traveling downward from the bottom leaf-layer. Likewise, `I_Up[1]` refers to flux traveling upward from the top leaf-layer (i.e., flux exiting the canopy), and so on through to `I_Up[layers]`, which

Program 7.2: iterate.awk

```
# Program to calculate the spectral reflectance of a        1
# plant canopy based on a simple two-stream model          2
# (upward and downward fluxes) for a user-specified        3
# number of leaf layers above a soil substrate. The        4
# program requires selected information to be provided      5
# via the command line:                                     6
#                                                           7
# Usage: gawk -f iterate.awk -v  R_Leaf=value \             8
#              -v T_Leaf=value -v R_Soil=value \            9
#              -v layers=value [ > outFile ]               10
#                                                          11
# Variables:                                               12
# R_Leaf       Spectral reflectance of each leaf layer     13
# T_Leaf       Spectral transmittance of each leaf layer   14
# R_Soil       Spectral transmittance of the soil substrate 15
# I_Down[z]    Flux traveling downards from layer z        16
# I_Up[z]      Flux traveling upwards from layer z         17
# layers       Number of leaf layers in canopy            18
# iteration    Number of iterations performed             19
                                                          20
BEGIN {                                                    21
   # Set the total incident solar radiation                22
   I_Down[0]=1;                                            23
                                                          24
   # Outer loop performs iterations                        25
   for (iteration=1;iteration<=20;iteration++){            26
                                                          27
      # Inner loop deals with each leaf layer in turn      28
      for (z=1; z<=layers; z++){                           29
         I_Up[z]=(I_Down[z-1]*R_Leaf)+(I_Up[z+1]*T_Leaf);  30
         I_Down[z]=(I_Up[z+1]*R_Leaf)+(I_Down[z-1]*T_Leaf); 31
      }                                                    32
                                                          33
      # ...and then the soil substrate                     34
      I_Up[layers+1]=I_Down[layers]*R_Soil;                35
                                                          36
   }                                                       37
                                                          38
   # Finally, print out the number of leaf layers (LAI) and 39
   # the spectral reflectance of the canopy as a whole      40
   printf("%-3i %5.4f\n", layers, I_Up[1]);                41
                                                          42
}                                                          43
```

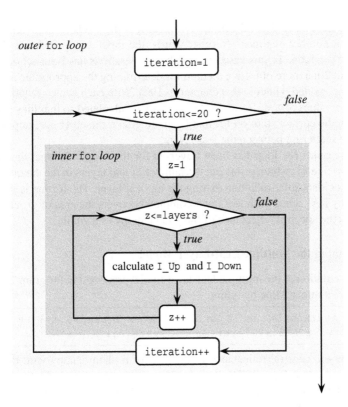

Figure 7.10: Flow-chart representation of the two nested `for` loops used in `iterate.awk` (Program 7.2).

denotes flux traveling upward from the bottom leaf-layer. Finally, `I_Up[layers+1]` indicates flux traveling upward from the soil substrate.

The majority of the code is placed in the `BEGIN` block (lines 21 to 43) because the program is intended to perform simulations, and hence to generate new data, rather than to process data contained in a file. In the `BEGIN` block, the total downwelling solar radiation incident on the top of the canopy is initialized to 1 (`I_Down[0]=1`) on line 23. The objective is to determine what fraction of this eventually exits the canopy (i.e., is reflected by it).

The remainder of the `BEGIN` block consists primarily of two `for` loops, one "nested" within the other (see also Figure 7.10). The inner `for` loop (lines 29 to 32) cycles through each of the leaf layers in turn, from the top (`z=1`) to the bottom (`z=layers`) of the canopy. In so doing it calculates the fraction of incident flux traveling upward and downward from each leaf layer (lines 30 and 31; Equations 7.9 and 7.10). After considering all of the leaf layers, the program moves on to deal with the special case of flux traveling upward from the soil substrate (line 35; Equation 7.11). This allows the incident flux (photons) to traverse its way down through the canopy to the soil substrate and eventually back up and out of the canopy via the top leaf-layer, in a manner analogous to that shown in Figure 7.3.

The outer `for` loop (lines 26 and 37) ensures that the work of the inner `for` loop is repeated a number of times. In other words, the inner `for` loop forms the body of the outer `for` loop. In this case, the number of iterations has been set, somewhat arbitrarily, to 20; a more objective method for determining the appropriate number of iterations is considered later in this chapter (p.190). With each cycle around the outer `for` loop, the values of `I_Up[z]` and `I_Down[z]` are recalculated so that they gradually converge on the correct solution for that layer. In effect, the outer `for` loop accounts for multiple scattering between the leaf layers.

After the outer `for` loop has been traversed for the twentieth time, the program progresses to line 41, which prints out the number of leaf layers in the canopy and the fraction of incident solar radiation exiting the top leaf layer. The former is equivalent to the canopy LAI, since each leaf layer completely covers the area of ground below. The latter is the spectral reflectance of the plant canopy as a whole.

7.3.4 Running the Multiple Leaf-Layer Model

The implementation of the multiple leaf-layer model presented in Program 7.2 can be run from the command line by typing

```
gawk  -f  iterate.awk  -v  R_Leaf=0.475  -v  T_Leaf=0.475  -v ↩     3
    ↪R_Soil=0.125  -v  layers=3  >  layers.nir
```

to determine the spectral reflectance in the NIR of a plant canopy with three leaf-layers, assuming that $R_L = 0.475$, $R_S = 0.125$ and $T_L = R_L$. Similarly, the following command line instruction can be used to calculate the canopy spectral reflectance at red wavelengths, assuming that $R_L = 0.04$, $R_S = 0.08$ and $T_L = R_L$:

```
gawk  -f  iterate.awk  -v  R_Leaf=0.04  -v  T_Leaf=0.04  -v ↩       4
    ↪R_Soil=0.08  -v  layers=3  >  layers.red
```

The revised model can be run numerous times, varying the value of `layers` on each occasion while keeping the other parameters constant. This allows the spectral reflectance of the plant canopy to be simulated for different numbers of leaf layers and, hence, different LAI. The results of each model run can be added to the end of the files `layers.nir` and `layers.red` by replacing the "greater than" symbol (>) in the example command lines shown above with two "greater than" symbols (>>), which means "append the output to the following file". For example, the following instruction adds the result to the end of the file `layers.nir`:

```
gawk  -f  iterate.awk  -v  R_Leaf=0.475  -v  T_Leaf=0.475  -v ↩     5
    ↪R_Soil=0.125  -v  layers=4  >>  layers.nir
```

Running the model in this way, and varying the number of leaf layers between 0 and 10, produces the data shown in Table 7.2.

7.3.5 Evaluating the Multiple Leaf-Layer Model

Figure 7.11 is a visualization of the data in Table 7.2, generated using the following gnuplot commands:

Table 7.2: Output from `iterate.awk` at red and NIR wavelengths.

LAI	Canopy Reflectance	
(Leaf Layers)	Red	NIR
0	0.0800	0.1250
1	0.0401	0.5050
2	0.0401	0.6249
3	0.0401	0.6755
4	0.0401	0.6994
5	0.0401	0.7112
6	0.0401	0.7169
7	0.0401	0.7196
8	0.0401	0.7208
9	0.0401	0.7212
10	0.0401	0.7214

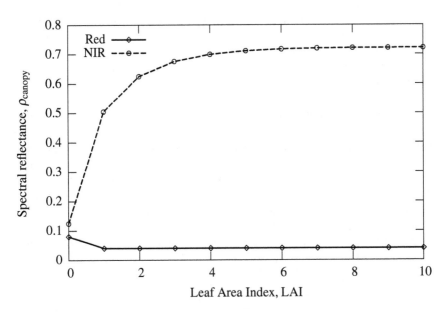

Figure 7.11: Relationship between LAI and canopy spectral reflectance at red and NIR wavelengths predicted by the numerical solution to the multiple leaf-layer model (`iterate.awk`; Program 7.2).

```
set ylabel "Spectral reflectance, rho_canopy"              1
set xlabel "Leaf Area Index, LAI"                          2
set style data linespoints                                 3
set key 1.25,0.75                                          4
plot 'layers.red' title "Red" lt 1 lw 2, \                 5
     'layers.nir' title "NIR" lt 2 lw 2                     6
```

These commands assume that the files layers.red and layers.nir are located in the working directory; otherwise, their full or relative pathnames must be provided.

Figure 7.11 suggests that the numerical solution to the multiple leaf-layer model reproduces the expected relationship between canopy spectral reflectance and LAI reasonably well (Figure 6.10 on page 156). Moreover, for the single leaf-layer case (LAI = 1), the results generated by iterate.awk (Program 7.2) are identical (at least to 4 decimal places) to those produced by the exact analytical solution, analytic.awk (Program 7.1), presented in the preceding section (Table 7.1 on page 178).

7.3.6   How Many Iterations are Required?

Program 7.2 performs 20 iterations (line 26) to calculate the fraction of incident solar radiation reflected by the multiple leaf-layer plant canopy. This value was selected rather arbitrarily; it could equally well have been 10 or 30. In this context, reducing the number of iterations from 20 to 10 would halve the time required to run the model. This is not a particularly significant consideration here because the time taken to run Program 7.2 is quite short anyway, even on a low-specification computer. There are many other instances, however, in which the model run times are very large (e.g., hours or days). It makes sense in such circumstances to limit the number of iterations, performing no more than is absolutely necessary. The question is, how many iterations are required? The obvious response is, a sufficient number to obtain an accurate answer. The challenge, though, is to establish an objective method against which to evaluate quantitatively this rather subjective criterion.

The effect of the number of iterations performed, when calculating the spectral reflectance of the multiple leaf-layer plant canopy, can be explored by making some minor modifications to the final few lines of code in Program 7.2. These are presented in Program 7.3. First, the printf statement (line 40 in Program 7.3) has been moved inside the outer for loop (note the closing brace ending the outer for loop on line 41 in Program 7.3; cf. line 37 in Program 7.2), so that it is performed once per iteration. Second, the printf statement has been altered so that the program prints out the number of iterations performed thus far, as well as the estimated fraction of incident solar radiation reflected by the plant canopy as a whole after that number of iterations.

The modified code can be run from the command line by typing

```
gawk -f iterate2.awk -v R_Leaf=0.475 -v T_Leaf=0.475 -v ↩      6
    ↪R_Soil=0.125 -v layers=10 > iterate2.nir
```

and

```
gawk -f iterate2.awk -v R_Leaf=0.04 -v T_Leaf=0.04 -v ↩        7
    ↪R_Soil=0.08 -v layers=10 > iterate2.red
```

Program 7.3: iterate2.awk

```
# Program to calculate the spectral reflectance of a          1
# plant canopy, based on a simple two-stream model           2
# (upward and downward fluxes) for a user-specified          3
# number of leaf layers above a soil substrate. The          4
# program requires selected information to be provided        5
# via the command line:                                      6
#                                                            7
# Usage: gawk -f iterate2.awk -v  R_Leaf=value \             8
#              -v T_Leaf=value -v R_Soil=value \             9
#              -v layers=value [ > outFile ]                 10
#                                                            11
# Variables:                                                 12
# R_Leaf       Spectral reflectance of each leaf layer       13
# T_Leaf       Spectral transmittance of each leaf layer     14
# R_Soil       Spectral transmittance of the soil substrate  15
# I_Down[z]    Flux traveling downwards from layer z         16
# I_Up[z]      Flux traveling upwards from layer z           17
# layers       Number of leaf layers in canopy              18
# iteration    Number of iterations performed               19
                                                             20
BEGIN {                                                      21
   # Set the total incident solar radiation                 22
   I_Down[0]=1;                                              23
                                                             24
   # Outer loop performs iterations                         25
   for (iteration=1;iteration<=20;iteration++){             26
                                                             27
      # Inner loop deals with each of the leaf layers in turn 28
      for (z=1; z<=layers; z++){                            29
         I_Up[z]=(I_Down[z-1]*R_Leaf)+(I_Up[z+1]*T_Leaf);   30
         I_Down[z]=(I_Up[z+1]*R_Leaf)+(I_Down[z-1]*T_Leaf); 31
      }                                                      32
                                                             33
      # ...and then the soil substrate                      34
      I_Up[layers+1]=I_Down[layers]*R_Soil;                 35
                                                             36
      # Print out the number of iterations performed thus far 37
      # and the estimated spectral reflectance of the canopy 38
      # as a whole after that number of iterations          39
      printf("%-3i %5.4f\n", iteration, I_Up[1]);           40
   }                                                         41
                                                             42
}                                                            43
```

Table 7.3: Output from `iterate2.awk` (Program 7.3) at red and NIR wavelengths, showing the predicted canopy spectral reflectance as a function of the number of iterations.

Iterations	1	2	3	4	5	10	20
Red	0.0400	0.0401	0.0401	0.0401	0.0401	0.0401	0.0401
NIR	0.4750	0.5822	0.6305	0.6578	0.6750	0.7091	0.7214

to determine, respectively, the red and NIR reflectance of a plant canopy consisting of 10 leaf-layers as a function of the number of iterations performed, based on the usual values for leaf reflectance, leaf transmittance and soil reflectance. The results obtained are summarized in Table 7.3.

It is evident from Table 7.3 that the 10 leaf-layer model reaches a stable value (0.0401) after only two iterations at red wavelengths. In other words, further iterations of the model do not produce an appreciable change in the predicted value of canopy reflectance in this part of the spectrum. As has already been noted, this is because the overwhelming majority of incident red radiation is reflected by the top leaf-layer, the contribution from which is modeled effectively after two iterations. Moreover, the impact of multiple scattering is negligible at red wavelengths because of the relatively strong absorption of radiation at these wavelengths by the leaf layers. By comparison, the predicted canopy reflectance at NIR wavelengths continues to increase as a function of the number of iterations performed, and does not appear to reach a stable value within the range examined. Nevertheless, these data describe an asymptotic relationship, which can be visualized in gnuplot as follows:

```
unset key                                                              7
set xlabel "Number of iterations"                                      8
plot 'iterate2.nir' w lp                                               9
```

Thus, the model predicts relatively small changes in the canopy reflectance after 10 or so iterations (Figure 7.12).

These results highlight two important issues: first, the number of iterations needed to reach a stable solution differs according to the input values presented to the model (i.e., as a function of wavelength in this instance); second, an objective method of determining the required number of iterations has yet to be established. These issues are, of course, interrelated. A solution to both of these problems is examined below.

7.3.7   Objective Determination of the Required Number of Iterations

Previously, the number of iterations employed to simulate the spectral reflectance of the multiple layer plant canopy was "hard wired" into the code (Program 7.2). Thus, the number of iterations performed was greater than strictly necessary at red wavelengths, while further iterations were required to reach a stable solution in the NIR. A much better approach is to allow the model to continue running while the result of the current iteration differs significantly from that of the preceding iteration because this implies that the model has not yet reached a stable solution. However, once the

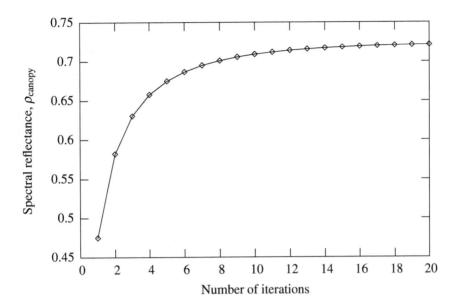

Figure 7.12: Estimated spectral reflectance of a 10 leaf-layer plant canopy at near-infrared wavelengths as a function of the number of model iterations.

difference between the results of successive iterations is less than a threshold value, defined by the user, the iterative process should cease. This is because further iterations are likely to result in only marginal changes to the model output. Put another way, assuming that the threshold value is small, the computational "cost" of performing extra iterations outweighs the potential "benefit" that may be obtained in terms of improvements to the accuracy of the model output.

The required number of iterations is generally not known in advance, so that a for loop, which performs a pre-determined number of iterations, is usually not the most appropriate control-flow structure to use. gawk provides two other control-flow structures that are much better suited to this type of problem: the while and do while loops (Aho *et al.* 1988, Robbins 2001). As their names suggest, these constructs are similar in terms of both syntax and function. The do while loop is used here. The general syntax of the do while statement is as follows (Robbins 2001, Figure 7.13):

```
do
    body
while (condition)
```

Thus, the lines of code in the body are performed repeatedly while the condition is true, otherwise the program moves on to the next line of code. The structure of the do while loop is such that the code in the body is performed at least once.

Program 7.4 presents an alternative implementation of the numerical solution to the multiple leaf-layer light-interaction model. This program uses a do while loop (lines 29 to 39) to control the number of iterations performed, replacing the outer for loop used in Program 7.2. Lines 30 to 38 represent the body of the do while loop.

Program 7.4: iterate3.awk

```
# Program to calculate the spectral reflectance of a        1
# plant canopy, based on a simple two-stream model          2
# (upward and downward fluxes) for a user-specified         3
# number of leaf layers above a soil substrate.             4
#                                                           5
# Usage: gawk -f iterate3.awk -v R_Leaf=value \             6
#                -v T_Leaf=value -v R_Soil=value \          7
#                -v layers=value -v threshold=value \       8
#                [ > outFile ]                              9
#                                                          10
# Variables:                                               11
# R_Leaf       Spectral reflectance of each leaf layer     12
# T_Leaf       Spectral transmittance of each leaf layer   13
# R_Soil       Spectral transmittance of the soil substrate 14
# I_Down[z]    Flux traveling downwards from layer z        15
# I_Up[z]      Flux traveling upwards from layer z          16
# I_Up_prev    Flux traveling upward from the top leaf-     17
#              layer during the previous iteration          18
# layers       Number of leaf layers in canopy             19
# iteration    Number of iterations performed thus far      20
# threshold    Reflectance threshold to terminate iterations 21
#                                                          22
BEGIN {                                                     23
   # Set the total incident solar radiation                24
   I_Down[0]=1;                                             25
                                                            26
   iteration=0;                                             27
   # Outer do-while loop controls number of iterations      28
   do{                                                      29
      I_Up_prev=I_Up[1];                                    30
      # Inner loop deals with each leaf layer in turn       31
      for (z=1; z<=layers; ++z){                            32
         I_Up[z]=(I_Down[z-1]*R_Leaf)+(I_Up[z+1]*T_Leaf);   33
         I_Down[z]=(I_Up[z+1]*R_Leaf)+(I_Down[z-1]*T_Leaf); 34
      }                                                     35
      # ...and then the soil substrate                      36
      I_Up[layers+1]=I_Down[layers]*R_Soil;                 37
      ++iteration;                                          38
   } while((I_Up[1]-I_Up_prev)>threshold)                   39
                                                            40
   # Print out threshold, iterations and canopy reflectance 41
   printf("%5.4f %3i %5.4f\n", threshold, iteration, I_Up[1]); 42
}                                                          43
```

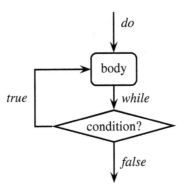

Figure 7.13: Flow-chart representation of the general structure of a do while loop.

Program 7.4 also introduces an additional command-line parameter, threshold (lines 8, 21, 39 and 42). This value, which is measured in terms of the difference in the predicted canopy spectral reflectance between successive iterations of the model, is used to decide when to halt the iterative procedure (line 39). At the beginning of each iteration, the fraction of incident solar radiation exiting the canopy via the top leaf-layer, estimated during the previous iteration, is stored in the variable I_Up_prev (line 30). The upward and downward fluxes from each layer in the canopy are then re-calculated (lines 32 through 37), after which the variable used to count the number of iterations performed thus far is incremented by one (++iteration, line 38). While the difference between the results from the current and previous iterations is larger than the user-defined threshold ((I_Up[1]-I_Up_prev)>threshold; line 39), further iterations of the model are performed (the program loops back from line 39 to line 30). As soon as the difference between successive iterations is equal to or smaller than the threshold value, the do while loop halts. The program then prints out the threshold value, the number of iterations performed and the fraction of incident solar radiation reflected by the canopy (i.e., the flux escaping the top leaf-layer; line 42).

7.3.8 Running and Evaluating the Revised Computational Model

Program 7.4 can be run from the command line by typing

```
gawk -f iterate3.awk -v R_Leaf=0.475 -v T_Leaf=0.475 -v ⟳          8
    ⟳R_Soil=0.125 -v layers=10 -v threshold=0.01 > iterate3⟳
    ⟳.nir
```

to determine the spectral reflectance of a plant canopy consisting of 10 leaf-layers, based on the usual values of leaf reflectance, leaf transmittance and soil reflectance at NIR wavelengths, and a threshold value of 0.01 (i.e., 1% reflectance). The following command line gives the equivalent result at red wavelengths:

```
gawk -f iterate3.awk -v R_Leaf=0.04 -v T_Leaf=0.04 -v ⟳          9
    ⟳R_Soil=0.08 -v layers=10 -v threshold=0.01 > iterate3.⟳
    ⟳red
```

Table 7.4: Output from `iterate3.awk` (Program 7.4) at red and NIR wavelengths, showing the number of iterations performed and the predicted canopy spectral reflectance, R_C.

		Wavelength			
Reflectance Threshold		Red		NIR	
Absolute	Percentage	Iterations	R_C	Iterations	R_C
0.1	10.00	1	0.0400	3	0.6305
0.05	5.00	1	0.0400	3	0.6305
0.01	1.00	2	0.0401	7	0.6950
0.001	0.10	2	0.0401	16	0.7191
0.0001	0.01	2	0.0401	28	0.7230

The results obtained are summarized in Table 7.4, along with those for further runs of the same model using different threshold values. Table 7.4 demonstrates that the number of iterations required increases as the reflectance threshold becomes more stringent (i.e., smaller). At NIR wavelengths, for instance, 7 iterations are needed for the model output to stabilize to within a threshold of 0.01 (1%) reflectance, while 16 iterations are required for a threshold of 0.001 (0.1%) reflectance.

> **Exercise 7.2**: Run Program 7.4 to simulate the spectral reflectance at NIR wavelengths of a plant canopy comprising 10 leaf-layers and a substrate comprising a layer of snow. Assume that $R_L = 0.475$, $R_L = R_T$ and $R_S = 0.75$. How many iterations are required for the model to stabilize using a reflectance threshold of 0.001?

7.4 BOUGUER'S LAW AND THE ATTENUATION COEFFICIENT

In this section, a further modification is made to the implementation of the multiple leaf-layer canopy reflectance model presented previously. This modification allows the model to be used to study the attenuation (extinction) of incident solar radiation as a function of the distance (i.e., the number of leaf layers) that it traverses into the canopy. In general, the amount of solar radiation penetrating a plant canopy varies inversely with depth because of the increasing probability that it is either absorbed or reflected upward by leaf layers higher up the canopy. Thus, for example, the amount of sunlight available to plants found on the forest floor is typically very much smaller than that available to the tallest trees, which form the canopy crown. The resulting vertical distribution of light within the forest has important implications for canopy photosynthesis and productivity, and different species of plants adopt contrasting strategies to compete for the available light resource (Jones 1983, Monteith and Unsworth 1995).

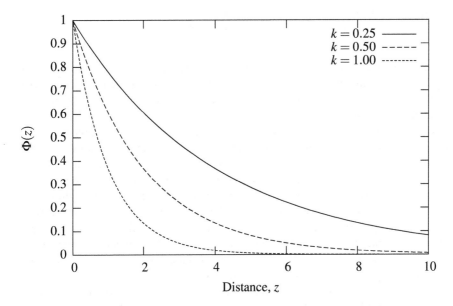

Figure 7.14: Attenuation of incident solar radiation as a function of the distance traversed through a homogeneous turbid medium, according to Bouguer's Law for various values of the attenuation coefficient, k.

The extinction of solar radiation within plant canopies is often modeled using Bouguer's Law, also known as Beer's Law or Lambert's Law of Absorption (Campbell and Norman 1998, Glickman 2000). This describes the attenuation of a parallel beam of monochromatic (single wavelength) radiation as it passes through a homogeneous turbid medium, and is given by

$$\Phi(z) = \Phi(0) \exp^{-kz} \tag{7.12}$$

where z is the distance traveled through the medium, $\Phi(0)$ is the flux density at $z = 0$ (before entering the medium), $\Phi(z)$ is the flux density remaining at distance z, and k is a constant of proportionality, known as the attenuation or extinction coefficient (Jones 1983, Monteith and Unsworth 1995). The value of the attenuation coefficient, k, therefore describes the overall optical density of the medium; that is, the extent to which it is either opaque or translucent. In the context of plant canopies, it provides a single value that can be used to compare the light regimes in different canopies or in the same canopy as it grows and develops over time. Equation 7.12 can be represented graphically in gnuplot using the following commands, which produce Figure 7.14:

```
set dummy z                          10
phi_0=1.0                            11
phi(k,z) = phi_0*exp(-k*z)           12
set key top right                    13
set xlabel "Distance, z"             14
set ylabel "Phi(z)"                  15
```

```
plot [0:10] phi(0.25,z) w l t "k=0.25", \                    16
            phi(0.50,z) w l t "k=0.50", \                    17
            phi(1.00,z) w l t "k=1.00"                       18
```

Note that line 10 instructs gnuplot to employ a "dummy" variable, z, to denote the
x-axis. This facility makes it possible to refer to the variable plotted on the *x*-axis (the
distance traversed into the canopy) by its conventional symbol, *z*, which is more convenient in this instance (Williams and Kelly 1998). Also note that the Bouguer Law
function is defined in terms of two parameters, k and z, which allows both parameters
to be varied when the function is plotted: the value of *z* varies continuously along the
x-axis; the value of *k* is specified explicitly on the command line (lines 16 to 18).

Strictly speaking, Bouguer's Law pertains to systems in which the radiation is absorbed without scattering as it passes through a homogeneous turbid medium (Monteith and Unsworth 1995). It is often applied empirically, though, in situations where
radiation is typically scattered once only; that is, where the effect of multiple scattering is negligible (Jones 1983, Monteith and Unsworth 1995). In terms of the multiple
leaf-layer model used here, this is a reasonable first approximation at red wavelengths
because the spectral reflectance and transmittance of leaves and soil are small in this
part of the electromagnetic spectrum; however, the assumption of single scattering
clearly does not hold at NIR wavelengths. In the remainder of this section, therefore,
Bouguer's Law is applied empirically to estimate the attenuation coefficient for the
multiple leaf-layer canopy at red wavelengths only. In doing so, it is assumed that the
number of leaf layers traversed down from the top of the canopy is a surrogate for the
distance parameter, *z*, in Bouguer's Law (Equation 7.12).

7.4.1   Implementing a Computational Model of Bouguer's Law

An implementation of Bouguer's Law is given in Program 7.5 (bouguer.awk). This
program is a very slightly modified version of Program 7.4. More specifically, lines
41 through 43 of Program 7.5 report the fraction of incident solar radiation remaining
at different levels within the canopy (I_Down[z]). A for loop is used to consider each
of the leaf layers in turn, starting at z=0 (at the top of the canopy; see Figure 7.9) and
finishing at z=layers (at the bottom of the canopy, beneath the lowest leaf-layer and
above the soil substrate).

7.4.2   Running and Evaluating the Modified Computational Model

Program 7.5 can be run from the command line by typing the following instruction
to determine the attenuation of incident solar radiation as a function of depth into
a 10 leaf-layer canopy at red wavelengths, based on a threshold value of 0.01 (1%
reflectance) and the usual values of leaf reflectance and transmittance and of soil
reflectance:

```
gawk -f bouguer.awk -v R_Leaf=0.04 -v T_Leaf=0.04 -v ↺       10
   ↺R_Soil=0.08 -v layers=10 -v threshold=0.01 > bouguer.↺
   ↺red
```

Program 7.5: bouguer.awk

```
# Program to calculate the extinction of incident solar          1
# radiation as a function if the number of leaf layers           2
# traversed into a vegetation canopy, based on a simple           3
# two-stream model (upward and downward fluxes).                  4
#                                                                 5
# Usage: gawk -f bouguer.awk -v  R_Leaf=value \                   6
#               -v T_Leaf=value -v R_Soil=value \                 7
#               -v layers=value -v threshold=value \              8
#               [ > outFile ]                                     9
#                                                                 10
# Variables:                                                      11
# R_Leaf        Spectral reflectance of each leaf layer           12
# T_Leaf        Spectral transmittance of each leaf layer         13
# R_Soil        Spectral transmittance of the soil substrate      14
# I_Down[z]     Flux traveling downards from layer z              15
# I_Up[z]       Flux traveling upwards from layer z               16
# I_Up_prev     Flux traveling upward from the top leaf-          17
#               layer during the previous iteration               18
# layers        Number of leaf layers in canopy                  19
# iteration     Number of iterations performed thus far          20
# threshold     Reflectance threshold to terminate iterations    21
                                                                 22
BEGIN {                                                          23
    # Set the total incident solar radiation                    24
    I_Down[0]=1;                                                 25
    iteration=0;                                                 26
    # Outer do-while loop controls number of iterations         27
    do{                                                         28
        I_Up_prev=I_Up[1];                                      29
        # Inner loop deals with each of the leaf layers in turn 30
        for (z=1; z<=layers; ++z){                              31
            I_Up[z]=(I_Down[z-1]*R_Leaf)+(I_Up[z+1]*T_Leaf);    32
            I_Down[z]=(I_Up[z+1]*R_Leaf)+(I_Down[z-1]*T_Leaf);  33
        }                                                       34
        # ...and then the soil substrate                        35
        I_Up[layers+1]=I_Down[layers]*R_Soil;                   36
        ++iteration;                                            37
    } while((I_Up[1]-I_Up_prev)>threshold)                      38
    # Print out the fraction of incident solar radiation        39
    # remaining at different depths (z) within the canopy.      40
    for(z=0; z<=layers; ++z){                                   41
        printf("%3i %5.4f\n", z, I_Down[z]);                    42
    }                                                           43
}                                                               44
```

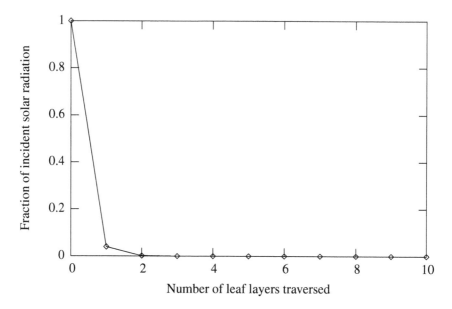

Figure 7.15: Attenuation of incident solar radiation at red wavelengths as a function of the number of leaf-layers traversed down from the top of the canopy.

7.4.3 Visualizing the Output

The results obtained using `bouguer.awk` (Program 7.5) can be visualized by typing the following commands in **gnuplot** assuming that the file `bouguer.red` is located in the working directory:

```
unset key                                                      19
set ylabel "Fraction of incident solar radiation"             20
set xlabel "Number of leaf layers traversed"                  21
plot 'bouguer.red' w lp                                       22
```

The resultant plot is presented in Figure 7.15. This shows the expected negative exponential relationship between the fraction of incident solar radiation remaining and increasing depth into the canopy; thus, the leaf layers at the bottom of the canopy receive considerably less solar radiation than those at the top. Note that over 90% of the incident radiation is attenuated after just one leaf-layer is traversed.

7.4.4 Function Fitting in gnuplot

The relationship between radiation extinction and depth into the model canopy can be expressed mathematically by fitting the equation for Bouguer's Law (Equation 7.12) to the data shown in Figure 7.15. The curve-fitting capabilities of **gnuplot** can be used for this purpose, as demonstrated below.

```
set fit logfile 'bouguer.fit'                                 23
fit phi(k,z) 'bouguer.red' via k                              24
```

```
FIT:       data read from 'bouguer.red'
           #datapoints = 11
           residuals are weighted equally (unit weight)

function used for fitting: phi(k,z)
fitted parameters initialized with current variable values




Iteration 0
WSSR          : 0.128181        delta(WSSR)/WSSR   : 0
delta(WSSR)  : 0               limit for stopping : 1e-05
lambda       : 0.146877

initial set of free parameter values

k                = 1

After 7 iterations the fit converged.
final sum of squares of residuals : 1.33222e-09
rel. change during last iteration : -5.84194e-07

degrees of freedom (ndf) : 10
rms of residuals        (stdfit) = sqrt(WSSR/ndf)       : 1.15422e-05
variance of residuals (reduced chisquare) = WSSR/ndf : 1.33222e-10

Final set of parameters            Asymptotic Standard Error
=======================            =========================

k               = 3.21639         +/- 0.0002874    (0.008935%)
```

Figure 7.16: Results reported by gnuplot when fitting the Bouguer Law function to the data in the file bouguer.red.

Note that the general form of the function phi(k,z) was defined previously (line 12 on page 197). Recall that this is a negative exponential function, specified in terms of two parameters, k and z, which correspond to the variables k and z in Equation 7.12. Line 23 instructs gnuplot to direct the results of the function-fitting procedure to the file bouguer.fit in the current directory. Line 24 fits the user-defined function phi(k,z) to the data contained in the file bouguer.red by varying the value of the k. Thus, this procedure finds the value of k that results in the best fit between the function phi(k,z) and the data in the file bouguer.red.

The output produced by these gnuplot commands is presented in Figure 7.16, which indicates that the estimated value of the attenuation coefficient, k, is 3.21639 at this wavelength. Another value given in the output is an estimate of the uncertainty associated with k (±0.0002874), which can be interpreted loosely as describing the confidence limits about k (Williams and Kelly 1998); the smaller this value is, the better the function fits the data. The result obtained here suggests a very good fit. This finding is also demonstrated in Figure 7.17, produced in gnuplot using the following command, which confirms how well the amount of solar radiation penetrating

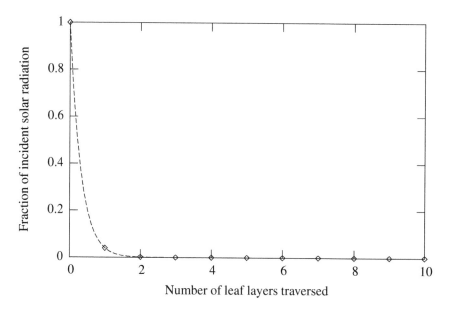

Figure 7.17: Attenuation of incident solar radiation at red wavelengths as a function of the number of leaf-layers traversed downward into a multi-layer plant canopy (points) and the negative exponential function (Bouguer's Law) fitted to these data (dashed lines).

to different depths in the model canopy is characterized by Bouguer's Law:

```
plot 'bouguer.red' w p, phi(k,z) w l                              25
```

   Relationships similar to the one observed in Figure 7.17 are often exploited by manufacturers of sensors designed to estimate, in a non-destructive way, the LAI of plant canopies, particularly arable crops. These instruments are used to measure the incident solar radiation above and below the canopy. By making certain assumptions about the angular distribution of leaves within the canopy, it is possible to derive fairly accurate estimates of LAI from such devices.

7.5 SUMMARY

This chapter explores two formulations, and several different implementations, of a model intended to represent the interaction of solar radiation with a plant canopy. The first formulation represents an analytical solution for the special case in which the canopy consists of a single leaf-layer that completely covers the soil substrate below. The second formulation employs a numerical solution to the problems of handling multiple leaf-layers and multiple scattering. Each of these two general approaches has its merits. It is not the intention of this chapter to establish which is "better". In any case, this will vary according to the nature of the problem under investigation. What it demonstrates, however, is that there is rarely only one way of solving an environmental modeling problem.

Much of this chapter is concerned with different computational implementations of Equation 7.9 through Equation 7.11, which underpin the numerical solution to the multiple leaf-layer model. These are employed in Program 7.2 to Program 7.5. It is instructive to note that these "two stream" equations are similar in form to the Kubelka-Munk equations, which describe the transport of solar radiation through a turbid medium that absorbs and scatters radiation (Monteith and Unsworth 1995, Campbell and Norman 1998). The Kubelka-Munk equations are commonly given as follows:

$$-dI^\uparrow = -(S+K)I^\uparrow dz + I^\downarrow Sdz \tag{7.13}$$

$$dI^\downarrow = -(S+K)I^\downarrow dz + I^\uparrow Sdz \tag{7.14}$$

where z is the distance traveled into the turbid medium, K is an absorption coefficient, S is a scattering coefficient, $I^\uparrow$ is the upward traveling flux and $I^\downarrow$ is the downward traveling flux (Monteith and Unsworth 1995). These equations form the basis of many of the more sophisticated models of light interaction with plant canopies that are widely used in studies of ecology, climatology and Earth observation (terrestrial remote sensing) (Goel 1987, Myneni *et al.* 1990, Pinty and Verstraete 1998). Readers wishing to pursue this aspect of environmental modeling further will find that the comprehensive review by Goel (1987) is an excellent starting point.

SUPPORTING RESOURCES

The following resources are provided in the `chapter7` sub-directory of the CD-ROM:

`analytic.awk`	gawk implementation of the exact analytical solution to the two-layer plant canopy model for the special case of $A = 1$
`iterate.awk`	gawk implementation of the iterative numerical solution to the multiple leaf-layer model
`iterate2.awk`	modified version of `iterate.awk`, which prints out the number of iterations taken to reach a stable solution
`iterate3.awk`	modified version of `iterate2.awk`, which uses a `do while` loop to control the number of iterations
`bouguer.awk`	modified version of `iterate3.awk`, which calculates the attenuation of solar radiation as a function of the number of leaf layers traversed into the canopy
`iterate.bat`	Command line instructions used in this chapter
`interate.plt`	gnuplot commands used to generate the figures presented in this chapter
`interate.gp`	gnuplot commands used to generate figures presented in this chapter as EPS files

Chapter 8

Population Dynamics

Topics

- Discrete and continuous models of population growth

- Difference equations and differential equations

- Constrained and unconstrained models of population growth

- Chaotic behavior in deterministic systems

- Inter-specific competition

- Predator-prey relationships

Methods and techniques

- Control-flow structures in **gawk**: the `while` loop

- Numerical integration using Euler and Runge-Kutta methods

8.1   INTRODUCTION

As a general rule, environmental systems tend to be highly dynamic; that is to say, the system state changes through time, sometimes dramatically so, in response to a set of forcing factors. A good example is the growth or decline of a population, where the system state is defined by the number of individuals of a particular species, typically within a finite geographical area. The species type is not a primary concern here. Instead, consideration is given to generic models that can be applied to populations of human beings, animals, plants, bacteria and a host of other organisms. This generality is desirable because it implies a degree of economy, such that a single model can be

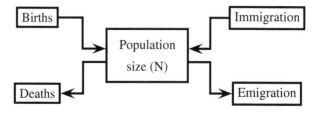

Figure 8.1: Factors leading to a change in the size of a population. The arrows denote the flow of individuals to and from the total population.

used to study several related problems. It is also beneficial because it enables parallels to be explored between environmental systems, so that knowledge of one can be used to help understand the functioning of others.

The models examined in this chapter can be divided into two broad categories: discrete and continuous. The distinction relates to the way in which each treats time as a variable. In discrete models, time proceeds in a series of finite steps that are typically, although not necessarily, of equal length; each step relates to a specific time interval such as an hour, a day, a month or a year. The system state is updated at the end of each time step to account for changes caused by processes operating on the system over the preceding period. In continuous models, by contrast, the system state is updated instantaneously in response to continuous changes in the forcing factors. Discrete and continuous models also differ in terms of the mathematical equations on which they are founded: the former are expressed in terms of difference equations; the latter employ differential equations. These are introduced informally throughout this chapter.

8.2 UNCONSTRAINED OR DENSITY-INDEPENDENT GROWTH

8.2.1 Development of the Conceptual Model

Various factors cause the population of a particular species in a given area to change in size over time. For instance, new individuals may be added to the population as a result of births or immigration from surrounding areas, while existing individuals may be lost through deaths or emigration to surrounding areas (Figure 8.1). Depending on the balance between these four demographic processes, the population may increase, remain stable or decline over time (Donovan and Welden 2002). Note that in studies of population dynamics it is common to refer to either the population size, which is the number of individuals of a given species, or the population density, which is the population size per unit area.

Following Alstad (2001), four assumptions are made to simplify the development of the initial model of population dynamics developed here, namely that (i) there is no migration of individuals between the study area and the surrounding regions (the study area is isolated in some way) or else the number of immigrants equals the number of emigrants, (ii) reproduction is asexual, (iii) every individual shares the same likelihood of producing offspring and of dying, and (iv) there is an infinite supply of the natural resources (food, water and shelter) required to support the population. The respective implications of these assumptions are that (i) changes to the population size (and density) are controlled solely by births and deaths, (ii) the relative abundance of males and females within the population can be ignored, (iii) the age structure of the population need not be taken into account, and (iv) there are no limits to the growth of the population (Alstad 2001). The last of these is a prerequisite for what is known as unconstrained or density-independent growth, which occurs when the population is unaffected by factors relating to sparsity (e.g., a lack of partners for mating) or over-crowding (e.g., through competition for a finite quantity of natural resources).

8.2.2 Formulation, Implementation and Evaluation of the Discrete Model

Formulation

In discrete models of population growth it is assumed that individuals are born in a series of cohorts at finite intervals in time (Alstad 2001). Models such as these are appropriate to species that reproduce on a seasonal or an annual basis; this category includes many types of plants, insects and mammals. With each successive time step, therefore, a number of individuals are born and a number die, depending on the size of the population at that point in time. This can be represented mathematically as follows:

$$N_{t+1} = N_t + BN_t - DN_t \tag{8.1}$$

where N_t is the population size (or density) during time step t, N_{t+1} is the population during the subsequent time step $(t+1)$, B is the average number of offspring produced by an individual in any one time step (the number of births per capita per time step, where $B \geq 0$) and D is the probability that an individual will die during any given time step (where $0 \leq D \leq 1$). For example, if the population size at time t is 100 ($N_t = 100$), if each individual produces on average two offspring per time step ($B = 2$) and if there is a one in two chance that an individual will die during any given time step ($D = 0.5$), then

$$N_{t+1} = 100 + (2 \times 100) - (0.5 \times 100) = 250 \tag{8.2}$$

Note that Equation 8.1 implicitly assumes that B and D are constant with respect to time, which means that they are independent of the population size (or density), N_t. Consequently, Equation 8.1 is known as a density-independent model of population growth (Donovan and Welden 2002).

Equation 8.1 is a form of finite difference equation (FDE), which describes a change (or difference) in the population over a finite interval of time. This is more obvious if Equation 8.1 is rearranged as follows:

$$\begin{aligned} N_{t+1} - N_t &= BN_t - DN_t \\ &= (B-D)N_t \end{aligned} \tag{8.3}$$

where $N_{t+1} - N_t$ is the change in the size of the population between consecutive time steps, t and $t+1$. The difference in the population size is often denoted by ΔN, while the corresponding interval of time is denoted by Δt. More generally, FDEs can be used to describe the change in any environmental property, not just population size, with respect to finite differences in a given dimension, not just time (e.g., the change in mean air temperature with distance along a geographical axis).

Equation 8.1 can be simplified as follows:

$$N_{t+1} = (1 + B - D)N_t \tag{8.4}$$

where the term $(1 + B - D)$, often replaced by the symbol λ, describes the discrete or finite rate of growth in the population. As a result, Equation 8.4 can be restated as

$$N_{t+1} = \lambda N_t \tag{8.5}$$

and hence

$$\lambda = \frac{N_{t+1}}{N_t} \tag{8.6}$$

Thus, λ describes the proportional change in the population size per unit interval of time (Alstad 2001). In some texts, λ is referred to as the Malthusian factor after the Reverend Thomas Malthus (1766–1834) whose *Essay on the Principle of Population*, first published in 1798, explored various issues concerning the growth of the human population and the ability to support it by means of agricultural production.

Assuming that B and D (and hence λ) remain constant over time and that the initial population size (N at $t = 0$; N_0) is known, it is possible to use Equation 8.5 to examine the growth of a population over a number of time steps. For instance, the population size at $t = 1$ is

$$N_1 = \lambda N_0 \tag{8.7}$$

while at $t = 2$ it is

$$N_2 = \lambda N_1 \tag{8.8}$$

Substituting Equation 8.7 into Equation 8.8 gives

$$\begin{aligned} N_2 &= \lambda(\lambda N_0) \\ &= \lambda^2 N_0 \end{aligned} \tag{8.9}$$

Similarly, the population at $t = 3$ is

$$\begin{aligned} N_3 &= \lambda N_2 \\ &= \lambda(\lambda N_1) \\ &= \lambda(\lambda(\lambda N_0)) \\ &= \lambda^3 N_0 \end{aligned} \tag{8.10}$$

Program 8.1: discrete.awk

```
# Discrete model of unconstrained (density-independent)     1
# population growth (see Eq. 8.11).                         2
#                                                           3
# Usage: gawk -f discrete.awk -v pop_init=value \           4
#                -v lambda=value -v period=value \          5
#                [ > outputFile ]                           6
#                                                           7
# Variables:                                                8
# ----------                                                9
# pop_init   initial population                             10
# lambda     discrete (finite) rate of population growth    11
# period     period (time steps) over which growth modeled  12
#                                                           13
BEGIN{                                                      14
    for(time_step=0;time_step<=period;++time_step){        15
        print time_step, pop_init*(lambda**time_step);     16
    }                                                       17
}                                                           18
```

and so on (Roughgarden 1998). Assuming that the time step is a year, that the initial population is $10,000$ ($N_0 = 10,000$) and that the population grows at a steady rate of 5% per annum ($B - D = 0.05$ and therefore $\lambda = 1.05$), after one year the population size will be $10,500$ ($N_1 = 1.05 \times 10,000 = 10,500$), after two years it will be $11,025$ ($N_2 = 1.05 \times 1.05 \times 10,000 = 11,025$), and so on. Note that this is a geometrical sequence, and populations that behave in this way are said to exhibit geometrical growth.

Equation 8.10 can be generalized to determine the size of a population after an arbitrary number of time steps, t,

$$N_t = \lambda^t N_0 \qquad (8.11)$$

where N_0 is the initial population (at $t = 0$), N_t is the population at time t, and t is a positive integer (a whole number). Equation 8.11 describes a population that grows without limit when $\lambda > 1$, that remains constant when $\lambda = 1$ and that declines toward extinction when $\lambda < 1$ (Alstad 2001).

Implementation

The next task is to convert the discrete density-independent model of population growth (Equation 8.11) into **gawk** code. Program 8.1 (discrete.awk) provides an example implementation. Apart from the comments (lines 1 through 12), the code in this program is placed entirely within the BEGIN block (lines 14 to 18) because the aim is to generate data (to perform a simulation) rather than to process data contained in a file. Note that values for each of the main variables — pop_init (the initial size of the population, N_0), lambda (the discrete rate of population growth, λ) and period (the number of time steps over which the population growth is to be simulated, t) — are not specified in the body of the code; they have to be supplied via the command line. This makes the code more flexible, permitting the user to experiment with different values of the variables without having to edit the program each time.

The code in the BEGIN block consists of a single for loop (lines 15 to 17). This is used to calculate the size of the population during each time step ($t = 0, 1, 2, \ldots$). The body of the loop (line 16) prints out the current time step, time_step, and the corresponding population size, pop_init*(lambda**time_step). Remember that any number, except zero, raised to the power of zero is equal to 1 ($x^0 = 1$, if $x \neq 0$; n.b. 0^0 is undefined). Thus, lambda**time_step=1 when time_step=0. As a result, line 16 prints out the initial population size (pop_init) the first time around the loop.

It is instructive to note that Program 8.1 and the mathematical equation on which it is based sometimes generate a fractional number of individuals (if $N_0 = 10$ and $\lambda = 1.05$, then $N_1 = 10 \times 1.05 = 10.5$). In reality, of course, fractional individuals do not exist, and Program 8.1 can be amended to prevent this outcome. The simplest way to do this is to use **gawk**'s int(x) function. This converts its argument to an integer (a whole number) by truncating the fractional component (everything after the decimal point). For example, the result of int(10.6) is 10. Thus, line 16 in Program 8.1 could be amended to read int(pop_init*(lambda^time_step)). Alternatively, the print statement could be replaced by a printf statement, such as printf("%i %.0f\n", time_step, pop_init*(lambda^time_step). The term %.0f in the printf format string instructs **gawk** to print out the corresponding item in the argument list (pop_init*(lambda^time_step)) as a floating-point number to zero decimal places. This rounds the value of the argument up or down to the nearest integer (values in the range 10.0 to 10.5 are reported as 10, while those in the range 10.51 to 10.99 are output as 11). In the remainder of this section, however, the code given in Program 8.1 is used unamended to perform various simulations.

Evaluation

Program 8.1 can be used to examine changes in the size of a population for which births and deaths take place in discrete cohorts at finite intervals in time. By providing the program with appropriate values on the command line, it is possible to explore the impact of different initial population sizes and discrete rates of growth. Assuming that $N_0 = 10$, for example, the growth of the population can be simulated over a period of 50 time steps for each of three different rates of growth ($\lambda = 1.05$, $\lambda = 1.06$ and $\lambda = 1.07$) by typing the following instructions on the command line:

```
gawk -f discrete.awk -v pop_init=10 -v lambda=1.05 -v ↺      1
   ↺period=50 > discr105.dat
gawk -f discrete.awk -v pop_init=10 -v lambda=1.06 -v ↺      2
   ↺period=50 > discr106.dat
gawk -f discrete.awk -v pop_init=10 -v lambda=1.07 -v ↺      3
   ↺period=50 > discr107.dat
```

Note that the result of each simulation is redirected to a separate data file in the working directory, namely discr105.dat, discr106.dat and discr107.dat.

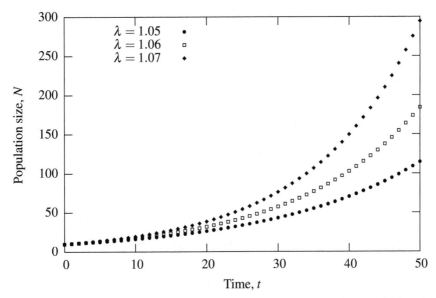

Figure 8.2: Population growth predicted by a discrete density-independent model based on three separate rates of growth ($\lambda = 1.05$, $\lambda = 1.06$ and $\lambda = 1.07$) and a common initial population ($N_0 = 10$).

Visualization

The results of the model runs listed above can be visualized by typing the following commands in gnuplot:

```
reset                                                               1
set style data points                                              2
set xlabel "Time, t"                                               3
set ylabel "Population size, N"                                    4
set key top left                                                   5
plot 'discr105.dat' t "lambda=1.05" lt 3 pt 7, \                   6
     'discr106.dat' t "lambda=1.06" lt 3 pt 6, \                   7
     'discr107.dat' t "lambda=1.07" lt 3 pt 5                      8
```

These commands assume that the relevant data files are located in the working directory. Note that the `points` data style is used here to emphasize the fact that the data pertain to discrete intervals in time (Figure 8.2).

Figure 8.2 demonstrates the geometric pattern of population growth that occurs when $\lambda > 1$. As expected, the population grows more rapidly as the value of λ increases. Figure 8.3 shows a similar plot generated using data produced by the same model for $N_0 = 100$ and $\lambda < 1$. In this case, the population declines over time and does so more rapidly as the value of λ decreases.

Figure 8.4 and Figure 8.5 show the same two data sets plotted on semi-logarithmic axes (i.e., a logarithmic scale on the y-axis and a linear scale on the x-axis) having issued the following commands in gnuplot:

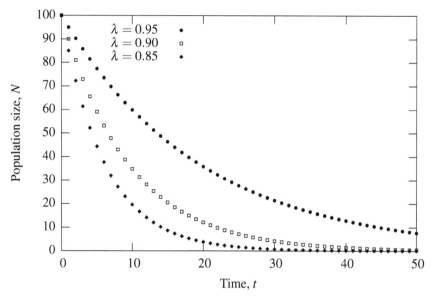

Figure 8.3: Population decline predicted by a discrete, density-independent model based on three separate rates of growth ($\lambda = 0.95$, $\lambda = 0.90$ and $\lambda = 0.85$) and a common initial population ($N_0 = 100$).

```
set logscale y                                                    9
replot                                                            10
```

These figures suggest that there is a simple linear relationship between the logarithm of N_t and t. Taking the natural logarithm (logarithm to the base e) of both sides of Equation 8.11 and simplifying, this relationship can be expressed as follows:

$$
\begin{aligned}
\ln(N_t) &= \ln(\lambda^t N_0) \\
&= \ln(\lambda)t + \ln(N_0)
\end{aligned}
\tag{8.12}
$$

(Roughgarden 1998). This equation has the general form $y = mx + c$. Thus, the slope of the lines in Figure 8.4 and Figure 8.5 is given by $\ln(\lambda)$. The significance of this result is examined in the next section.

Finally, before proceeding any further, the following gnuplot command should be issued to turn off the logarithmic scaling on the y-axis in all subsequent plots:

```
unset logscale y                                                 11
```

> **Exercise 8.1**: Reproduce the results shown in Figure 8.3 by running Program 8.1 for $N_0 = 100$, $\lambda = 0.95$, $\lambda = 0.90$ and $\lambda = 0.85$, visualizing the output in gnuplot. Plot the same data on a semi-logarithmic scale by typing set logscale y and then replot in gnuplot to reproduce Figure 8.5. Remember to unset logscale y before producing any further plots.

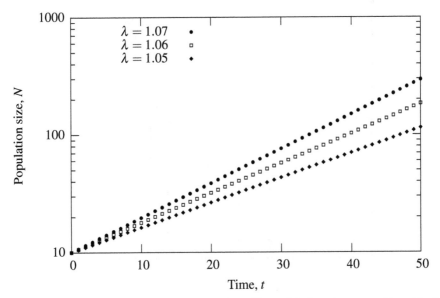

Figure 8.4: Discrete population growth curves from Figure 8.2 plotted on a semi-logarithmic scale.

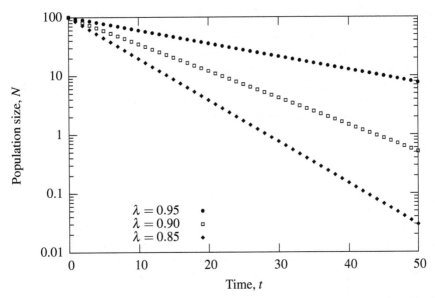

Figure 8.5: Discrete population growth curves from Figure 8.3 plotted on a semi-logarithmic scale.

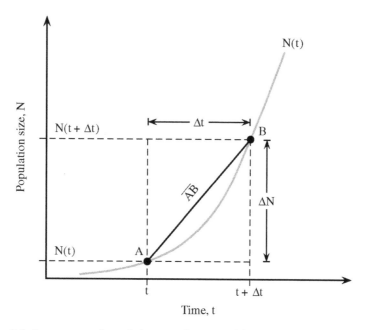

Figure 8.6: Average rate of population growth measured between two census points, *A* and *B*,
expressed as the slope of the secant line $\overline{AB}$. The underlying continuous growth
curve, $N(t)$, is shown by the curved gray line.

8.2.3   Formulation, Implementation and Evaluation of the Continuous Model

The populations of certain species, including *Homo sapiens* and some bacteria grown
in vitro, vary in size almost continuously through time, and it is appropriate to use
continuous models to describe their growth or decline. Although the mathematical
formulation of such models can be derived from first principles (Alstad 2001), this
requires knowledge of calculus, which some readers may not possess. Consequently,
a less formal, more graphical, approach is adopted here.

Formulation

Imagine that, instead of having been produced by a discrete model of population
growth, the data in Figure 8.2 come from a series of censuses conducted at regular
intervals in time for three continuously varying populations, each of which grows at
a different rate. The underlying trends in population growth are therefore continuous,
and each census provides a "snapshot" of the population sizes at a particular moment
in time. The challenge is to derive a continuous mathematical function that describes
the size of each population, and its rate of growth, at any point in time. This is
addressed informally, below.

   Let Δt denote the time interval between successive censuses *A* and *B* conducted
at times t and $t + \Delta t$, respectively (Figure 8.6). This interval might, for example,
be a year, a month or a day. Similarly, let ΔN denote the amount by which the

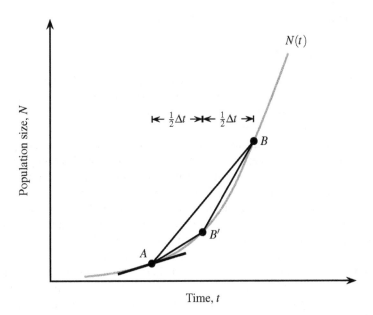

Figure 8.7: Effect of reducing Δt on the representation of the continuous population growth curve, $N(t)$. The tangent to $N(t)$ at point A is shown by the bold black line. This represents the rate of change in the population size at that instant in time.

population changes in size over this period of time. The average rate of change in the size of the population between census A and census B is therefore ΔN divided by Δt ($\Delta N / \Delta t$). This is represented graphically by the slope of the (solid black) secant line $\overline{AB}$ shown in Figure 8.6. In the absence of any other information about the way in which the population varies in size between census A and census B, the secant line, $\overline{AB}$, represents a first approximation to the shape of the (solid gray) continuous growth curve, $N(t)$, over the corresponding interval of time. It is evident, however, that points along the secant line, $\overline{AB}$, tend to overestimate the population size; that is, they lie above the curve, $N(t)$. This is because the slope of the secant line, $\overline{AB}$, is greater than that of the continuous growth curve, $N(t)$, at point A. The latter defines the instantaneous rate of change in the size of the population at time t.

Now imagine that another census B' is conducted midway in time between A and B; that is, the time interval (Δt) between successive censuses is halved (Figure 8.7). The resulting secant lines, $\overline{AB'}$ and $\overline{B'B}$, are shorter than $\overline{AB}$, and they provide a better approximation to the shape of the continuous growth curve $N(t)$. Furthermore, the slope of the secant line $\overline{AB'}$ is nearer to that of the continuous growth curve at point A. It should be evident that halving the time interval between censuses once again will produce an even more accurate representation of the continuous growth curve. Taking this to the limit, as Δt approaches zero ($\Delta t \rightarrow 0$), the resulting secant lines become infinitesimally small so that they effectively define the continuous growth curve. Moreover, as $\Delta t \rightarrow 0$ each secant line is tangent to the continuous growth curve at that point in time, and its slope therefore defines the rate at which the population is growing at that instant. This is expressed more formally as

$$\lim_{\Delta t \to 0} \frac{\Delta N}{\Delta t} = \frac{dN}{dt} = rN \tag{8.13}$$

where N is the population size, $\frac{dN}{dt}$ ("dN by dt") is known as the derivative of N with respect to t, and r is known as the intrinsic or instantaneous rate of population growth. Thus, $\frac{dN}{dt}$ describes the rate of change in the size of the population at any given moment in time (i.e., the instantaneous rate of change). This is equivalent to the slope of the curve $N(t)$. Note that Equation 8.13 is a form of differential equation.

Equation 8.13 indicates that the rate of change in the size of the population at any given moment in time, $\frac{dN}{dt}$, is equal to the population size, N, at that time, t, multiplied by the instantaneous rate of growth, r. The value of r, however, is as yet unknown. To resolve this, recall that the data in Figure 8.2 were transformed into a straight-line relationship by plotting them on semi-logarithmic axes (Figure 8.4). Now imagine, once again, that the data points in Figure 8.4 were obtained from censuses conducted at regular intervals in time (Δt) for each of the three continuously varying populations. At the limit, when $\Delta t \to 0$, there is an infinite number of census points for each population, such that they effectively describe continuous straight lines. In the preceding section, it was shown that the slope of these lines is $\ln(\lambda)$. It follows, therefore, that

$$r = \ln(\lambda) \tag{8.14}$$

and, hence, that

$$\lambda = e^r \tag{8.15}$$

Substituting Equation 8.15 into Equation 8.11 and modifying the notation slightly to present N as a continuous function of t therefore gives

$$N(t) = e^{rt} N(0) \tag{8.16}$$

where $N(t)$ is the population size at an arbitrary point in time, t, $N(0)$ is the initial size of the population (at $t = 0$), and r is the instantaneous rate of population growth. Equation 8.16 indicates that the population size grows (or declines) exponentially with respect to time, depending on the instantaneous rate of growth, r.

Implementation

Program 8.2 (continue.awk) is an implementation of the continuous model of density-independent population growth presented in Equation 8.16. The code is very similar to that used in Program 8.1. The main difference occurs on line 16, where the formula for the continuous model is implemented. Although Program 8.2 uses the same looping structure as Program 8.1 (discrete.awk; the discrete model), stepping forward in integer units of time, it is possible to use the continuous model code to predict the population at any point in time. For example, the variable time could be incremented in steps of 0.1 by replacing the statement ++time on line 15 with time+=0.1. Finally, note that $\exp(0) = 1$. The first time around the for loop, when time=0, the program prints out the initial population size given on the command line.

Program 8.2: continue.awk

```
# Continuous model of unconstrained (density-independent)    1
# population growth (see Eq. 8.16).                          2
#                                                            3
# Usage: gawk -f continue.awk -v pop_init=value \            4
#                -v growth_rate=value -v period=value \      5
#                [ > outputFile ]                            6
#                                                            7
# Variables:                                                 8
# ----------                                                 9
# pop_init       initial population                         10
# growth_rate  instantaneous (intrinsic) growth rate        11
# period         period of time over which growth is modeled 12
#                                                           13
BEGIN{                                                      14
    for(time=0;time<=period;++time){                       15
        print time, pop_init*exp(growth_rate*time);        16
    }                                                       17
}                                                           18
```

Evaluation

Program 8.2 can be used to simulate continuous population growth over a period of
50 units of time for three different instantaneous rates of growth ($r = 0.05$, $r = 0.06$
and $r = 0.07$), based on an initial population of 10 individuals ($N(0) = 10$), by typing
the following instructions on the command line:

```
gawk -f continue.awk -v pop_init=10 -v growth_rate=0.05 -v↺   4
    ↺ period=50 > cont005.dat
gawk -f continue.awk -v pop_init=10 -v growth_rate=0.06 -v↺   5
    ↺ period=50 > cont006.dat
gawk -f continue.awk -v pop_init=10 -v growth_rate=0.07 -v↺   6
    ↺ period=50 > cont007.dat
```

Recall from Equation 8.14 that $r = \ln(\lambda)$, so that the instantaneous growth rates
used here are roughly equivalent to the finite growth rates used in the discrete model
simulations reported in the preceding section ($\ln(1.05) \approx 0.05$, $\ln(1.06) \approx 0.06$ and
$\ln(1.07) \approx 0.07$). Note that the result of each simulation is redirected to a separate
data file, namely cont005.dat, cont006.dat and cont007.dat.

Visualization

The results of the model runs described above can be visualized by issuing the following set of commands in **gnuplot**:

```
set style data lines                                        12
plot 'cont005.dat' t "r=0.05", \                            13
```

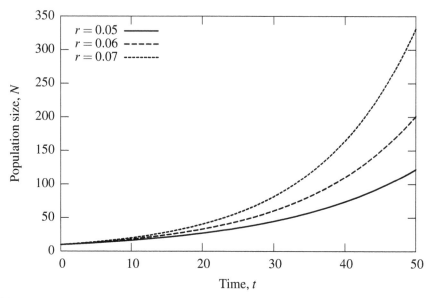

Figure 8.8: Results from the continuous model of density-independent population growth (Program 8.2; continue.awk) for $N_0 = 10$ and $r = 0.05$, $r = 0.06$ and $r = 0.07$.

```
'cont006.dat' t "r=0.06", \                                          14
'cont007.dat' t "r=0.07"                                             15
```

Note that the lines data style is used here to emphasize the fact that population size is a continuous function of time. The results are presented in Figure 8.8, which can be compared directly to the discrete model case presented in Figure 8.2. Before moving on, it is worth emphasizing once again the difference between the discrete and continuous models of density-independent growth: the latter produces a value of N for any value of t ($t = 1.25$), whereas the former reports values of N for integer values of t only ($t = 1, 2, 3, \ldots$).

Exercise 8.2: Run Program 8.2 for $N = 100$ and three different values of r (-0.05, -0.1 and -0.15). Visualize the output in gnuplot and compare the resulting plot with Figure 8.3

8.3 CONSTRAINED OR DENSITY-DEPENDENT GROWTH

Few populations grow geometrically or exponentially without limit for a long period of time. There are many reasons why this is the case; prime among these are the constraints imposed by the finite pools of natural resources, such as food, water, light and shelter, that are available to support the population. In general, the competition for resources between individuals of the same species, known as intra-specific competition, becomes ever more intense as the population grows in size (Donovan and

Welden 2002). Consequently, the birth and death rates per capita vary according to the size of the population (Alstad 2001). This phenomenon is known as constrained or density-dependent growth because the population growth rate is affected by factors relating to sparsity and over-crowding.

It is possible to derive both discrete and continuous models of density-dependent population growth. In this section, the continuous model is examined first, before moving on to consider a remarkable outcome of the discrete model.

8.3.1 Developing the Conceptual Model

One of the earliest attempts to model density-dependent population growth was made by the Belgian mathematician Pierre François Verhulst (1804–1849). In essence, Verhulst hypothesized that there is a limit to the number of individuals that can be supported on a continuing basis by any given environment (Whittaker 1982). He reasoned that this is because of the constraints imposed by factors such as the finite amount of food available for consumption. The theoretical maximum population that can be supported by an environmental system in steady state is known as the carrying capacity (Roughgarden 1998). Verhulst suggested that the rate of growth of a population is dependent on both the population size and the carrying capacity of the environment that it inhabits. For instance, when a population is small in number relative to the carrying capacity, the resources available to each individual are plentiful; as a result, the population is able to grow at a rate largely determined by the intrinsic birth and death rates for that species (i.e., at a rate approximately equal to that of the density-independent model). By contrast, when the population size is equal to the carrying capacity of the environment, the available resources are sufficient to sustain that number of individuals and no more; further growth in the population is therefore inhibited by resource limitations. In these circumstances, the population growth rate tends to zero (Alstad 2001).

8.3.2 Continuous Logistic Model

Formulation

Verhulst's model of density-dependent growth, also known as the logistic growth model, is usually presented as

$$\frac{dN}{dt} = rN\left(\frac{K-N}{K}\right) \tag{8.17}$$

or, equivalently, as

$$\frac{dN}{dt} = rN\left(1 - \frac{N}{K}\right) \tag{8.18}$$

where N is the population size, r is the instantaneous rate of population growth and K is the carrying capacity (Roughgarden 1998). Equation 8.18 differs from its density-independent counterpart (Equation 8.16) through inclusion of a multiplicative term,

$(K - N)/K$. Alstad (2001) interprets $K - N$ as a measure of the unused resources in the environment and $(K - N)/K$ as the unused fraction of the carrying capacity. Thus, $(K - N)/K \approx 1$, and hence $\frac{dN}{dt} \approx rN$, when $N \approx 0$ (the population is very small); that is to say, the rate of change in the population size is approximately equal to the unconstrained or density-independent case (Equation 8.16). So, when small, the population experiences a period of almost exponential growth. By contrast, $(K - N)/K \approx 0$, and hence $\frac{dN}{dt} \approx 0$, when $N \approx K$ (the population is very large). Consequently, as the population size approaches the carrying capacity, the rate of growth slows to zero. Note that it is theoretically possible for the population size to exceed the carrying capacity $(N > K)$ for short periods of time, whether as a result of immigration or of over-stocking. In these circumstances, $(K - N)/K$, and hence $\frac{dN}{dt}$, becomes negative. As a result, the population decreases in size over time. In effect, therefore, the term $(K - N)/K$ in Equation 8.17 introduces negative feedback between the size of a population and its rate of growth (Alstad 2001).

It is possible to solve Equation 8.18 analytically to derive a formula that expresses N as a continuous function of t. The formal derivation of this solution is beyond the scope of this book — interested readers should consult Giordano *et al.* (1997) — but it can be shown that

$$N(t) = \frac{K}{1 + [(K - N(0))/N(0)]e^{-rt}} \tag{8.19}$$

where $N(t)$ is the population at time t, $N(0)$ is the initial population (at $t = 0$), r is the instantaneous rate of growth and K is the carrying capacity (Roughgarden 1998).

Implementation

The continuous logistic model of population growth (Equation 8.19) is implemented in Program 8.3, `cntlogst.awk`. The structure of this code is very similar to that of Program 8.2, `continue.awk`, in that it makes use of a `for` loop contained within the `BEGIN` block. Two additional variables are employed, namely `carry_cap`, which is used to store the value of the carrying capacity (lines 14 and 19), and `pop_now`, which is used to store the population calculated for time t (lines 12, 19 and 21). This calculation (implementation of Equation 8.19) is performed on lines 19 and 20. These lines of code constitute a single **gawk** statement, which has been split over two lines using the line continuation symbol (\) at the end of line 19. The current time and population size are printed out on line 21. Thus, Program 8.3 reports the size of a continuously varying population at regular intervals in time, controlled by the parameters of the `for` loop.

Evaluation

Program 8.3 can be used to simulate the density-dependent growth of a continuously varying population, initially consisting of 10 individuals ($N(0) = 10$), over a period of 200 units of time and for three intrinsic rates of growth ($r = 0.05$, $r = 0.06$ and $r = 0.07$) by issuing the following instructions on the command line:

Program 8.3: cntlogst.awk

```
# Continuous-time model of density-dependent (constrained)     1
# population growth, based on the Verhulst (logistic)          2
# equation (see Eq. 8.19).                                     3
#                                                              4
# Usage: gawk -f cntlogst.awk -v pop_init=value \              5
#               -v growth_rate=value -v carry_cap=value \      6
#               -v period=value [ > output_file ]              7
#                                                              8
# Variables:                                                   9
# ----------                                                   10
# pop_init      initial population                             11
# pop_now       current population                             12
# growth_rate   instantaneous (intrinsic) growth rate          13
# carry_cap     carrying capacity                              14
# period        period of time over which growth is modeled    15
                                                               16
BEGIN{                                                         17
  for(time=0;time<=period;++time){                             18
    pop_now=carry_cap/(1+((carry_cap-pop_init)/pop_init)* \    19
      exp(-growth_rate*time));                                 20
    print time, pop_now;                                       21
  }                                                            22
}                                                              23
```

```
gawk -f cntlogst.awk -v pop_init=10 -v growth_rate=0.05 -v↩   7
   ↩ carry_cap=1000 -v period=200 > c_log005.dat
gawk -f cntlogst.awk -v pop_init=10 -v growth_rate=0.06 -v↩   8
   ↩ carry_cap=1000 -v period=200 > c_log006.dat
gawk -f cntlogst.awk -v pop_init=10 -v growth_rate=0.07 -v↩   9
   ↩ carry_cap=1000 -v period=200 > c_log007.dat
```

Note that values for the initial population size, carrying capacity and period of time over which the population growth is to be modeled are specified on the command line, and that the result of each simulation is redirected to a separate data file.

Visualization

The output from the simulations described above can be visualized in gnuplot by issuing the following set of commands:

```
plot 'c_log005.dat' t "r=0.05" lw 2, \               16
     'c_log006.dat' t "r=0.06" lw 2, \               17
     'c_log007.dat' t "r=0.07" lw 2                  18
```

The resultant population growth curves (Figure 8.9) are sigmoidal (S-shaped). To begin with, each population grows at an almost exponential rate (i.e., at a rate that

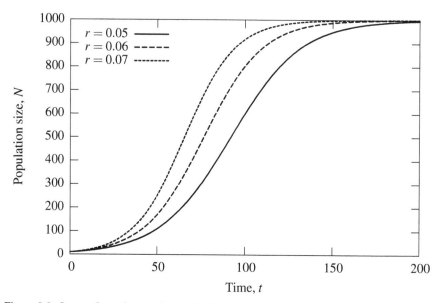

Figure 8.9: Output from the continuous logistic model of population growth (Program 8.3; `cntlogst.awk`) for $N(0) = 10$, $K = 1000$ and various values of r.

is approximately equal to the unconstrained growth model), but as it increases in size the competition for resources becomes more intense and hence its rate of growth decreases. Later, as the size of the population approaches the carrying capacity for the environment, the growth rate decreases to zero, and the population size gradually approaches its steady-state value (Haefner 1996).

The relationship between population size (N) and population growth ($\frac{dN}{dt}$) in the logistic model is shown in Figure 8.10. Note that population growth is small when N is small because, despite the fact that the growth rate ($r((K-N)/K)$) is high, the number of individuals available to reproduce is limited. Population growth is also small when N is large because, despite the fact that the population is large, the growth rate is small. Population growth is maximized when $N = K/2$.

Exercise 8.3: What happens if the initial population exceeds the carrying capacity? Test this by running Program 8.3 for $N(0) = 1000$, $K = 100$ ($N(0) > K$) and $r = 0.07$. Visualize the output in gnuplot.

8.3.3 Discrete Logistic Model

It is possible to construct a version of the logistic model for species that are born in discrete cohorts separated by finite intervals of time (i.e., a discrete logistic growth model), and one such model is examined here. The primary intention in doing so is to demonstrate the remarkable behavior that this model exhibits under certain con-

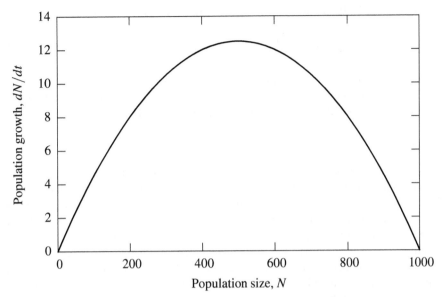

Figure 8.10: Relationship between population size (N) and population growth (dN/dt) in the continuous density-dependent model of population growth for $N(0) = 10$, $K = 1000$ and $r = 0.05$.

ditions. More specifically, it illustrates how a relatively simple deterministic equation can generate apparently random, and hence unpredictable, output (Gleick 1987, Hall 1991). This type of "chaotic" behavior in population growth models was first demonstrated by Sir Robert May (1974).

Formulation

Numerous different formulations of the discrete logistic growth model are reported in the literature. Perhaps the most obvious example is

$$\Delta N_t = RN_t \left(\frac{K - N_t}{K} \right) \tag{8.20}$$

where ΔN_t is the amount by which the population grows (or declines) during time step t, N_t is the population size during time step t, $R = B - D$ (the per capita birth rate minus the per capita death rate; see Section 8.2.2) and K is the carrying capacity (Donovan and Welden 2002). This is simply the product of Equation 8.3 (i.e., the discrete version of the unconstrained growth model) and the logistic growth term, $(K - N_t)/K$. The corresponding finite difference equation is, therefore,

$$N_{t+1} = N_t + RN \left(\frac{K - N_t}{K} \right) \tag{8.21}$$

(Roughgarden 1998). A different solution, suggested by May (1976), is

$$N_{t+1} = N_t \exp\left[r\left(\frac{K - N_t}{K}\right)\right] \tag{8.22}$$

where r is the instantaneous or intrinsic rate of population growth. This formulation is directly equivalent to the continuous logistic model (Equation 8.17) at the limit, where $\Delta t \rightarrow 0$ (Alstad 2001), and is the solution employed here.

Implementation

Equation 8.22 is implemented in **gawk** code in Program 8.4, dsclogst.awk. After the usual set of explanatory comments (lines 1 to 14), the program employs the BEGIN block and a for loop to simulate the growth of the population with each successive time step. Each time the program traverses the for loop, the value of the current time step and the corresponding population size are printed out (line 18), after which the population size in the next time step is calculated (line 19). This is repeated until the requisite number of time steps (steps) have elapsed. Note that values must be supplied on the command line for the initial population size (pop), instantaneous rate of population growth (growth_rate), carrying capacity (carry_cap) and number of time steps over which the simulation is to be performed (steps).

Evaluation

Program 8.4 can be run from the command line as follows:

```
gawk -f dsclogst.awk -v pop=10 -v growth_rate=0.05 -v ↪          10
    ↪carry_cap=1000 -v steps=200 > d_log005.dat
gawk -f dsclogst.awk -v pop=10 -v growth_rate=0.06 -v ↪          11
    ↪carry_cap=1000 -v steps=200 > d_log006.dat
gawk -f dsclogst.awk -v pop=10 -v growth_rate=0.07 -v ↪          12
    ↪carry_cap=1000 -v steps=200 > d_log007.dat
```

These commands assume that the population consists of 10 individuals initially and that the growth of the population is to be modeled over a total period of 200 time steps. Three separate simulations are performed, each for a different (instantaneous) rate of population growth ($r = 0.05$, $r = 0.06$ and $r = 0.07$). The results are redirected into separate data files (d_log005.dat, d_log006.dat and d_log007.dat) in the working directory.

Visualization

The output from the simulations described above can be visualized in **gnuplot** by issuing the following set of commands, which produce Figure 8.11:

```
set style data points                                            19
plot 'd_log005.dat' t "r=0.05" w p, \                            20
    'd_log006.dat' t "r=0.06" w p, \                             21
    'd_log007.dat' t "r=0.07" w p                                22
```

Program 8.4: dsclogst.awk

```
# Discrete-time model of density-dependent (constrained)      1
# population growth, based on the Verhulst (logistic)          2
# equation (see Eq. 8.22).                                     3
#                                                              4
# Usage: gawk -f dsclogst.awk -v pop=value \                   5
#                 -v growth_rate=value -v carry_cap=value \    6
#                 -v steps=value [ > output_file ]             7
#                                                              8
# Variables:                                                   9
# ----------                                                   10
# pop            population at time step                       11
# growth_rate    instantaneous (intrinsic) growth rate         12
# carry_cap      carrying capacity                             13
# steps          total number of time steps                    14
                                                               15
BEGIN{                                                         16
  for(time_step=0;time_step<=steps;++time_step){              17
    print time_step, pop;                                     18
    pop=pop*exp(growth_rate*(1-pop/carry_cap));               19
  }                                                            20
}                                                              21
```

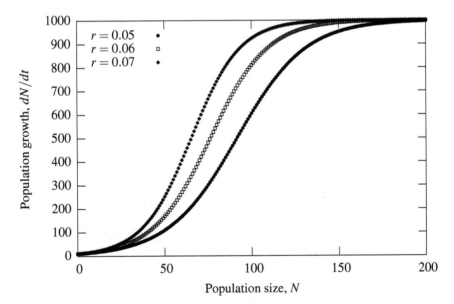

Figure 8.11: Output from the discrete logistic model of population growth (Program 8.4; dsclogst.awk) for $N_0 = 10$, $K = 1000$ and three different values of r (0.05, 0.06 and 0.07).

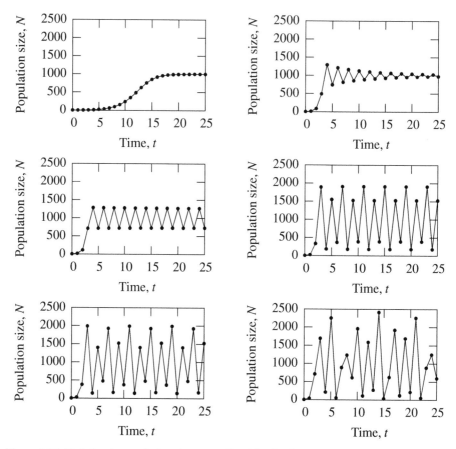

Figure 8.12: Variation in population size as predicted by the discrete logistic growth model as a function of r for $N_0 = 2$ and $K = 1000$. Key: $r = 0.5$ (top left), $r = 1.9$ (top right), $r = 2.05$ (middle left), $r = 2.6$ (middle right), $r = 2.67$ (bottom left) and $r = 3.0$ (bottom right).

Note that the `points` data style has been used to emphasize the fact that the data relate to discrete time steps. It should be evident from Figure 8.11 that the resultant curves are identical in form and magnitude to those in Figure 8.9; that is, the discrete version of the logistic model behaves in the same way as its continuous counterpart for the values of r examined here.

The behavior of the discrete logistic model alters dramatically, however, as the value of r increases. For example, Figure 8.12 presents the results of six different runs of the discrete logistic model (Program 8.4; `dsclogst.awk`) using the same initial population size ($N_0 = 2$; `pop`) and carrying capacity ($K = 1000$; `carry_cap`) on each occasion, but varying the rate of population growth ($r = 0.5, 1.90, 2.05, 2.6, 2.7$ and 3.0; `growth_rate`). To make it easier to interpret the resulting patterns, lines have been drawn between consecutive data points. For $r = 0.5$, the population size follows the normal sigmoidal growth curve, reaching a steady-state value (i.e., the carrying

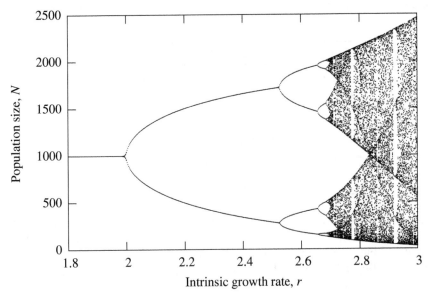

Figure 8.13: "Bifurcation diagram" showing the population sizes predicted by the discrete version of the density-dependent population growth model as a function of the intrinsic rate of increase, *r*.

capacity) after about 20 time steps (Figure 8.12, top left). When the growth rate is increased to $r = 1.9$ the population grows dramatically at first (Figure 8.12, top right), so that by the fourth time step the population consists of almost 1300 individuals. This exceeds the carrying capacity of the environment, so that there are insufficient resources to support all of the individuals. Consequently, the population decreases to around 750 by the fifth time step. This is less than the carrying capacity, however, so the population grows once more to around 1200 by the sixth time step. During subsequent time steps, the population goes through a series of these "boom and bust" cycles, which gradually diminish in magnitude so that the population size stabilizes around a steady-state value (the carrying capacity).

The behavior of the system changes once again when $r > 2$. Here, the population increases in size very dramatically over the first few time steps and then settles into an oscillatory pattern centered around the value of the carrying capacity. For instance, the system state oscillates between two population sizes when $r = 2.05$ (Figure 8.12, middle left), four when $r = 2.6$ (Figure 8.12, middle right) and eight when $r = 2.67$ (Figure 8.12, bottom left). The behavior of the system becomes chaotic when $r \geq 3$ (Figure 8.12, bottom right), with seemingly random fluctuations in the population size from one time step to the next, and the population does not reach an equilibrium size. This behavior is remarkable because it is produced by a simple deterministic equation. May (1991) refers to this as "*deterministic chaos*".

The behavior of the discrete logistic model can be visualized over a range of values of *r*, not just the few selected above. Figure 8.13, for example, was produced by running the model repeatedly for $r = 1.8$ to $r = 3.0$ in steps of 0.0025. On each

Table 8.1: Approximate threshold values for different types of behavior in the discrete logistic population growth model.

Growth Rate	Model Behavior
$0 < r \leq 2.0$	Reaches stable equilibrium
$2.0 < r \leq 2.52$	Two-point cycle
$2.52 < r \leq 2.65$	Four-point cycle
$2.65 < r \leq 2.6825$	Eight-point cycle
$2.6825 < r < 3.0$	16-, 32-, 64-point cycles, and so on
$r \geq 3.0$	Chaotic

occasion the model was run for a total of 500 time steps, and the results for the last 100 of these were saved for each model run in a single data file (Roughgarden 1998). Figure 8.13 indicates that the population size tends toward a single steady-state value for $r < 2.0$. Beyond this it oscillates, first in a two-point cycle, then four-point, eight-point, sixteen-point and so on, as r increases, before breaking into chaotic behavior when $r \geq 3$ (Roughgarden 1998). The values of r that mark the break-points between these different types of behavior are shown in Table 8.1. Note that, despite the apparently increasing disorder above $r = 2.7$, there is a considerable amount of fine structure hidden within this part of Figure 8.13. This is evident in the enlarged section shown in Figure 8.14, which clearly demonstrates that the bifurcation patterns evident in the range $2.0 \leq r \leq 2.8$ are repeated in miniature in the range $2.92 \leq r \leq 2.93$. The fine structure is also present for $r > 3.0$. Thus, although the population size varies chaotically when $r > 3.0$, the fluctuations are not random (Gleick 1987).

One consequence of the chaotic behavior exhibited by the discrete logistic model when $r > 3.0$ is that the model output is highly sensitive to the initial population size, N_0 (Roughgarden 1998). This is illustrated in Figure 8.15, which presents the results of two model runs, one using $N_0 = 99$ and the other using $N_0 = 101$, for $r = 3.0$ and $K = 500$. Despite the fact that the two populations differ in size by only two individuals to begin with, which is small in both absolute and relative terms, their trajectories diverge rather rapidly (Figure 8.15). This is significant because there is often great uncertainty over the values that should be used to initialize a model; here, such uncertainty can affect dramatically the nature of the model output.

8.4 NUMERICAL INTEGRATION (OR STEPPING) METHODS

In the preceding section, analytical solutions to both the continuous and the discrete versions of the logistic population growth model were obtained from the literature (Equation 8.19 and Equation 8.21, respectively). This was done for the sake of convenience and brevity. Normally, however, analytical solutions have to be derived from the corresponding differential or difference equations. This can be challenging, and it is often the case that an analytical solution cannot be found or else does not exist (Wainwright and Mulligan 2004). In such circumstances, the model has to be solved

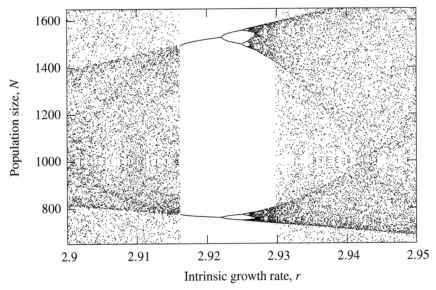

Figure 8.14: Enlarged section of the "bifurcation diagram" presented in Figure 8.13 showing the fine structure present.

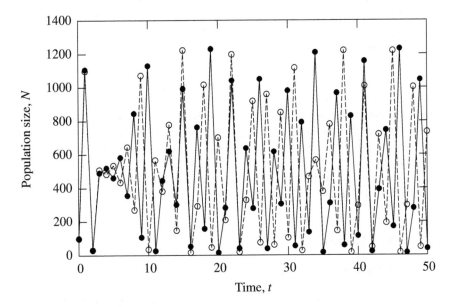

Figure 8.15: Effect of a small difference in N_0, $N_0 = 99$ (●) and $N_0 = 102$ (○), on the population dynamics of the discrete logistic model for $r = 3.0$ and $K = 500$.

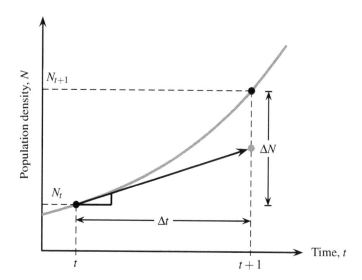

Figure 8.16: Diagrammatic representation of Euler's method of numerical integration used to estimate N_{t+1} based on N_t and $\frac{dN}{dt}$, where $\Delta t = 1$. The solid gray line represents the underlying continuous function $N(t)$.

numerically, using methods similar to those introduced in Chapter 7. In this section, a set of numerical techniques, known formally as numerical integration (Harris and Stocker 1998) and informally as stepping techniques (Borse 1997), is introduced. These methods are used to provide a numerical solution to the differential equations of the continuous logistical population growth model.

Consider, for the moment, Figure 8.16. Imagine that the continuous gray curve represents the underlying, but as yet unknown and unspecified, relationship between population size and time. Suppose that, for some reason, it is not possible to derive by analytical means a continuous mathematical function that describes the form of this relationship, but that the size of the population at time t (N_t) and its rate of growth at that instant ($\frac{dN}{dt}$) are known. The challenge is to use this information to predict the population size at other points in time, such as at $t + 1$ (N_{t+1}). This is sometimes known as an initial value problem. Two approaches to this type of problem, based on the methods of Euler and Runge-Kutta, are demonstrated below.

8.4.1   Euler's Method

Conceptual Basis and Mathematical Formulation

The simplest solution to the initial value problem, known as Euler's method (Harris and Stocker 1998), is to take the rate of population growth ($\frac{dN}{dt}$) at time t and to project this forward for a finite period of time (Δt) to estimate the change in the size of the population (ΔN) over that interval. The continuous function is, thus, approximated

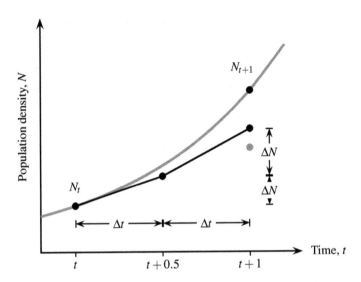

Figure 8.17: Euler's method of numerical integration used to estimate N_{t+1} based on N_t and $\frac{dN}{dt}$, where $\Delta t = 0.5$. The gray dot is the value of N_{t+1} estimated using $\Delta t = 1$.

by a straight line segment (Figure 8.16). This can be expressed formally as

$$\Delta N \approx \frac{dN}{dt} \cdot \Delta t \tag{8.23}$$

and hence

$$N_{t+1} \approx N_t + \frac{dN}{dt} \cdot \Delta t \tag{8.24}$$

where N_t is the population size at time t, $\frac{dN}{dt}$ is the rate of population growth at that moment and N_{t+1} is the population size at time $t+1$.

The weakness of this very basic approach is immediately apparent from Figure 8.16. Specifically, if Euler's method is used to project the population size forward over the entire period between time t and time $t+1$ ($\Delta t = 1$) it often yields a poor estimate of the actual population at time $t+1$ (N_{t+1}). In this particular example it underestimates substantially the true value. This is because the derivative, $\frac{dN}{dt}$, of the underlying function, $N(t)$, varies continuously with time (the slope of the curve $N(t)$ varies as a function of t).

The situation can often be improved by reducing Δt, the time period over which the population growth is projected. Figure 8.17, for example, illustrates what happens when Δt is halved ($\Delta t = 0.5$), so that the population size at time $t+1$ (N_{t+1}) is estimated in two steps. First, the rate of population growth at time t is used to estimate the change in the size of the population up to time $t+0.5$. The predicted population size at time $t+0.5$ is then employed to calculate the rate of population growth at that instant, and this is subsequently used to estimate the population at time $t+1$.

In the example shown in Figure 8.17, halving Δt increases the accuracy with which the population size at $t+1$ is estimated. Despite this, the result is still a poor approximation to the actual population size at that moment in time (N_{t+1}). The accuracy of the estimate can be improved by reducing Δt further, increasing the number of steps used to calculate N_{t+1}. Theoretically, at the limit, $\Delta t \rightarrow 0$, Euler's method yields the exact analytical solution (Borse 1997). There are, however, two important practical constraints. First, decreasing Δt and increasing the number of steps affects the computational load; the more steps, the longer it takes to compute the result. Second, computers do not store numbers with infinite precision, and the resulting rounding errors ultimately limit the level of accuracy that can be attained (Borse 1997). The most appropriate value of Δt to use for any given model is often, therefore, a matter of trial-and-error. This value can be established by progressively reducing Δt until further reductions no longer yield significant changes in the model output.

Implementation

Program 8.5 illustrates how Euler's method can be applied to the continuous logistic model of population growth. Apart from the comments (lines 1 to 14), the code is entirely contained in the BEGIN block (lines 16 to 31), because the intention is to generate data as part of a simulation. In this particular example, values for most of the main parameters — that is, the initial population size (N_0, pop), the intrinsic growth rate (r, growth_rate), the carrying capacity (K, carry_cap) and the time period over which the population growth is to be modeled (t, stop_time) — are hard-wired into the code (lines 17 to 21), rather than being provided on the command line, for the sake of convenience. A while loop (lines 23 to 30) is used to perform a sequence of actions (lines 24 to 29) repeatedly while the current time (time; initialized to zero on line 20) is less than or equal to the time at which the simulations are meant to stop (stop_time). The first of these actions is to print out the values of current time and the population size (line 24). Line 27 then increments the current time by Δt, the value of which is entered via the command line. The estimated change in the size of the population over this finite interval of time (ΔN; delta_pop) is calculated on line 28. This represents an implementation of Equation 8.23, where $\frac{dN}{dt} = rN\left(1 - \frac{N}{K}\right)$ (Equation 8.18). Finally, the updated population size is calculated on line 29 (Equation 8.24).

Program 8.5 can be run from the command line as follows, for three different values of Δt (1.0, 0.5 and 0.1):

```
gawk  -f  euler.awk  -v  dt=1.0  >  euler1.dat              13
gawk  -f  euler.awk  -v  dt=0.5  >  euler05.dat             14
gawk  -f  euler.awk  -v  dt=0.1  >  euler01.dat             15
```

Note that the results are redirected to three separate data files, namely euler1.dat, euler05.dat and euler01.dat, which are located in the working directory.

Visualization and Evaluation

The results obtained by running Program 8.5 for $\Delta t = 1.0$, 0.5 and 0.1 are visualized in Figure 8.18, which was produced by typing the following commands in **gnuplot**:

Program 8.5: euler.awk

```
# Continuous-time model of density-dependent (constrained)    1
# population growth, based on the Verhulst (logistic)          2
# equation (see Eq. 22), solved using Euler's method of        3
# numerical integration.                                       4
#                                                              5
# Usage: gawk -f euler.awk -v dt=value [ > output_file ]       6
#                                                              7
# Variables:                                                   8
# ----------                                                   9
# pop            population at time step                      10
# growth_rate    instantaneous (intrinsic) growth rate        11
# carry_cap      carrying capacity                            12
# time           current time                                 13
# stop_time      time at which simulation should stop         14
# dt             integration time step (1/steps)              15
#                                                             16
BEGIN{                                                        17
  pop=10.0;                                                   18
  growth_rate=0.5;                                            19
  carry_cap=1000;                                             20
  time=0;                                                     21
  stop_time=10;                                               22
                                                              23
  while(time<=stop_time){                                     24
    printf("%.1f\t%.3f\n", time, pop);                        25
                                                              26
    # Numerical integration using Euler's method              27
    time+=dt;                                                 28
    delta_pop=growth_rate*pop*(1-(pop/carry_cap))*dt;         29
    pop+=delta_pop;                                           30
  }                                                           31
}                                                             32
```

```
N=10.0                                                        23
r=0.5                                                         24
K=1000.0                                                      25
set dummy t                                                   26
pop(t)=K/(1+((K-N)/N)*exp(-r*t))                              27
plot pop(t) t "Analytical solution" w 1 lw 2, \              28
     'euler1.dat' t "Euler: dt=1.0" w lp pt 7, \             29
     'euler05.dat' every 2 t "Euler: dt=0.5" w lp pt 8, \    30
     'euler01.dat' every 10 t "Euler: dt=0.1" w lp pt 10     31
```

These commands assume that the data files euler1.dat, euler05.dat and euler01.dat are located in the working directory; otherwise, their full or relative pathnames must be provided. Lines 23 to 27 define the parameters and the equation of the analytical

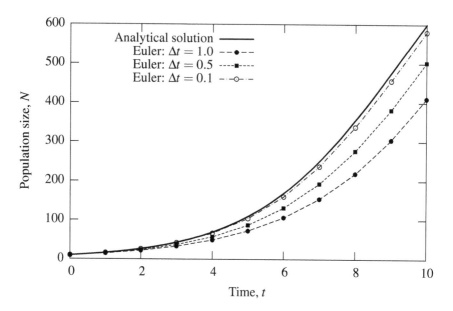

Figure 8.18: Analytical and numerical (Euler's method) solutions to the continuous logistic model of population growth for $r = 0.5$, $N_0 = 10$ and $\Delta t = 1, 0.5$ and 0.1.

solution to the continuous logistic model (Equation 8.19). These commands should be familiar from the preceding chapter (Section 7.4). The plot is created on lines 28 to 31 using a single **gnuplot** command, which is split over four lines of code by the line continuation symbol (\) on lines 28 through 30. Note that the keyword every on lines 30 and 31 instructs **gnuplot** to plot every n^{th} value in the named data set. For example, the file euler05.dat contains estimates of the population size at $t = 0$, 0.5, 1.0, 1.5, and so on. By selecting every second record of data from this file (every 2), only the results for $t = 0$, 1.0, and so on, are plotted. This allows the results of the three model runs to be directly compared.

It is clear from Figure 8.18 that the numerical solution gets closer to the analytical solution as Δt becomes smaller. Table 8.2 presents similar results, examining the difference at $t = 10$ between the exact analytical solution ($N_{10} = 599.86$) and the numerical solution based on Euler's method, for different values of Δt. Note that the percentage difference between the analytical and numerical solutions decreases by an order of magnitude as Δt is reduced by a factor of 10. At the same time, however, the total number of computational steps required to calculate N_{10} increases by an order of magnitude. Thus, there is a trade-off between computation time and accuracy.

8.4.2 Runge-Kutta Methods

Conceptual Basis and Mathematical Formulation

The main advantages of Euler's method are that it is relatively simple to program and fast to compute (Harris and Stocker 1998). It is unreliable in many instances,

Table 8.2: Difference at $t = 10$ between the analytical solution to the continuous logistic model for $r = 0.5$, $K = 1000$ and $N_0 = 10$ and the numerical solution based on Euler's method for different values of Δt.

Δt	Steps	N_{10}	% Difference
1.0	1	410.548	31.559
0.5	2	502.277	16.267
0.1	10	580.571	3.216
0.01	100	597.945	0.319
0.001	1000	599.668	0.032
0.0001	10000	599.840	0.003

however, and so is rarely used in practice. Nevertheless, it is important because it forms the basis for understanding virtually all other methods of numerical integration (Borse 1997). Among the most widely used of these are the mid-point method (also known as the modified Euler method) and the method of Runge-Kutta (Borse 1997, Harris and Stocker 1998, Wainwright and Mulligan 2004). The mid-point method is typically more accurate than the classical Euler method; this is because it uses the slope of the continuous function $N(t)$ in the middle of the interval ($\frac{dN}{dt}$ at $t + 0.5$; Figure 8.19), which is normally a better approximation to the average rate of change over the whole interval (Borse 1997, Harris and Stocker 1998). The method of Runge-Kutta takes this a stage further; the slope that is used to predict the population size at the end of the time interval is a weighted average of the slope at the start of the interval (at t) and at several other points between t and $t + 1$.

The fourth-order Runge-Kutta method is used in this section. This method offers a good compromise between programming effort, computation time and numerical accuracy, and can be expressed as

$$N_{t+1} \approx N_t + \frac{1}{6} (k_1 + 2k_2 + 2k_3 + k_4) \Delta t \qquad (8.25)$$

where

$$
\begin{aligned}
k_1 &= f'(t, N_t) \\
k_2 &= f'(t + 0.5\Delta t, N_t + 0.5k_1\Delta t) \\
k_3 &= f'(t + 0.5\Delta t, N_t + 0.5k_2\Delta t) \\
k_4 &= f'(t + \Delta t, N_t + k_3\Delta t)
\end{aligned}
\qquad (8.26)
$$

and where $f'(t_i, N_i)$ is the derivative of the function $N(t)$ at t, which is simply a different way of writing $\frac{dN}{dt}$ (Harris and Stocker 1998). Despite the rather daunting appearance of these equations, the calculations involved are quite simple and their subsequent implementation in gawk code is relatively straightforward.

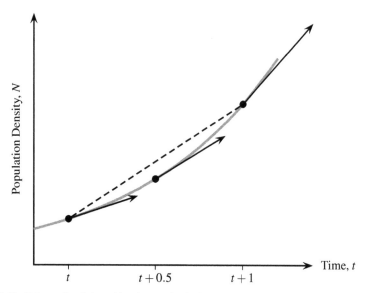

Figure 8.19: Schematic of the mid-point method of numerical integration. The arrows indicate
the slope of the continuous function $N(t)$ (solid gray curve) at $t, t+0.5$ and $t+1$;
the dashed line represents the average rate of change between t and $t+1$.

Implementation

Program 8.6, rk4.awk, illustrates the use of the fourth-order Runge-Kutta method to
solve numerically the continuous logistic model of population growth. The first 21
lines of this program are identical to those of Program 8.5, euler.awk, which used
Euler's method of numerical integration, and they perform a similar function. The
values of the Runge-Kutta parameters, k_1, k_2, k_3 and k_4 (Equation 8.26), are calculated
on lines 29 to 35. Note that lines 29, 31, 33 and 35 use the differential equation form
of the continuous logistic model (Equation 8.18). The k parameters are used, in turn,
to calculate values for the variables pop_k1, pop_k2 and pop_k3, which are estimates
of the population size at intermediate points between time t and time $t+1$. These
variables are employed on line 36 to estimate the change in the population size over
the period t to $t+1$ (Equation 8.25). Finally, the estimated population size at time
$t+1$ is calculated on line 37.

Program 8.6 can be run from the command line as follows:

```
gawk  -f  rk4.awk  -v  dt=0.1  >  rk4.dat                        16
```

assuming that $\Delta t = 0.1$. The results are redirected to the file rk4.dat.

Visualization

The output from Program 8.6 can be visualized in gnuplot as follows (Figure 8.20):

```
plot  pop(t)  t  "Analytical  solution"  w  l  lw  2,  \        32
      'rk4.dat'  every  10    t  "rk4:  dt=0.1"  w  p  pt  7  ps  2    33
```

Program 8.6: rk4.awk

```
# Continuous-time model of density-dependent (constrained)    1
# population growth, based on the Verhulst (logistic)          2
# equation (see Eq. 22), solved using fourth-order            3
# Runge-Kutta numerical integration.                          4
#                                                             5
# Usage: gawk -f rk4.awk -v dt=value [ > output_file ]        6
#                                                             7
# Variables:                                                  8
# ----------                                                  9
# pop            population at time step                      10
# growth_rate    instantaneous (intrinsic) growth rate        11
# carry_cap      carrying capacity                            12
# time           current time                                 13
# stop_time      time at which simulation should stop         14
# dt             integration step size                        15
#                                                             16
BEGIN{                                                        17
  pop=10.0;                                                   18
  carry_cap=1000;                                             19
  growth_rate=0.5;                                            20
  time=0;                                                     21
  stop_time=10;                                               22
                                                             23
  while(time<=stop_time){                                     24
    printf("%f\t%f\n", time, pop);                            25
                                                             26
    # Fourth-order Runge-Kutta method                        27
    #(after Harris and Stocker, 1998)                         28
    k1=growth_rate*pop*(1-(pop/carry_cap));                   29
    pop_k1=pop+k1*dt/2.0;                                     30
    k2=growth_rate*pop_k1*(1-(pop_k1/carry_cap));             31
    pop_k2=pop+k2*dt/2.0;                                     32
    k3=growth_rate*pop_k2*(1-(pop_k2/carry_cap));             33
    pop_k3=pop+k3*dt;                                         34
    k4=growth_rate*pop_k3*(1-(pop_k3/carry_cap));             35
    pop+=(1.0/6.0)*(k1+2.0*k2+2.0*k3+k4)*dt;                  36
    time+=dt;                                                 37
  }                                                           38
}                                                             39
```

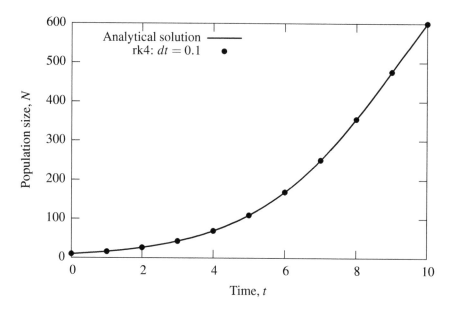

Figure 8.20: Results of the analytical (solid line) and fourth-order Runge-Kutta numerical (dots) solutions to the continuous logistic model of population growth for $r = 0.05$, $N(0) = 10$ and $dt = 0.1$.

Recall that the function pop(t) on line 32 was defined previously and is the analytical solution to the continuous logistic model. It is evident that the fourth-order Runge-Kutta method provides a close approximation to the analytical solution in this case.

8.5   INTER-SPECIFIC COMPETITION

8.5.1   Conceptual Basis and Mathematical Formulation

The models that have been considered thus far have been concerned with the growth or decline of a single species taken in isolation. It is extremely rare, however, for just one species to have exclusive access to the resources of a particular environment; it is much more common for these resources to be shared, often unequally, between two or more species. Individuals must therefore compete for the available resources, not only with members of their own species (intra-specific competition), but also with members of other species (inter-specific competition). Ultimately, the competitive interaction between two species produces one of two outcomes: sustained coexistence of the two species or the demise of the less competitive species (Alstad 2001).

The competition between two species can be represented in simple terms by extending the continuous logistic model of population growth, which accounted for intra-specific competition (Section 8.3; Equation 8.18), as follows:

$$\frac{dN_1}{dt} = r_1 N_1 \left(\frac{K_1 - (N_1 + N_2)}{K_1} \right) \tag{8.27}$$

where N_1 is the population size of species 1, N_2 is the population size of species 2, r_1 is the intrinsic growth rate of species 1 and K_1 is the carrying capacity of the environment for that species. Thus, an increase in the population of species 2 (N_2) will decrease the amount of resources that remains available to species 1 because the combined populations of species 1 and 2 ($N_1 + N_2$) are subtracted from K_1 in the negative feedback term of Equation 8.27. An equivalent formulation can be derived for species 2:

$$\frac{dN_2}{dt} = r_2 N_2 \left(\frac{K_2 - (N_1 + N_2)}{K_2} \right) \tag{8.28}$$

where r_2 is the intrinsic growth rate of species 2 and K_2 is the carrying capacity of the environment for that species.

Equation 8.27 and Equation 8.28 suggest that the competition between the two species is simple, direct and brutal. An increase in the population of species 1 means that the resources remaining available to support the growth of species 2 are reduced accordingly. It is highly unusual, however, for two species to make identical demands on an environment; for instance, their diets may differ (Alstad 2001). If the type of food that the two species consume differs only slightly, the competition between them for this resource will be intense. If their diets differ substantially, however, the level of inter-specific competition will be much less pronounced. This can be represented by introducing a scaling factor, or competition coefficient, into Equation 8.27 as follows:

$$\frac{dN_1}{dt} = r_1 N_1 \left(\frac{K_1 - (N_1 + \alpha N_2)}{K_1} \right) \tag{8.29}$$

where α is a factor that describes the effect of competition on a member of species 1 caused by members of species 2 (Roughgarden 1998). For example, if $\alpha = 1$ an extra individual of species 2 has exactly the same effect on species 1 as another member of species 1 (i.e., inter-specific competition is as intense as intra-specific competition); if $\alpha > 1$, an extra individual of species 2 has a greater effect on species 1 than the introduction of another member of species 1 (i.e., inter-specific competition is more intense than intra-specific competition); if $\alpha < 1$, an extra individual of species 2 has a smaller effect on species 1 than the introduction of another member of species 1 (i.e., inter-specific competition is less intense than intra-specific competition); if $\alpha = 0$, there is no inter-specific competition and the two populations grow independently, according to separate continuous logistic equations (Roughgarden 1998).

A similar competition coefficient, β, can be introduced to Equation 8.28:

$$\frac{dN_2}{dt} = r_2 N_2 \left(\frac{K_2 - (N_2 + \beta N_1)}{K_2} \right) \tag{8.30}$$

where β describes the effect of competition on a member of species 2 caused by members of species 1. In most cases, the values of α and β differ so that the effect of competition between species 1 and 2 is unequal and not reciprocal (Alstad 2001).

Equation 8.29 and Equation 8.30 are known as the Lotka-Volterra competition equations, named after the two scientists, Alfred Lotka and Vito Volterra, who developed them independently in the 1920s (Lotka 1925, Volterra 1926). Note that there

pop_1	10		*1*
pop_2	20		*2*
carry_1	1000		*3*
carry_2	750		*4*
alpha	0.7		*5*
beta	0.6		*6*
growth_1	0.5		*7*
growth_2	0.75		*8*
stop_time	100		*9*
delta_t	0.1		*10*

Figure 8.21: Example data file, params1.dat, containing the set of parameter values that are
required as input to Program 8.7, compete.awk.

is no closed-form analytical solution to these differential equations and so numerical
integration techniques must be employed.

8.5.2 Implementation

Program 8.7 presents an implementation of the Lotka-Volterra competition equations
(Equation 8.29 and Equation 8.30) using Euler's method of numerical integration.
The first 18 lines of this code are comments, which outline the purpose of the program, explain how it should be run via the command line, and list the primary variables that it employs. Lines 20 to 29 employ a group of pattern-action rules to read
the parameter values that are required by the model (N_1, N_2, K_1, K_2, α, β, r_1, r_2, t
and Δt) from an input data file. Given the number of parameter values that have to be
specified, this approach is more convenient than entering the values via the command
line and more flexible than stating them explicitly within the program itself. An example of the corresponding data file is given in Figure 8.21. Note that the first field of
each record in the data file contains the name of the variable in Program 8.7 associated with that parameter; the second field specifies its value. The pattern-action rules
read these values from the data file and store them using the appropriate variables.

Lines 31 through 47 of the program contain an END block, which is used here to
perform the model simulations (i.e., to generate data) after the parameter values have
been read from the input data file. A while loop (lines 33 through 46) is employed
inside the END block to model the growth of the two populations over the specified
period of time (i.e., while time is less than or equal to stop_time). The current time
(time) is initially set to zero (line 32). The first action of the while loop is to print out
the current time and the population size for species 1 and 2 (line 34). The current time
is then increased by a small amount Δt (delta_t; line 36), and the change in the size
of each population over this interval of time is calculated using Equation 8.29 and
Equation 8.30 (lines 38 to 39 and 41 to 42). These values are then used to calculate
the two population sizes at the new point in time (lines 44 and 45). The updated time
and population sizes are printed out at the start of the next iteration of the while loop.

Program 8.7: compete.awk

```
# Simple model of population growth with inter-specific      1
# competition described by the Lokta-Volterra equations      2
# (Eqs. 4 and 5), solved using Euler's method of             3
# numerical integration.                                     4
#                                                            5
# Usage: gawk -f compete.awk params.dat [ > outputFile ]     6
#                                                            7
# Variables                                                  8
# -----------                                                9
# pop_1, pop_2 Initial population density, species 1 and 2   10
# carry_1      Carrying capacity, species 1                  11
# carry_2      Carrying capacity, species 2                  12
# alpha, beta  Competition coefficients, species 1 and 2     13
# growth_1     Growth rate, species 1                        14
# growth_2     Growth rate, species 2                        15
# time         Current time                                  16
# stop_time    Time period after which simulation should stop 17
# dt           Time step for numerical integration           18
                                                             19
(NR==1){pop_1=$2;}                                           20
(NR==2){pop_2=$2;}                                           21
(NR==3){carry_1=$2;}                                         22
(NR==4){carry_2=$2;}                                         23
(NR==5){alpha=$2;}                                           24
(NR==6){beta=$2;}                                            25
(NR==7){growth_1=$2;}                                        26
(NR==8){growth_2=$2;}                                        27
(NR==9){stop_time=$2;}                                       28
(NR==10){dt=$2;}                                             29
                                                             30
END{                                                         31
    time=0;                                                  32
    while(time<=stop_time){                                  33
       print time, pop_1, pop_2;                             34
       # Increment time by dt                                35
       time+=dt;                                             36
       # Calculate change in population species 1 (Eq. 4)    37
       delta_pop_1=dt*growth_1*pop_1* \                      38
           ((carry_1-pop_1-(alpha*pop_2))/carry_1);          39
       # Calculate change in population species 2 (Eq. 5)    40
       delta_pop_2=dt*growth_2*pop_2* \                      41
           ((carry_2-pop_2-(beta*pop_1))/carry_2);           42
       # Calculate new population sizes species 1 and 2      43
       pop_1+=delta_pop_1;                                   44
       pop_2+=delta_pop_2;                                   45
    }                                                        46
}                                                            47
```

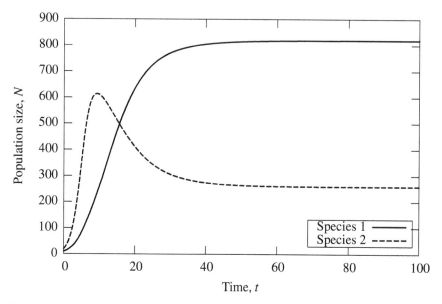

Figure 8.22: Sustained coexistence: inter-specific competition modeled using Program 8.7, compete.awk, for $N_1 = 10$ and $N_2 = 20$ initially, $r_1 = 0.5$, $r_2 = 0.75$, $\alpha = 0.7$, $\beta = 0.6$, $K_1 = 1000$ and $K_2 = 750$.

8.5.3   Running the Model

Program 8.7 can be run from the command line as follows:

```
gawk -f compete.awk params1.dat > compete1.dat
```
17

This command assumes that $N_1 = 10$ and $N_2 = 20$ initially, and that $r_1 = 0.5$, $r_2 = 0.75$, $\alpha = 0.7$, $\beta = 0.6$, $K_1 = 1000$ and $K_2 = 750$ (Figure 8.21). Thus, species 2 has the higher intrinsic rate of growth (0.75 as opposed to 0.5 for species 1), but a lower carrying capacity in this environment (750 as opposed to 1000 for species 1). The competition coefficients α and β are similar, although the relative impact of inter-specific competition is greater on species 1 than it is on species 2 ($\alpha > \beta$). Finally, the growth of both populations is simulated over a period of 100 units of time and for $\Delta t = 0.1$. The output from the model is redirected to the file compete1.dat.

8.5.4   Visualization

The output from Program 8.7 can be visualized by typing the following commands in gnuplot (Figure 8.22):

```
set style data lines
set key bottom right box
plot 'compete1.dat' u 1:2 t "Species 1" lw 2, \
     'compete1.dat' u 1:3 t "Species 2" lw 2
```
34
35
36
37

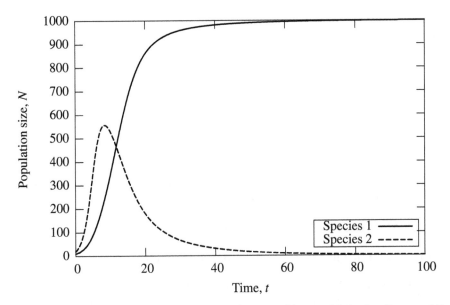

Figure 8.23: Demise of one species: inter-specific competition modeled using Program 8.7, compete.awk, for $N_1 = 10$ and $N_2 = 20$ initially, and $r_1 = 0.5$, $r_2 = 0.75$, $\alpha = 0.5$, $\beta = 0.8$, $K_1 = 1000$ and $K_2 = 750$.

These commands assume that the data file compete1.dat is located in the working directory; otherwise, its full or relative pathname must be supplied.

Figure 8.22 indicates that the population of species 2 initially grows in number faster than that of species 1. This is because of the higher intrinsic rate of growth of species 2 ($r_2 > r_1$) and the competitive impact that it has on species 1 ($\alpha = 0.7$). Despite this, the population of species 1 continues to grow in size and exceeds that of species 2 after $t = 16$, primarily because of the environment's greater carrying capacity for species 1. The increased competition resulting from the large number of individuals of species 1 causes the population of species 2 to decrease in number. By $t = 60$, however, the two populations settle into a state of sustained coexistence in which $N_1 \approx 819$ and $N_2 \approx 259$. Note that the size of each population under these steady-state conditions is smaller than the carrying capacity of the environment for either species in isolation ($K_1 = 1000$ and $K_2 = 750$).

The Lotka-Volterra equations do not always produce conditions of sustained co-existence. Figure 8.23, for example, shows the results of a further simulation using Program 8.7 in which species 2 eventually becomes extinct as a result of competition with species 1. This simulation was performed using slightly different values of the competition coefficients ($\alpha = 0.5$ and $\beta = 0.8$) compared to the preceding example, but the values of the other variables are unchanged (Figure 8.24).

```
gawk -f compete.awk params2.dat > compete2.dat
```
18

```
pop1            10                                          1
pop2            20                                          2
carry1          1000                                        3
carry2          750                                         4
alpha           0.5                                         5
beta            0.8                                         6
growth1         0.5                                         7
growth2         0.75                                        8
stop_time       100                                         9
delta_t         0.1                                         10
```

Figure 8.24: Second example data file, `params2.dat`, containing the set of parameter values that are required as input to Program 8.7, `compete.awk`.

8.6 PREDATOR-PREY RELATIONSHIPS

8.6.1 Conceptual Basis and Mathematical Formulation

The final model of population growth examined in this chapter considers the case in which one species preys upon another. All other things being equal, one might expect that the predator species will generally become more numerous as the population of the prey species increases in number because of the greater level of resources available to support the predator population. As the number of predators increases, however, more prey are likely to be consumed, which may lead to a decline in the population of the prey species. This may, in turn, lead to a reduction in the number of predators and eventually, perhaps, to the recovery of the prey species population. There is, thus, a fine balance and interrelationship between the populations of the two species, which is another form of inter-specific competition.

One of the simplest mathematical models of predator-prey relationships is based, once again, on the pioneering work of Lotka (1925) and Volterra (1926), and is given by the following pair of differential equations:

$$\frac{dN_1}{dt} = rN_1 - aN_1N_2 \tag{8.31}$$

and

$$\frac{dN_2}{dt} = abN_1N_2 - dN_2 \tag{8.32}$$

where N_1 is the population size of the prey species, N_2 is the population size of the predator species, r is the intrinsic rate of growth of the prey species, d is the death rate of the predators, a is a coefficient that describes the rate of success with which predators capture prey and b is a coefficient relating the number of prey consumed by each predator to the number of predator births (Roughgarden 1998). There are several important assumptions that underpin these equations. The first is that, in the absence of predation, the growth of the prey species is unconstrained (density-independent), so that $dN_1/dt = rN_1$ when $N_2 = 0$ (the prey species increases in number exponentially over time when there are no predators; Equation 8.31). The second assumption, which is implicit, is that the predators encounter the prey in a random fashion and that the number of encounters increases in proportion to the number of predators and prey ($N_1 N_2$). Only a fraction of encounters results in a predator capturing its prey, and hence in the death of the prey ($a N_1 N_2$; Equation 8.31). The third assumption is that the predators are never satiated; that is, they consume as many prey as they can capture. Finally, it is assumed that the number of predator births is proportional to the number of prey consumed ($ab N_1 N_2$; Equation 8.32), and that a fixed proportion of the predator population dies at any given point in time ($d N_2$; Equation 8.32).

8.6.2 Implementation

Program 8.8, `predprey.awk`, presents an implementation in **gawk** of the Lotka-Volterra predator-prey model (Equation 8.31 and Equation 8.32). The program employs fourth-order Runge-Kutta numerical integration to solve this pair of differential equations because a closed-form analytical solution does not exist.

The first six lines of the program consist of comments that outline the purpose of the code and that show how it should be run via the command line. The bulk of the code is contained within the BEGIN block (lines 8 to 43) because the intention is to generate data via a simulation model. For the sake of simplicity, the values of the main variables have been hard-wired into the program (lines 9 to 16), although they could equally well have been entered on the command line or via an input data file. A `while` loop is used to simulate the growth of the predator and prey populations over time, continuing while the current time (`time`) is less than or equal to the specified point in time when the simulation should stop (`stop_time`). The first action within the `while` loop is to print out the current time and the sizes of the prey and predator populations (line 19). The current time is then incremented by a small amount, Δt (`delta_t`; line 21). The size of each population at the new point in time is recalculated (lines 24, 26, 28, 30, 32, 34, 36 and 40 for the prey species and lines 25, 27, 29, 31, 33, 35, 37 and 41 for the predator species) using fourth-order Runge-Kutta numerical integration to solve Equation 8.31 and Equation 8.32. The updated population sizes are printed out in the next iteration of the `while` loop.

8.6.3 Running the Model

Program 8.8 can be run from the command line as follows:

```
gawk -f predprey.awk -v dt=0.01 > predprey.dat
```
19

Remember that the values of the main variables are hard-wired in the code and hence do not have to be supplied on the command line. Thus, the population of each species initially consists of 20 individuals ($N_1 = 20$ and $N_2 = 20$), the prey species exhibits a relatively high intrinsic rate of growth ($r = 0.9$) and 60% of the predator population dies in each time step ($d = 0.6$). Moreover, on average the predators are successful in capturing the prey once in every 10 encounters ($a = 0.1$) and produce an offspring for every two prey that they consume ($b = 0.5$). Note that output from the computational model is redirected to the file `predprey.dat` in the working directory.

Program 8.8: predprey.awk

```
# A simple model of predator-prey interaction based        1
# on the Lotka-Volterra equations (see Equations 8.31 and  2
# 8.32), solved using fourth-order Runge-Kutta numerical   3
# integration techniques.                                  4
#                                                          5
# Usage: gawk -f predprey.awk -v dt=value [ > outputFile ] 6
                                                           7
BEGIN{                                                     8
    prey=20.0;          # Population density, prey (N_1)    9
    pred=20.0;          # Population density, predator (N_2) 10
    death=0.6;          # Death rate, predator species (d)  11
    p_coeff=0.1;        # Coefficient of predation (a)      12
    p_effic=0.5;        # Predator efficiency (b)           13
    growth_rate=0.9;    # Growth rate, prey species (r)     14
    time_stop=60;       # Time at which simulation should stop (t)15
    time=0;             # Current time                      16
                                                           17
    while(time<=time_stop){                                18
       print time, prey, pred;                             19
                                                           20
       time+=dt;                                           21
                                                           22
       # Prey species (Eq. 8.31), predator species (Eq. 8.32) 23
       k1_prey=(growth_rate*prey)-(p_coeff*prey*pred);     24
       k1_pred=(p_effic*p_coeff*prey*pred)-(death*pred);   25
       prey_1=prey+k1_prey*dt/2;                           26
       pred_1=pred+k1_pred*dt/2;                           27
       k2_prey=(growth_rate*prey_1)-(p_coeff*prey_1*pred_1); 28
       k2_pred=(p_effic*p_coeff*prey1*pred_1)-(death*pred_1); 29
       prey_2=prey+k2_prey*dt/2;                           30
       pred_2=pred+k2_pred*dt/2;                           31
       k3_prey=(growth_rate*prey_2)-(p_coeff*prey_2*pred_2); 32
       k3_pred=(p_effic*p_coeff*prey_2*pred_2)-(death*pred_2); 33
       prey_3=prey+k3_prey*dt;                             34
       pred_3=pred+k3_pred*dt;                             35
       k4_prey=(growth_rate*prey_3)-(p_coeff*prey_3*pred_3); 36
       k4_pred=(p_effic*p_coeff*prey_3*pred_3)-(death*pred_3); 37
                                                           38
       # Calculate revised populations                     39
       prey+=(1/6)*(k1_prey+2*k2_prey+2*k3_prey+k4_prey)*dt; 40
       pred+=(1/6)*(k1_pred+2*k2_pred+2*k3_pred+k4_pred)*dt; 41
    }                                                      42
}                                                          43
```

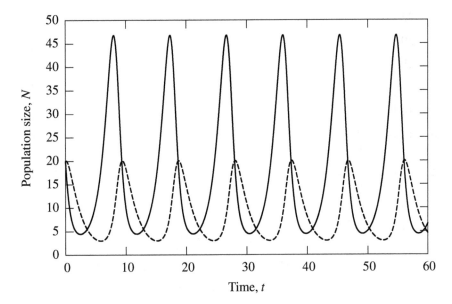

Figure 8.25: Output from the Lotka-Volterra predator-prey equations (solid line = prey species; dashed line = predator species).

8.6.4 Visualization

The output produced by Program 8.8, `predprey.awk`, can be visualized (Figure 8.25) by typing the following commands in **gnuplot**:

```
unset key                                        38
plot 'predprey.dat' u 1:2 lw 2, \                39
     'predprey.dat' u 1:3 lw 2                    40
```

These commands assume that the file `predprey.dat` is located in the working directory; otherwise, its full or relative pathname must be supplied. Note that this file contains three fields of data: the current time, the population size of the prey species and the population size of the predator species.

Figure 8.25 demonstrates that the Lotka-Volterra predator-prey model produces regular oscillations in the sizes of the predator and prey populations, based on the parameter values used in Program 8.8. The oscillations are slightly out of phase, so that an increase in the size of the prey population is followed, after a short lag, by a corresponding increase in the predator population. As the predator population grows, the number of prey consumed rises and hence the prey population starts to fall. The prey population eventually becomes too small to support the number of predators and, after another short lag, the predator population begins to decrease. The reduced number of predators means that the prey population flourishes once more, and the whole cycle begins again.

The same information can be represented in the form of a phase-plane diagram (Alstad 2001), where the sizes of the prey and predator populations are plotted on the

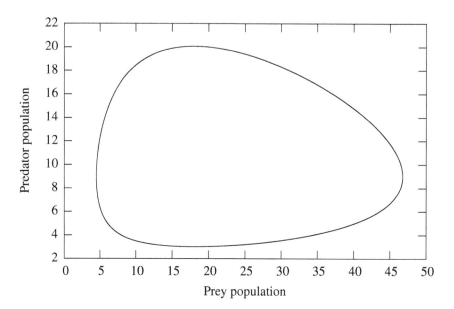

Figure 8.26: Phase-plane diagram showing the hysteresis loop between the sizes of the predator and prey populations predicted by the Lotka-Volterra model (Program 8.8).

x- and y-axes, respectively. This can be achieved in **gnuplot** as follows:

```
set xlabel "Prey population"                                    41
set ylabel "Predator population"                                42
plot 'predprey.dat' u 2:3 lw 2                                  43
```

These commands produce Figure 8.26. Notice that the repeated oscillations seen in Figure 8.25 trace a closed loop in Figure 8.26. This is sometimes known as a hysteresis loop because the change in the predator population lags somewhat behind the change in the prey population that causes it. Figure 8.26 and Figure 8.25 also demonstrate that the predator and prey species continue to coexist in this realization of the model, with neither forcing the other into extinction (Roughgarden 1998); other values of the model parameters may, however, produce very different outcomes.

> **Exercise 8.4**: Experiment with Program 8.8, changing the values of the main variables (N_1, N_2, r, d, a and b) in the code. Which conditions lead to coexistence of the predator and the prey populations and which, if any, lead to the demise of one or both of them? Plot the results in **gnuplot**.

8.7 SUMMARY

This chapter examines a range of models of population growth, starting with a simple model of unconstrained (density-independent) growth, progressing through to models of constrained (density-dependent) growth, to ones that represent competition for resources between species (inter-specific competition) and the relationships between predator and prey species. Each is comparatively simple, in the sense that few real populations behave exactly in the way that the models describe. Nevertheless, each provides important insights that help to understand the functioning of environmental systems. Readers who wish to explore population growth models further are encouraged to consult the following sources: Roughgarden (1998), Wilson (2000), Alstad (2001), May (2001), and Donovan and Welden (2002).

In the following chapter, a population growth model, similar to the ones explored here, is used to simulate the growth of two species of daisy on an imaginary planet, known as Daisyworld. The two types of daisy compete with one another for the available resources. Their growth rates are also controlled by the temperature of the planet, which is, in turn, a function of the amount of solar radiation that the planet receives.

SUPPORTING RESOURCES

The following resources are provided in the `chapter8` sub-directory of the CD-ROM:

`discrete.awk`	**gawk** implementation of the discrete model of unconstrained (density-independent) population growth.
`continue.awk`	**gawk** implementation of the continuous model of unconstrained (density-independent) population growth.
`cntlogst.awk`	**gawk** implementation of the continuous model of constrained (density-dependent) population growth, based on the Verhulst (logistic) equation.
`dsclogst.awk`	**gawk** implementation of the discrete model of constrained (density-dependent) population growth, based on the Verhulst (logistic) equation.
`euler.awk`	**gawk** implementation of the continuous model of constrained (density-dependent) population growth, based on the Verhulst (logistic) equation, solved using Euler's method.
`rk4.awk`	**gawk** implementation of the continuous model of constrained (density-dependent) population growth, based on the Verhulst (logistic) equation, solved using the fourth-order Runge-Kutta method.
`compete.awk`	**gawk** implementation of the continuous model of inter-specific competition, described by the Lotka-Volterra equations, solved using Euler's method.
`predprey.awk`	**gawk** implementation of the continuous model of predator-prey interaction, described by the Lotka-Volterra equations, solved using the fourth-order Runge-Kutta method.

`params1.dat`	Example parameter value file required as input to `compete.awk`.
`params2.dat`	Example parameter value file required as input to `compete.awk`.
`populate.bat`	Command line instructions used in this chapter.
`populate.plt`	gnuplot commands used to generate the figures presented in this chapter.
`populate.gp`	gnuplot commands used to generate figures presented in this chapter as EPS files.

Chapter 9

Biospheric Feedback on Daisyworld

Topics

- Daisyworld model and the Gaia hypotheses

- Biospheric feedback and steady-state conditions

- Perturbing Daisyworld by increasing solar luminosity

- Exploring the impact of biodiversity

Methods and techniques

- Modularizing gawk code with user-defined functions

- Sensitivity analysis

9.1 INTRODUCTION

This chapter examines a model of the feedback mechanisms that exist between the biota (the living organisms) and the abiotic environment of an imaginary planet, known as Daisyworld. The model combines many of the elements examined in the preceding four chapters. For instance, it explores the growth in the populations of two species of daisy over time (Chapter 8) as a function of planetary temperature. The latter is, in turn, partly controlled by changes in the amount of solar radiation incident on the planet's surface (Chapter 5). The two species of daisy differ solely with respect to color: one is dark, the other is light, compared to the soil substrate in which they

Table 9.1: Multiple Gaia hypotheses (after Kirchner 2002).

Hypothesis	Description
Influential Gaia	Biota collectively significantly affect planet's abiotic environment
Co-evolutionary Gaia	Evolution of biota and abiotic environment are closely coupled
Homeostatic Gaia	Biota act to stabilize abiotic environment (negative feedback loops dominate)
Geophysiological Gaia	Biosphere operates as a single, giant, organism
Optimizing Gaia	Biota optimize the abiotic environment for their own benefit
Gaia as a metaphor	—

grow. Thus, the model is also concerned with the interaction between solar radiation and plant canopies, or, more specifically, the fractions of incident radiation that are reflected or absorbed by the daisies (Chapter 6 and Chapter 7).

The Daisyworld model was developed by Watson and Lovelock (1983) largely in response to criticism of Lovelock's Gaia hypothesis (named after the goddess of the Earth in classical Greek mythology), which suggests that Earth behaves like a self-regulating super-organism in which the biota and abiotic environment interact to maintain conditions that are suitable for, and adapted to, the continued existence of life (Lovelock 1995b). One of the main criticisms of the Gaia hypothesis is that it is teleological; that is to say, it implies that the biota are imbued with foresight or a sense of purpose, which they employ to achieve a specific goal (Doolittle 1981, Dawkins 1982). Lovelock (1995a) strongly contests this assertion and, together with Andrew Watson, he developed the Daisyworld model to show how the biota might regulate their abiotic environment without recourse to foresight or planning, through a combination of positive and negative feedback mechanisms (Saunders 1994, Lovelock 1995a). Despite this, a number of criticisms of the Gaia hypothesis remain, the most serious of which are that it is untestable (Kirchner 1989) and unfalsifiably vague (Kirchner 1990). The second of these two criticisms arises partly because, over time, Lovelock and co-workers have expressed the Gaia hypothesis in a number of different ways (Kirchner 2002). These variants are summarized in Table 9.1.

Setting aside the merits and deficiencies of the Gaia hypothesis (or hypotheses), which it was originally developed to defend, the Daisyworld model has considerable value in its own right as a heuristic tool. While the model does not purport to show how the biotic and abiotic components of Earth's biosphere are actually linked, it suggests a way in which they might interact so that the biota exert a degree of control over their abiotic environment (Hardisty *et al.* 1993). The model also displays certain properties that may be relevant to the way in which Earth's climate and biosphere have evolved. For these reasons, it is examined in detailed in this chapter.

9.2 DESCRIPTION AND ASSUMPTIONS OF THE CONCEPTUAL MODEL

Daisyworld is an imaginary planet illuminated by a distant sun (Watson and Lovelock 1983, Hardisty *et al.* 1993). Its atmosphere is cloudless and contains a negligible quantity of greenhouse gases. The only forms of life on the planet are two species of daisy, which differ solely in terms of color: one species is dark, the other is light, compared to the soil substrate in which they grow. For the sake of convenience, the dark and light species are referred to as "black" and "white" daisies, respectively, hereafter; by implication the soil is "gray". As a consequence of their physical characteristics (i.e., their color), the black daisies reflect less of the incident solar radiation than does the soil substrate, and hence they absorb more; in contrast, the white daisies reflect more and absorb less solar radiation than does the soil substrate.

To simplify matters still further, Daisyworld is assumed to be completely flat. This characteristic has an important ramification: the amount of solar radiation that is incident on the planet is uniform across the whole of its surface; by contrast, for a spherical planet, such as Earth, it varies from equator to pole (Watson and Lovelock 1983). It is also implicitly assumed that Daisyworld traces a circular orbit around its sun and that the planet's surface is always normal to the incident solar radiation (i.e., the angle of obliquity is zero). Thus, Daisyworld experiences neither diurnal variation nor seasonal changes in the amount of incident solar radiation (Hardisty *et al.* 1993). Over time, however, Daisyworld's sun gradually becomes brighter (its luminosity increases), which is typical of a main sequence star, such as Earth's sun (Saunders 1994). As a result, there is a concomitant increase in the amount of solar radiation incident on the planet's surface, which has important consequences for the climate of Daisyworld.

A fraction of the planet's surface is suitable for daisy growth. The availability of this fertile land is one of three factors that control the rates at which the daisy populations grow; the other two factors are the fraction of daisies that die during a given period of time and the rate at which new daisies appear (are born) over the same interval of time. The death rate is assumed to be fixed, independent of the population size, and the daisies that die decompose very rapidly to become soil. By contrast, the birth rate varies according to temperature: the optimum temperature for daisy growth is 22.5 °C, but growth can occur at any temperature in the range of 5 °C to 40 °C (22.5 °C ± 17.5 °C). The amount of fertile land available for daisy growth is, of course, limited; hence, there is both intra-specific and inter-specific competition among the black and white daisies for this resource.

The temperature of Daisyworld as a whole (i.e., the average global temperature) is dependent on the amount of incident solar radiation and the albedo of the planet surface. The latter is, in turn, the sum of the albedo values of the various materials that cover the planet's surface (black daisies, white daisies and bare soil) weighted by the fraction of ground covered by each material type (its relative areal extent). The surface temperature differs somewhat from one location to another, however, due to local differences in the planet's albedo. Thus, the temperature is lower than the global average over patches of white daisies because they reflect more and, hence, absorb less of the incident solar radiation than either the soil substrate or the black daisies.

Similarly, the temperature is higher than the global average over patches of black daisies because they reflect less and, hence, absorb more of the incoming radiation than either the soil substrate or the white daisies.

9.3 FORMULATING THE MATHEMATICAL MODEL

The processes and interactions that are described verbally in the preceding section are represented graphically in Figure 9.1. This diagram provides a useful overview of how Daisyworld functions as a system; it highlights the various feedback loops, which cause the system to behave in a complex non-linear fashion. The next step is to develop a mathematical model of the system. This model is formulated below.

It is convenient to start by representing the amount of solar energy incident on Daisyworld. As noted previously, this quantity varies with time and is independent of the other state variables in the system (it is an exogenous variable). Consequently, it is drawn outside the main system box in Figure 9.1, which is indicated by the dashed line. Watson and Lovelock (1983) express the amount of solar energy incident on the surface of Daisyworld as the product of the solar constant, S (see also Section 5.4.1), and a factor, L, which describes the relative luminosity (or brightness) of the sun. They give the value of S as 9.17×10^5 erg·cm^{-2}·s^{-1}, which is equivalent to 917 W·m^{-2}. The luminosity factor of Daisyworld's sun would therefore need to be about 1.5 to produce the same level of exo-atmospheric irradiance as Earth's sun ($S \times L = 917$ W·m$^{-2} \times 1.5 \approx 1380$ W·m^{-2}; see also Equation 5.7 on page 123).

The amount of solar radiation incident on Daisyworld is critically important to the system as a whole because of the direct effect that it has on the average temperature of the planet, T_{global}. T_{global} is also dependent on the fraction of incident solar radiation that is reflected from the planet surface and, hence, the fraction that it absorbs: the greater the fraction that is reflected, the less is absorbed; the smaller the fraction that is absorbed, the less the planet surface heats up. The fraction of incident solar radiation that is reflected by the planet as a whole is given by the global albedo, A_{global}. By implication, the fraction of incident solar radiation that is absorbed by Daisyworld is $1 - A_{global}$ (i.e., everything except that which is reflected). Thus, the total amount of radiation absorbed by Daisyworld is $S \times L \times (1 - A_{global})$.

Assuming that Daisyworld behaves like a blackbody radiator, the amount of solar energy that it absorbs and the amount that it emits must be equal. Consequently, the Stefan-Boltzmann equation, which was introduced in Chapter 5 (Equation 5.3 on page 122), can be used to estimate the average temperature of the planet. Remember that the Stefan-Boltzmann equation relates the energy emitted by a blackbody radiator to its temperature, as follows:

$$M = \sigma T^4 \qquad (9.1)$$

where M is the radiation emitted by the object (W·m^{-2}), T is its temperature in kelvin (K) and σ is the Stefan-Boltzmann constant (5.67×10^{-8} W·m^{-2}·K). Equation 9.1 can be rearranged to express the temperature of a blackbody in terms of the amount of

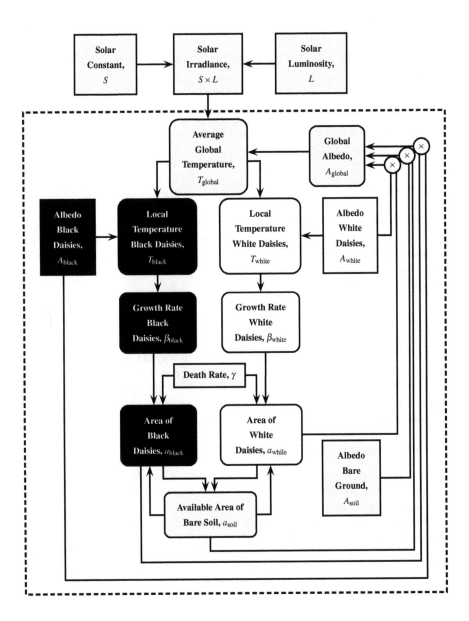

Figure 9.1: Diagrammatic representation of the Daisyworld model.

energy that it emits, as follows:

$$T = \left(\frac{M}{\sigma}\right)^{0.25} \tag{9.2}$$

It has already been shown that the amount of solar radiation absorbed by Daisyworld and hence, as a blackbody radiator, the amount that it emits is

$$M = S \times L \times (1 - A_{global}) \tag{9.3}$$

Substituting Equation 9.3 into Equation 9.2, and subtracting 273.2 to convert from kelvin into degrees Celsius, gives

$$T_{global} = \left(\frac{SL(1 - A_{global})}{\sigma}\right)^{0.25} - 273.2 \tag{9.4}$$

which describes the average global temperature of Daisyworld.

As noted earlier, the average global albedo of Daisyworld, A_{global}, varies as a function of the fraction of the planet surface that is covered by black daisies, by white daises and by bare soil, as well as their respective albedo values. This can be expressed as follows:

$$A_{global} = (a_{soil}A_{soil}) + (a_{black}A_{black}) + (a_{white}A_{white}) \tag{9.5}$$

where a_{soil}, a_{black} and a_{white} are the fractional areas of the planet surface covered by bare soil, black daisies and white daisies, respectively, and where A_{soil}, A_{black} and A_{white} are the corresponding albedo values. Watson and Lovelock (1983) use the following initial values for these variables: $A_{soil} = 0.5$, $A_{black} = 0.25$, $A_{white} = 0.75$, $a_{black} = 0.01$ and $a_{white} = 0.01$. Finally, the fraction of the planet surface that is not covered by daisies, but which is suitable for daisy growth (i.e., fertile bare soil), is

$$a_{soil} = a_{suit} - (a_{black} + a_{white}) \tag{9.6}$$

where a_{suit} is the fraction of the planet surface that is suitable for daisy growth. In this chapter, it is assumed that $a_{suit} = 1.0$. If a value less than one is used for a_{suit}, Equation 9.5 must be modified accordingly.

Watson and Lovelock (1983) suggest that the local temperature of an area covered by either black daises or white daisies can be expressed as a function of the difference between the local albedo and the average global albedo, which causes local variations in the rate of surface heating, and the rate at which heat energy is redistributed from warmer to cooler areas. The relevant formulation is as follows:

$$T_{black} = \left(q'(A_{global} - A_{black}) + T_{global}\right) \tag{9.7}$$

$$T_{white} = \left(q'(A_{global} - A_{white}) + T_{global}\right) \tag{9.8}$$

where T_{black} and T_{white} are the local temperatures over areas of black daisies and white daisies, respectively, and q' is a factor that describes the redistribution (i.e., the

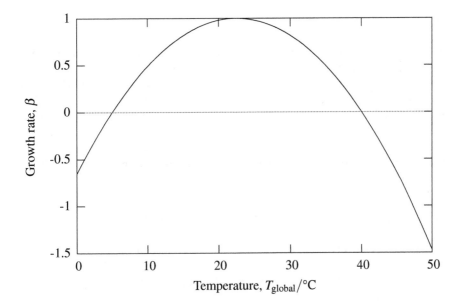

Figure 9.2: Parabolic relationship between daisy growth rate and local temperature.

conduction) of thermal energy from warmer to cooler areas. Thus, if $q' = 0$, the local temperatures are always equal to the global average temperature because there is perfect conduction of thermal energy from warmer locations to cooler ones. Watson and Lovelock (1983) suggest that the value of q' should be less than $0.2SL/\sigma$, and they employ $q' = 20$ in their own simulations.

The intrinsic (instantaneous) growth rates of the black daisies and the white daisies are assumed to be simple parabolic functions of local temperature, as follows:

$$\beta_{black} = 1 - 0.003265 \left(22.5 - T_{black}\right)^2 \tag{9.9}$$

$$\beta_{white} = 1 - 0.003265 \left(22.5 - T_{white}\right)^2 \tag{9.10}$$

where β_{black} and β_{white} are the growth rates for the black daisies and the white daisies, respectively (Figure 9.2). The numerical constants (1, 0.003265 and 22.5) used in these equations are such that daisy growth occurs when the local temperature lies in the range of 5 °C to 40 °C, and is maximized when the local temperature is 22.5 °C. Outside this range, Equation 9.9 and Equation 9.10 result in a negative growth rate; that is, a decline in the population for that species of daisy.

As the populations of the two species of daisy grow (or decline) over time, the fractional area of the planet surface that each occupies changes. This is expressed by the following pair of differential equations:

$$\frac{da_{black}}{dt} = \left(a_{black}\left(a_{soil}\beta_{black} - \gamma\right)\right) \tag{9.11}$$

$$\frac{da_\text{white}}{dt} = \left(a_\text{white}\left(a_\text{soil}\beta_\text{white} - \gamma\right)\right) \tag{9.12}$$

where γ is the death rate for both types of daisy. These equations encapsulate both inter-specific and intra-specific competition. Thus, the growth rate for each species is multiplied by the amount of fertile land that remains unoccupied ($a_\text{soil}\beta_\text{black}$ and $a_\text{soil}\beta_\text{white}$), for which daisies from each species must compete with daisies of the same species (inter-specific competition) and with daisies of the other species (intra-specific competition). Note, however, that the death rate, γ, is assumed to be identical for both species and is independent of the population size. Watson and Lovelock (1983) employ $\gamma = 0.3$ in their simulations; that is, 30% of the population dies in each time step.

9.4 IMPLEMENTING THE COMPUTATIONAL MODEL

In view of the relatively large number of variables involved in the Daisyworld model, the first step taken here is to summarize each of the variables in a table, together with their corresponding mathematical symbols, the names used to identify them in the gawk code and, where appropriate, their units of measurement (Table 9.2).

Four separate implementations of the computational model are examined below. In the first, the luminosity, L, of Daisyworld's sun is held constant through time. The intention is to examine how Daisyworld responds to a fixed level of external forcing and, in particular, to explore the steady-state behavior of the system under these conditions. In the second implementation, the solar luminosity is increased in steps of 0.025 from $L = 0.5$ to $L = 1.7$. At each step, the system is allowed to run until it reaches steady state before proceeding to the next value of L. The objective of this implementation is to examine how the system responds to a gradual change in external forcing. In the third implementation, an extra species of daisy is added to the model to investigate the effect of increasing biodiversity on Daisyworld. Finally, the fourth implementation replaces elements of the code from the third implementation with user-defined functions. The intention here is to show how the code can be made modular so that, in this particular example, the model can be extended to simulate the effect of having a very large number of daisy species, each with a different albedo.

9.4.1 Implementation 1: Constant Solar Luminosity

Implementation

Program 9.1, `daisy1.awk`, is an implementation of the Daisyworld model for a fixed (user-defined) value of solar luminosity. The code is contained entirely within the BEGIN block (lines 1 to 46) because the intention is to generate data from the model. Inside the BEGIN block, lines 4 to 13 initialize the values of the main variables used in the program, rather than requiring them to be entered via the command line. Thus, the fraction of the planet's surface that is initially covered by black daisies is set to 0.01 (i.e., 1% of the planet surface; line 4); the same value is used to initialize the fractional area covered by white daisies (line 5). These values provide a starting point from

Table 9.2: List of parameters and variables in the Daisyworld model and their implementation in the corresponding **gawk** code (Program 9.1).

Entity	Symbol	Variable name	Units
Emitted flux density	M	–	$W \cdot m^{-2}$
Stefan-Boltzmann constant	σ	`Stefan`	$W \cdot m^{-2} \cdot K^{-4}$
Solar constant	S	`solar_const`	$W \cdot m^{-2}$
Solar luminosity	L	`luminosity`	–
Global albedo	A_{global}	`global_albedo`	–
Soil albedo	A_{soil}	`albedo_soil`	–
Albedo of black daisies	A_{black}	`albedo_black`	–
Albedo of white daisies	A_{white}	`albedo_white`	–
Fraction of planet suitable for daisy growth	a_{suit}	`area_suit`	–
Fraction of planet covered by bare soil	a_{soil}	`area_soil`	–
Fraction of planet covered by black daisies	a_{black}	`area_black`	–
Fraction of planet covered by white daisies	a_{white}	`area_white`	–
Global temperature (average)	T_{global}	`global_temp`	°C
Local temperature (black daisies)	T_{black}	`temp_black`	°C
Local temperature (white daisies)	T_{white}	`temp_white`	°C
Energy diffusion factor	q'	`qfactor`	–
Growth rate of black daisies	β_{black}	`growth_black`	–
Growth rate of white daisies	β_{white}	`growth_white`	–

Program 9.1: daisy1.awk

```
BEGIN {                                                                    1
                                                                           2
  # Initialize main variables                                             3
  area_black   = 0.01;                                                     4
  area_white   = 0.01;                                                     5
  area_suit    = 1.0;                                                      6
  solar_const  = 917.0;                                                    7
  Stefan_Boltz = 5.67E-08;                                                 8
  albedo_soil  = 0.5;                                                      9
  albedo_black = 0.25;                                                    10
  albedo_white = 0.75;                                                    11
  death_rate   = 0.3;                                                     12
  q_factor     = 20.0;                                                    13
                                                                          14
  # Run model for 100 time steps                                         15
  for(time=0;time<=100;++time){                                          16
    # Equation 9.6                                                       17
    area_soil=area_suit-(area_black+area_white);                         18
    # Equation 9.5                                                       19
    albedo_global=(area_soil*albedo_soil)+ \                             20
      (area_black*albedo_black)+(area_white*albedo_white);               21
    # Equation 9.4                                                       22
    temp_global=(((solar_const*luminosity* \                             23
      (1-albedo_global))/Stefan_Boltz)**0.25)-273;                       24
    # Equations 9.7 and 9.8                                              25
    temp_black=((q_factor*(albedo_global-albedo_black))+ \               26
      temp_global);                                                      27
    temp_white=((q_factor*(albedo_global-albedo_white))+ \               28
      temp_global);                                                      29
    # Print current time, global temperature and daisy areas             30
    printf("%3i %7.4f %6.4f %6.4f\n", \                                  31
      time, temp_global, area_black, area_white);                        32
    # Equations 9.9 and 9.10                                             33
    growth_black=(1-(0.003265*((22.5-temp_black)**2)));                  34
    growth_white=(1-(0.003265*((22.5-temp_white)**2)));                  35
    # Update daisy area for next time step                               36
    # Equations 9.11 and 9.12                                            37
    area_black+=area_black*((area_soil*growth_black)- \                  38
      death_rate);                                                       39
    area_white+=area_white*((area_soil*growth_white)- \                  40
      death_rate);                                                       41
    # Do not allow daisies to become extinct                            42
    if(area_black<0.01){area_black=0.01};                               43
    if(area_white<0.01){area_white=0.01};                               44
  }                                                                       45
                                                                          46
}                                                                         47
```

which the daisies can grow. In this implementation, the entire planet is assumed to be suitable for daisy growth (line 6). The solar constant is given as $917\,\text{W·m}^{-2}$ (line 7) and the value of the Stefan-Boltzmann constant (σ; Equation 9.1) is set to 5.67×10^{-8} (line 8); note that the latter could have been coded as Stefan_Boltz = 0.0000000567, but it is more convenient when dealing with very large or very small numbers, such as this, to use the equivalent scientific notation (Stefan_Boltz = 5.67E-8). The albedo values of the soil, black daisies and white daisies are initialized on lines 9 through 11; the values employed here are the same as those used by Watson and Lovelock (1983). Thus, the soil reflects half of the incident solar radiation (and, hence, absorbs the other half), while the black daisies and white daisies reflect one quarter and three quarters, respectively. This means that the black daisies are not perfectly black. Instead, they are a dark shade of gray. Similarly, the white daisies are not perfectly white, but are a pale shade of gray. Finally, the death rate of the daisies, γ, is set to 0.3 (line 12) and the heat diffusion factor, q', is fixed at 20.0 (line 13), in common with Watson and Lovelock (1983).

The bulk of the code is contained within a for loop (lines 16 to 45). This is used to run the model for 101 iterations; each iteration representing a discrete time-step ($t = 0$ to $t = 100$ inclusive). This is intended to allow the model to reach steady-state conditions, if they exist. The lines of code inside the for loop implement the mathematical equations presented in the preceding section. Thus, for example, line 18 corresponds to Equation 9.6. In many cases, the gawk statements are quite long; they have therefore been split over two lines of code using the line continuation symbol (\), and the continuation line is indented by one tab stop. For instance, lines 20 and 21 represent a single gawk statement, implementing Equation 9.5.

In general, the translation from mathematical equations to gawk code is fairly straightforward and self-evident. There are, however, two particular points that are worth noting. First, in Equation 9.4 a number is raised to the power of 0.25; this is implemented in Program 9.1 using one of gawk's exponentiation symbols, ** (line 24). Second, the multiplication operator, which is implicit in the mathematical equations (the term $a_{soil}A_{soil}$ in Equation 9.6), must be made explicit in the gawk code (area_soil*albedo_soil; line 20).

The order in which the equations are implemented in Program 9.1 is different from the sequence in which they were introduced in the previous section. The reason for this is that the result of one equation typically forms the input to another. Thus, the average global albedo (albedo_global; lines 20 and 21) cannot be established until the area available for daisy growth (area_soil) has been determined (line 18). Similarly, the average global temperature (lines 23 and 24) can only be calculated once the average global albedo is known. The average global temperature is, in turn, required to calculate the local temperatures over the patches of black daisies and white daisies (lines 26 to 29).

A printf statement is used on lines 31 and 32 of the program to print out the average global temperature, and the fractional areas covered by the black daisies and the white daisies, at the current point in time. The growth rates of the black daisies and the white daisies are calculated on lines 34 and 35, respectively, based on the corresponding local temperature values. The rates of growth are used, in turn, to

```
 0  26.8742  0.0100  0.0100
 1  26.7692  0.0140  0.0168
 2  26.5566  0.0195  0.0280
 3  26.1617  0.0273  0.0462
 4  25.4810  0.0380  0.0749
 5  24.4002  0.0530  0.1182
 6  22.8705  0.0735  0.1782
 7  21.0655  0.1002  0.2507
 8  19.4890  0.1312  0.3211
 9  18.6609  0.1617  0.3720
10  18.5866  0.1874  0.3995
11  18.9154  0.2074  0.4115
12  19.3701  0.2227  0.4155
13  19.8215  0.2344  0.4160
14  20.2216  0.2432  0.4149
```

Figure 9.3: First 15 lines of the file daisy1.dat.

determine the change in the fractional area of the planet's surface covered by each species of daisy (lines 38 to 41), and hence to calculate revised values of the relevant variables (area_black and area_white) for the start of the next time step (the next iteration around the for loop). Finally, lines 42 and 43 of the code ensure that neither species of daisy becomes extinct. Thus, the fractional area covered by either species is reset to 0.01 (1% of the planet's surface) if its actual area falls below this value. These lines of code do not have direct counterparts in the mathematical model, but they provide a simple constraint, suggested by Watson and Lovelock (1983), on the dynamics of the Daisyworld system. They imply that there are always small refugia within which the daisy species can survive, even under the harshest conditions.

Evaluation

Program 9.1 can be used to examine the behavior of Daisyworld under conditions of constant solar luminosity, as follows:

```
gawk -f daisy1.awk -v luminosity=1.0 > daisy1.dat
```
1

In this example, the variable that records the solar luminosity is assigned a value of 1.0 on the command line and the output from the model is redirected to the file daisy1.dat in the working directory (Figure 9.3). The model can be run using other values of solar luminosity by changing the above command line accordingly.

Visualization

The data produced by daisy1.awk (Program 9.1) can be visualized by typing the following commands in gnuplot, assuming that the file daisy1.dat is located in the working directory:

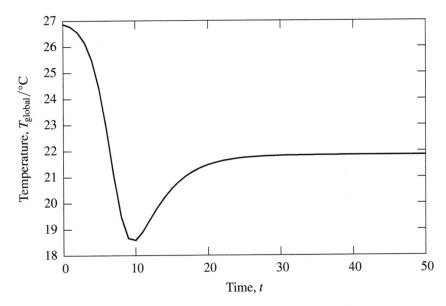

Figure 9.4: Average global temperature of Daisyworld as a function of time, assuming a constant solar luminosity ($L = 1.0$).

```
reset                                                      1
set style data lines                                       2
set ylabel "Temperature, T_global/deg C"                   3
set xlabel "Time, t"                                       4
unset key                                                  5
set xrange [0:50]                                          6
plot 'daisy1.dat' u 1:2 lw 2                               7
```

These commands, which should be familiar from the preceding chapters, produce a plot of the average global temperature of Daisyworld as a function of time, under conditions of constant solar luminosity ($L = 1.0$). The resulting plot (Figure 9.4) indicates that the temperature of the planet starts at a little less than 27 °C, declines to about 18.5 °C by $t = 10$, and ultimately reaches a steady state of just under 22 °C (i.e., very close to the optimum temperature for daisy growth) after $t = 30$.

To understand why the planet behaves in this way, it is instructive to examine how the fractional area covered by each type of daisy varies as a function of time. This can be explored in gnuplot by entering the following commands:

```
set ylabel "Fractional area, a"                            8
set format y "%4.2f"                                       9
set key bottom right box                                  10
plot 'daisy1.dat' u 1:3 t "Black daisies" lw 2, \         11
     'daisy1.dat' u 1:4 t "White daisies" lw 2            12
```

Note that line 9, above, specifies the format used for the tic-mark labels on the y-axis; that is, four digit floating-point numbers reported to two decimal places (e.g., 0.25).

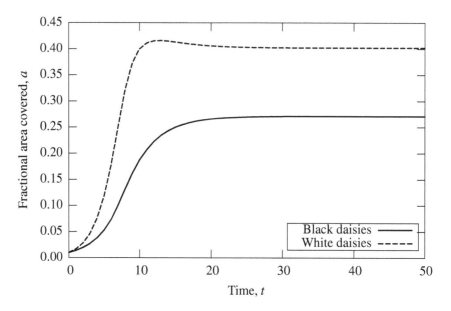

Figure 9.5: Fractional area of Daisyworld covered by black daisies and white daisies as a function of time, assuming a constant solar luminosity ($L = 1.0$).

The resulting plot indicates that both species of daisy increase in area very rapidly during the first 10 or so time steps (Figure 9.5). Initially, the white daisies grow faster because they reflect more of the incident solar radiation; consequently, the local temperature over these patches is cooler than over the rest of the planet, and hence it is closer to the optimum for daisy growth. As the fractional area covered by white daisies increases, the global average temperature of Daisyworld decreases (i.e., the white daisies exert a negative feedback effect on the temperature of the planet). As the planet becomes cooler, the growth rate of the black daisies increases and the fractional area covered by this species increases correspondingly. This has the effect of warming the planet because the black daisies absorb more of the incident solar radiation (i.e., the black daisies exert a positive feedback effect on the temperature of the planet). By about $t = 30$, the relative warming effect of the black daisies and the relative cooling effect of the white daisies counterbalance one another, so that the average global temperature and the fractional areas covered by the black daisies and white daisies stabilize with respect to time (i.e., the planet enters a steady state).

Program 9.1 can be run using different values of solar luminosity to demonstrate that steady-state behavior is the norm, rather than the exception, on Daisyworld (see Figure 9.6). Note that the average temperature of the planet in steady state decreases as the solar luminosity increases over the range of the values considered here.

Exercise 9.1: Show how the fractional areas of black daisies and white daisies vary as a function of time for $L = 0.8, 0.9, 1.0, 1.1$ and 1.2.

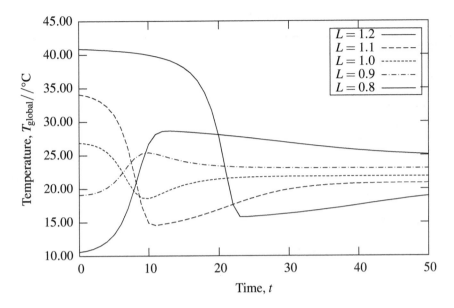

Figure 9.6: Global average temperature versus time for $L = 0.8, 0.9, 1.0, 1.1$ and 1.2.

9.4.2 Implementation 2: Increasing Solar Luminosity

The main focus of the paper by Watson and Lovelock (1983) is the way in which Daisyworld responds to a gradual increase in solar luminosity as a function of time. This behavior mimics that of main sequence stars, such as Earth's sun, which tend to increase in brightness with age (Saunders 1994), and it has important consequences for the climate of Daisyworld. Note that the planet is allowed to reach steady state each time, before the luminosity is increased again (Watson and Lovelock 1983).

Implementation

Program 9.2, `daisy2.awk`, is an implementation of the Daisyworld model in which the solar luminosity, L, is gradually increased in steps of 0.025 from 0.5 to 1.7. As in the previous implementation, the code is contained entirely within the BEGIN block (lines 1 to 46), because the intention is to generate data (i.e., to perform a simulation) using the model. For the sake of convenience, lines 3 to 13 initialize the values of the main variables rather than requiring them to be entered via the command line. An additional variable, `threshold`, is initialized on line 13, the purpose of which is outlined below. A for loop (lines 15 to 45) is used to run the model for several different values of solar luminosity. Inside this outer loop, a do-while loop (lines 18 to 42) is used to run the model through time until the system approaches steady state for the current value of solar luminosity. In this example, steady state is deemed to have been reached when the global average temperature of Daisyworld changes by less than a user-specified amount (line 42) between successive iterations of the do while loop; here, the threshold value is set to 0.01 °C (line 13). The absolute temperature

Program 9.2: daisy2.awk

```
BEGIN {                                                           1
  # Initialize main variables                                     2
  area_black   = 0.01;                                            3
  area_white   = 0.01;                                            4
  area_suit    = 1.0;                                             5
  solar_const  = 917;                                             6
  Stefan_Boltz = 5.67E-08;                                        7
  albedo_soil  = 0.5;                                             8
  albedo_black = 0.25;                                            9
  albedo_white = 0.75;                                            10
  death_rate   = 0.3;                                             11
  q_factor     = 20;                                              12
  threshold    = 0.01;                                            13
                                                                  14
  for(luminosity=0.5;luminosity<=1.7;luminosity+=0.025){          15
    # Run model until it reaches steady state for                 16
    # current solar luminosity                                    17
    do{                                                           18
      prev_temp_global=temp_global;                               19
      area_avail=area_suit-(area_black+area_white);               20
      albedo_global=(area_avail*albedo_soil)+ \                   21
        (area_black*albedo_black)+(area_white*albedo_white);      22
      temp_global=(((solar_const*luminosity* \                    23
        (1-albedo_global))/Stefan_Boltz)**0.25)-273;              24
      temp_black=((q_factor*(albedo_global-albedo_black))+ \      25
        temp_global);                                             26
      temp_white=((q_factor*(albedo_global-albedo_white))+ \      27
        temp_global);                                             28
      growth_black=(1-(0.003265*((22.5-temp_black)**2)));         29
      growth_white=(1-(0.003265*((22.5-temp_white)**2)));         30
      area_black+=area_black*((area_avail*growth_black)- \        31
        death_rate);                                              32
      area_white+=area_white*((area_avail*growth_white)- \        33
        death_rate);                                              34
      # Do not allow daisies to become extinct                    35
      if(area_black<0.01){area_black=0.01};                       36
      if(area_white<0.01){area_white=0.01};                       37
      # Check whether global average temperature for current      38
      # and previous model run differ by more than threshold      39
      temp_difference=temp_global-prev_temp_global                40
      if(temp_difference<=0){temp_difference*=-1}                 41
    } while(temp_difference>threshold)                            42
    printf("%5.3f %7.4f %6.4f %6.4f\n", \                         43
      luminosity, temp_global, area_black, area_white);           44
  }                                                               45
}                                                                 46
```

difference between successive iterations of the do while loop is calculated on lines 40 and 41; the temperature of the previous iteration is recorded on line 19. Further iterations are performed for the current solar luminosity value while the absolute temperature difference remains greater than the threshold value (line 42). Once the temperature difference between two successive iterations is less than or equal to the threshold value, the program exits the do while loop and prints out the current values of the solar luminosity, global average temperature and the fractional area of the planet surface covered by black daisies and by white daisies, respectively (lines 43 and 44). The program then returns to the for loop, incrementing the solar luminosity by 0.025, before entering the do while loop again.

Evaluation

Program 9.2 can be run from the command line as follows:

```
gawk -f daisy2.awk > daisy2.dat
```
2

Note that the output from the model is redirected to the file daisy2.dat in the working directory. It is also instructive to run the same model a second time without daisies, to examine the effect that the biota have on the evolution of Daisyworld's temperature as a function of increasing solar luminosity. This can be achieved most simply by commenting out (i.e., placing a # symbol at the start of) lines 3, 4, 36 and 37 in Program 9.2, which has the effect of setting the area covered by each species of daisy to zero. The revised program can then be saved in a new file, nolife.awk, and run as follows:

```
gawk -f nolife.awk > nolife.dat
```
3

Visualization

The two data sets produced above, daisy2.dat and nolife.dat, can be visualized together by typing the following commands in **gnuplot**:

```
set format y "%g"
set xrange [0.5:1.7]
set xzeroaxis
set key top left Left box
set ylabel "Temperature, T_global/deg C"
set xlabel "Solar luminosity, L"
plot 'nolife.dat' u 1:2 t "No biota" lw 2, \
     'daisy2.dat' u 1:2 t "With biota" lw 2
```
13
14
15
16
17
18
19
20

These commands assume that both data files are located in the working directory; otherwise, their full or relative pathnames must be provided. Note that line 13 sets the format of the labels placed beside the tic-marks on the *y*-axis to standard scientific notation, line 14 restricts the values displayed on the *x*-axis (i.e., solar luminosity) to the range 0.5 to 1.7, and line 15 draws a line at $y = 0$ marking the *x*-axis. This set of commands produces the plot shown in Figure 9.7.

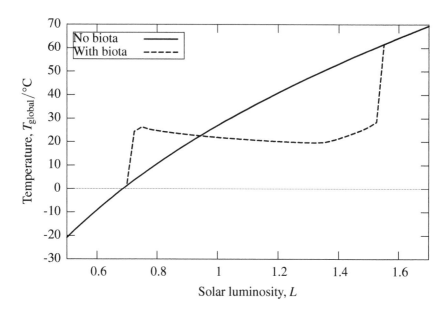

Figure 9.7: Global average temperature of Daisyworld, with and without biota, as a function of solar luminosity.

Figure 9.7 demonstrates the behavior of the Daisyworld model with and without biota (daisies). In the absence of biota (solid line), the temperature of the planet rises monotonically with increasing solar luminosity. With the biota present in the model, the planet exhibits a different response (dashed line). The daisies begin to exert an influence on the climate of Daisyworld when the solar luminosity rises above 0.7. At this point, the daisies cause the planet to warm much more rapidly than is the case when the biota are absent, and the global average temperature rises to a peak of just over 26 °C at $L = 0.75$. Subsequent increases in solar luminosity up to $L = 1.325$ are accompanied by a gradual reduction in the global average temperature; that is, the presence of daisies causes the planet to cool slightly. Thus, the effect of the daisies is to stabilize the temperature of the planet over a wide range of solar luminosity values. When the solar luminosity exceeds 1.325, however, the temperature of Daisyworld starts to increase gradually once more until $L = 1.55$ when it rises very suddenly, eventually merging with the temperature curve for the planet without life.

To understand why Daisyworld behaves in this way, it is instructive to examine how the fractional area of the planet's surface covered by each type of daisy varies as a function of solar luminosity. This can be achieved by entering the following commands in gnuplot, assuming that daisy2.dat is located in the working directory:

```
unset key                                                21
set ylabel "Fractional area covered, a"                  22
plot 'daisy2.dat' u 1:3 lw 2, \                          23
      'daisy2.dat' u 1:4 lw 2                             24
```

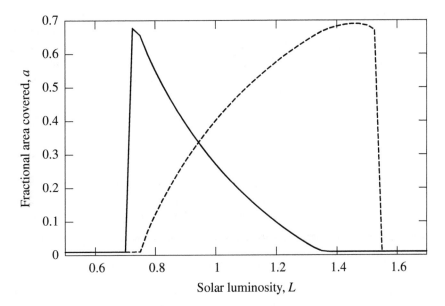

Figure 9.8: Solar luminosity versus the fractional area of Daisyworld covered by black (solid line) and white (dashed line) daisies.

The resulting plot (Figure 9.8) indicates that the black daisies grow very rapidly at first, covering approximately two-thirds of the planet's surface by $L = 0.725$; they then gradually die back, declining almost to zero by $L = 1.375$. The white daisies, by contrast, grow more slowly at first, eventually reaching a peak at $L = 1.45$, before dying back very rapidly by $L = 1.55$.

The results obtained using this implementation of the Daisyworld model can be interpreted as follows. Initially, when Daisyworld's sun is very weak (i.e., the solar luminosity is low; $L \leq 0.7$), the planet is cold and neither species of daisy is able to grow. Gradually, though, as the solar luminosity increases, the temperature of the planet rises to the point where daisy growth commences. At this stage, Daisyworld's climate favors the growth of the black daisies. This is because they absorb more of the incident solar radiation and, as a result, the local temperature over the patches of black daisies is somewhat higher than the global average. The temperature of the planet as a whole, however, remains too low to permit the growth of white daisies and, in the absence of inter-specific competition, the black daisies expand to cover the majority of the planet's surface. In doing so, they raise the temperature of the planet to the point at which the white daisies start to grow and, beyond this, the black daisies begin to die back. The white daisies grow slowly at first, partly because they have to compete with the black daisies for the limited amount of fertile land that remains unoccupied. As the area covered by white daisies increases so does the planetary albedo. This has the effect of cooling the planet slightly because a greater fraction of the incident solar radiation is reflected back out to space. For a while the two species of daisy co-exist. Moreover, the delicate balance between the relative warming effect of the black daisies and the corresponding cooling effect of the white ones causes the temperature of the planet to stabilize, despite the continuing increases in solar luminosity. Watson and Lovelock (1983) describe this situation as one of biological homeostasis. Eventually, though, the increasing solar luminosity causes the temperature of Daisyworld to rise above the maximum that the daisies can tolerate, and the populations of both species crash.

> **Exercise 9.2**: Modify Program 9.2, `daisy2.awk`, so that there is only one species of daisy present (black or white). Save the file as `daisy2x.awk`. Run the revised program from the command line. Create plots of the global average temperature of Daisyworld and the fractional area covered by daisies as a function of solar luminosity. Compare the results with those of the two-species model. Repeat the exercise for the other species.

9.4.3   Implementation 3: Exploring the Impact of Biodiversity

In the third implementation of the Daisyworld model examined here, an additional species of daisy is introduced. This minor modification forms the basis of a simple experiment that is designed to evaluate the impact of biodiversity on the behavior of the model system. Specifically, the experiment tests the extent to which the system becomes more or less stable when a third species of daisy is added.

Implementation

Adding an extra species of daisy to the computational model is a relatively straightforward task; it involves a small number of modifications to the previous program, `daisy2.awk`. These modifications are presented in Program 9.3, `daisy3.awk`.

One change is the inclusion of two further variables, `area_gray` and `albedo_gray`, which are used to store the fractional area of ground covered by gray daisies and their albedo, respectively (lines 4 and 11). Note that the albedo of the gray daisies is given as 0.5 in this case, which is the same as that of the soil substrate. The equations used to calculate the area of land available for further daisy growth (Equation 9.6; lines 22 and 23) and to determine the global albedo (Equation 9.5; lines 24 to 27) have been amended to take into account the gray daisies. Further lines of code have also been added to calculate the local temperature of the gray daisies (lines 32 and 33), their rate of growth (line 37), the resulting change in the fractional area of the planet's surface that they cover (lines 41 and 42), and to prevent the gray daisies from becoming extinct (line 47). Finally, the `print` statement (lines 54 to 56) has been amended so that it also outputs the fractional area covered by the gray daisies.

Evaluation

Program 9.3 can be run from the command line as follows:

```
gawk -f daisy3.awk > daisy3.dat
```
4

Program 9.3: daisy3.awk (lines 1 to 44)

```
BEGIN {                                                              1
  # Initialize main variables                                       2
  area_black   = 0.01;                                              3
  area_gray    = 0.01;                                              4
  area_white   = 0.01;                                              5
  area_suit    = 1.0;                                               6
  solar_const  = 917;                                               7
  Stefan_Boltz = 5.67E-08;                                          8
  albedo_soil  = 0.5;                                               9
  albedo_black = 0.25;                                              10
  albedo_gray  = 0.5;                                               11
  albedo_white = 0.75;                                              12
  death_rate   = 0.3;                                               13
  q_factor     = 20;                                                14
  threshold    = 0.01;                                              15
                                                                    16
  for(luminosity=0.5;luminosity<=1.7;luminosity+=0.025){            17
    # Run model until it reaches steady state for                   18
    # current solar luminosity                                      19
    do{                                                             20
      prev_temp_global=temp_global;                                 21
      area_avail=area_suit- \                                       22
        (area_black+area_white+area_gray);                          23
      albedo_global= (area_avail*albedo_soil)+ \                    24
        (area_black*albedo_black)+ \                                25
        (area_white*albedo_white) + \                               26
        (area_gray*albedo_gray);                                    27
      temp_global=(((solar_const*luminosity* \                      28
        (1-albedo_global))/Stefan_Boltz)**0.25)-273;                29
      temp_black=((q_factor*(albedo_global-albedo_black))+ \        30
        temp_global);                                               31
      temp_gray=((q_factor*(albedo_global-albedo_gray))+ \          32
        temp_global);                                               33
      temp_white=((q_factor*(albedo_global-albedo_white))+ \        34
        temp_global);                                               35
      growth_black=(1-(0.003265*((22.5-temp_black)**2)));           36
      growth_gray=(1-(0.003265*((22.5-temp_gray)**2)));             37
      growth_white=(1-(0.003265*((22.5-temp_white)**2)));           38
      area_black+=area_black*((area_avail*growth_black)- \          39
        death_rate);                                                40
      area_gray+=area_gray*((area_avail*growth_gray)- \             41
        death_rate);                                                42
      area_white+=area_white*((area_avail*growth_white)- \          43
        death_rate);                                                44
```

Program 9.3 (continued): daisy3.awk (lines 45 to 58)

```
      # Do not allow daisies to become extinct              45
      if(area_black<0.01){area_black=0.01};                 46
      if(area_gray<0.01){area_gray=0.01};                   47
      if(area_white<0.01){area_white=0.01};                 48
      # Check whether global average temperature for current 49
      # and previous model run differ by more than threshold 50
      temp_difference=temp_global-prev_temp_global          51
      if(temp_difference<0){temp_difference*=-1}            52
    } while(temp_difference>threshold)                       53
    printf("%5.3f %7.4f %6.4f %6.4f %6.4f\n", \            54
      luminosity, temp_global, area_black, \                55
      area_gray, area_white);                               56
  }                                                          57
}                                                            58
```

Visualization

The output produced by the three-species version of the Daisyworld model (Program 9.3; daisy3.awk) can be visualized in gnuplot, where it can also be compared to the results obtained using the two-species model (Program 9.2; daisy2.awk) and the no biota model (nolife.awk), by typing the following commands (Figure 9.9):

```
set key top left box                                        25
set ylabel "Temperature, T_global/deg C"                    26
plot 'nolife.dat' u 1:2 t "No biota" lt 0 lw 2, \          27
     'daisy2.dat' u 1:2 t "2 species" lt 1 lw 2, \         28
     'daisy3.dat' u 1:2 t "3 species" lt 2 lw 2            29
```

These commands assume that all three of the data files, nolife.dat, daisy2.dat and daisy3.dat, are located in the working directory. Similarly, the fractional area of the planet's surface that is covered by each of the three daisy species can be visualized as follows:

```
unset key                                                   30
set ylabel "Fractional area, a"                             31
plot 'daisy3.dat' u 1:3 t "Black daisies" lw 2, \          32
     'daisy3.dat' u 1:4 t "Gray daisies" lw 2, \           33
     'daisy3.dat' u 1:5 t "White daisies" lw 2             34
```

These commands produce the plot shown in Figure 9.10.

Figure 9.9 indicates that the relationship between global average temperature and solar luminosity predicted by the three-species model is similar to that predicted by the two-species model. The temperature of Daisyworld is arguably very slightly more stable for solar luminosity values between 0.725 and 1.5 in the three-species model, with the exception of a small perturbation around $L = 1.0$, but the difference is small and the evidence is, at best, inconclusive. The impact of an increase in biodiversity

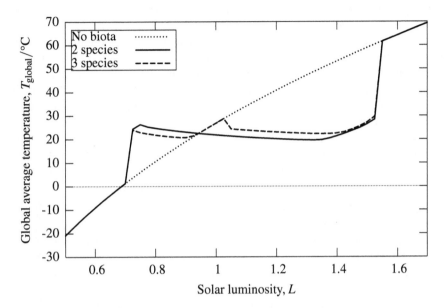

Figure 9.9: Variation in the global average temperature of Daisyworld as a function of solar luminosity. Comparison of results obtained using the two-species and three-species versions of the model.

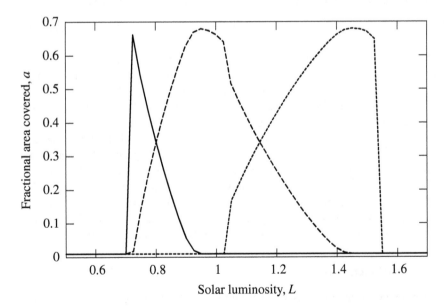

Figure 9.10: Fractional area of the planet surface covered by black (solid line), gray (long-dash line) and white (short-dash line) daisies as a function of relative solar luminosity.

on the global average temperature of Daisyworld (i.e., from two to three species of daisy) therefore appears to be minimal. However, close inspection of Figure 9.10 reveals that there are only ever two species of daisy that cover more than the specified minimum fraction (0.01) of the planet's surface for any given value of solar luminosity (i.e., the white daisies and gray daisies when $0.725 \leq L \leq 0.925$, and the gray daisies and black daisies when $1.025 \leq L \leq 1.525$). It is possible that this is an artifact of the specific value used for the albedo of the gray daisies in this simulation, which is the same as that of the soil substrate; this is left as an exercise for the reader to investigate (see Exercise 9.3, below). Alternatively, it is possible that the full impact of biodiversity becomes apparent only when a larger number of daisy species is modeled. This possibility is examined in the following section.

Exercise 9.3: Modify Program 9.3, `daisy3.awk`, so that the albedo of the gray daisies is specified on the command line. Save the file as `daisy3x.awk`. Run the revised program from the command line using a different value for the albedo of the gray daisies ($0.3 \leq A_g \leq 0.7$) each time. Redirect the output to separate data files. Create plots of the global average temperature of Daisyworld, and of the fractional areas covered by each species of daisy, as a function of solar luminosity. Compare the results with those from `daisy3.awk`.

9.4.4 Implementation 4: Modularizing the Code with User-Defined Functions

Modifying Program 9.2 (`daisy2.awk`) so that it incorporates an extra species of daisy is a relatively straightforward task. It requires only a few minor changes to the code, including additional lines to calculate the local temperature and the growth rate of the gray daisies, and the fractional area of Daisyworld that they cover (Program 9.3; `daisy3.awk`). This process is readily achieved using a text editor, by copying and pasting the equivalent lines of code for the black daisies and then editing these lines so that occurrences of the word black are replaced by the word gray. This approach is reasonably acceptable for small programs, but it is problematic for longer and more complex ones. Imagine, for example, having to modify Program 9.3 so that it simulates 10 or more species of daisy. There is considerable potential for making mistakes in these circumstances and "bugs" (coding errors) inevitably creep in. User-defined functions represent a much more elegant solution to this problem. They also make the code modular; that is to say, they help to divide the code into a number of smaller modules, which are more manageable individually and can be re-used.

User-Defined Functions

A user-defined function consists of a number of **gawk** statements, which perform a specific task, encapsulated in a named block of code. User-defined functions offer several important benefits: they help to modularize a program by breaking up the code into a collection of discrete building blocks; they can reduce the length of a program and improve its readability because they can be used many times in the same program without having to repeat the corresponding lines of code; and they foster the re-use of common functions in different programs, either by cutting and pasting the function definitions between program files or, more effectively, through the use of a function "library" (a separate file containing a number of frequently employed user-defined functions).

A user-defined function typically receives a number of inputs, which are known as the "arguments" to the function, and it manipulates these to generate an output. The general syntax of a function definition, which commences with the keyword function, is as follows:

```
function name_of_function (list_of_parameters)
{
    body_of_function
}
```

Function definitions can appear before, after or between the BEGIN, main and END pattern-action blocks, but may not be placed inside them. The name of a user-defined function can comprise any sequence of letters, digits and underscores, except that it may not start with a digit (Robbins 2001). The convention used here is to begin function names with the prefix fn_ to distinguish them from variables and arrays; for instance, a function to calculate the growth rate of daisies might be called fn_growth. Note that, in general, the same name cannot be used for a variable and a function, a function and an array, or an array and a variable in a single gawk program. Moreover, there must not be any blank spaces between the name of the function and the left parenthesis of its parameter list. The parameter list specifies the items of data on which the function operates; there may be zero or more items in this list. Where there is more than one item in the list, the individual items are separated by commas. The number of items in the list and the order in which they are listed are significant, as is demonstrated later in this section. The body of the function contains one or more gawk statements, which specify the actions to be performed; these define what the function does.

A user-defined function is said to be "called" when it is used within the BEGIN, main or END pattern-action blocks of a program. Note that it is not necessary to define a function prior to the point at which it is called in a program, because gawk parses all of the code before executing it (Robbins 2001). When a function is called, the values given in the parameter list are passed to, and stored in, the corresponding variables used in the body of the function; this is sometimes referred to as "calling by value". The actions specified in the body of the function are then performed on these values. In many instances, the result of a user-defined function is returned to the main part of the gawk program so that, for example, it can be employed in subsequent calculations. On other occasions, a user-defined function may simply perform an action, such as printing out a number, rather than returning a value; this is the equivalent of a void function in the C programming language or a procedure in Pascal (Robbins 2001).

Figure 9.11 provides a schematic representation of how a user-defined function is specified and employed in a gawk program. Note that in this example the function, fn_name, is defined after the END block, and that the parameter list consists of two

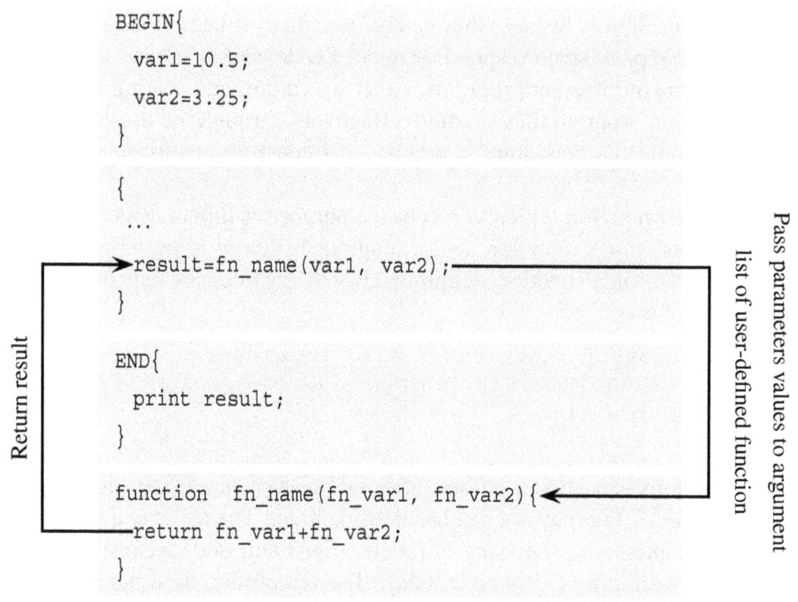

```
BEGIN{
  var1=10.5;
  var2=3.25;
}

{
  ...
  result=fn_name(var1, var2);
}

END{
  print result;
}

function  fn_name(fn_var1, fn_var2){
  return fn_var1+fn_var2;
}
```

Return result

Pass parameters values to argument list of user-defined function

Figure 9.11: Schematic representation of a user-defined function in gawk. The function fn_name is defined at the end of the program, and is "called" from the main block.

arguments, fn_var1 and fn_var2, separated by a comma. The function is called in the main pattern-action block of the program, where it is passed a copy of the values held by the variables var1 and var2; these variables are initialized in the BEGIN block (var1=10.5 and var2=3.25). The values passed to the function are stored in the variables fn_var1 and fn_var2, respectively; that is, fn_var1 receives the value of var1 (fn_var1=10.5) and fn_var2 receives the value of var2 (fn_var2=3.25). Note that fn_var1 and fn_var2 are referred to as local variables, since their values are only known within the user-defined function and they cease to exist once the function is exited, in contrast to global variables, such as var1 and var2, whose values are known globally and persist throughout the program (Robbins 2001). The values of the variables fn_var1 and fn_var2 are manipulated by the body of the function, which sums their values in this example. The result is returned to the main pattern-action block of the program where it is stored in the variable result (result=13.75). The value of result is printed out in the END block.

Re-Implementation of the Daisyworld Model with User-Defined Functions

Program 9.4, daisy4.awk, re-implements the three-species Daisyworld model with a number of user-defined functions, fn_temp, fn_growth and fn_area; this program is otherwise identical to Program 9.3, daisy3.awk. Each function is called three times, once for each species of daisy (white, gray and black). The first function, fn_temp, which is defined on lines 63 to 67, calculates the local temperatures over each of the

Program 9.4: daisy4.awk (lines 1 to 43)

```
BEGIN {                                                          1
                                                                 2
  # Initialize main variables                                   3
  area_black   = 0.01;                                          4
  area_gray    = 0.01;                                          5
  area_white   = 0.01;                                          6
  area_suit    = 1.0;                                           7
  solar_const  = 917;                                           8
  Stefan       = 5.67E-08;                                      9
  albedo_soil  = 0.5;                                          10
  albedo_black = 0.25;                                         11
  albedo_gray  = 0.5;                                          12
  albedo_white = 0.75;                                         13
  death_rate   = 0.3;                                          14
  q_factor     = 20;                                           15
  threshold    = 0.01;                                         16
                                                                17
  for(luminosity=0.5;luminosity<=1.7;luminosity+=0.025){       18
    # Run model until it reaches steady state for               19
    # current solar luminosity                                 20
    do{                                                        21
      prev_temp_global=temp_global;                            22
                                                                23
      area_avail=area_suit- \                                  24
        (area_black+area_white+area_gray);                     25
      albedo_global=(area_avail*albedo_soil)+ \                26
        (area_black*albedo_black)+ \                           27
        (area_white*albedo_white)  + \                         28
        (area_gray*albedo_gray);                               29
      temp_global=(((solar_const*luminosity* \                 30
        (1-albedo_global))/Stefan)**0.25)-273;                 31
                                                                32
      temp_black=fn_temp(albedo_black);                        33
      temp_gray=fn_temp(albedo_gray);                          34
      temp_white=fn_temp(albedo_white);                        35
                                                                36
      growth_black=fn_growth(temp_black);                      37
      growth_gray=fn_growth(temp_gray);                        38
      growth_white=fn_growth(temp_white);                      39
                                                                40
      area_black+=fn_area(area_black,growth_black);            41
      area_gray+=fn_area(area_gray,growth_gray);               42
      area_white+=fn_area(area_white,growth_white);            43
```

Program 9.4 (continued): daisy4.awk (lines 44 to 80)

```
                                                                          44
    # Do not allow daisies to become extinct                             45
    if(area_black<0.01){area_black=0.01};                                46
    if(area_gray<0.01){area_gray=0.01};                                  47
    if(area_white<0.01){area_white=0.01};                                48
                                                                          49
    # Check whether global average temperature for current               50
    # and previous model run differ by more than threshold               51
    temp_difference=temp_global-prev_temp_global                         52
    if(temp_difference<0){temp_difference*=-1}                           53
  } while(temp_difference>threshold)                                      54
                                                                          55
  printf("%5.3f %7.4f %6.4f %6.4f %6.4f\n", \                           56
    luminosity, temp_global, area_black, \                              57
    area_gray, area_white);                                             58
  }                                                                       59
}                                                                         60
                                                                          61
# Function to calculate local temperatures                               62
function fn_temp(albedo_daisy)                                            63
{                                                                         64
  return ((q_factor*(albedo_global-albedo_daisy))+ \                    65
    temp_global);                                                        66
}                                                                         67
                                                                          68
# Function to calculate daisy growth rates                               69
function fn_growth(temp_daisy)                                            70
{                                                                         71
  return (1-(0.003265*((22.5-temp_daisy)**2)));                         72
}                                                                         73
                                                                          74
# Function to calculate change in daisy area                             75
function fn_area(area_daisy,growth_daisy)                                76
{                                                                         77
  return (area_daisy*((area_avail*growth_daisy)- \                      78
    death_rate));                                                        79
}                                                                         80
```

three species of daisy; it accepts a single argument, the albedo of the selected species, albedo_daisy, and returns the local temperature over that species; it also employs three global variables, q_factor, albedo_global and temp_global, which are either initialized or calculated in the BEGIN block. The function is called on lines 33 to 35, in the BEGIN block. On each occasion, the function is passed the albedo value of a particular species of daisy and the result is stored in the corresponding variable for the local temperature of that species. The second function, fn_growth, which is defined on lines 70 to 73, operates in a similar manner; it accepts a single argument, the temperature of the selected species of daisy, temp_daisy, and returns the rate of growth for that species. This function is called on lines 37 to 39, in the BEGIN block. The third function, fn_area, which is defined on lines 76 to 80, accepts two arguments, the fractional area of Daisyworld covered by the chosen species of daisy, area_daisy, and the growth rate of that species, growth_daisy; it also uses two global variables, area_avail and death_rate, which are either initialized or calculated elsewhere in the program. The values of these variables are used to calculate the amount by which the area covered by that species of daisy changes during the current time step. This function is called on lines 41 to 43, in the BEGIN block.

Evaluation

Program 9.4 can be run from the command line as follows:

```
gawk -f daisy4.awk > daisy4.dat
```
6

Note that the output from the model is redirected to a file, daisy4.dat, in the working directory. The contents of this file should be identical to those of daisy3.dat because the programs used to generate them are functionally equivalent.

Exercise 9.4: Verify that Program 9.4, daisy4.awk, produces the same results as Program 9.3, daisy3.awk, by plotting the output from both in gnuplot. Modify daisy4.awk so that it simulates conditions on Daisyworld in the presence of five species of daisy, each with a different albedo in the range $0.25 \leq A \leq 0.75$. Save the revised program as daisy4x.awk and run this from the command line, redirecting the output to a data file, daisy4x.dat. Create a plot of the global average temperature of Daisyworld versus solar luminosity. Compare the results with the three-species, two-species and no biota models.

Revised Implementation using Separate Program Files

Earlier, it was noted that user-defined functions can be placed in a file separate from the program that calls them. This approach allows commonly used functions to be employed in many different programs without having to incorporate the corresponding code explicitly in each one. This feature helps to keep the main program files short and manageable; it also encourages the re-use of tried and tested code. While a

Program 9.5: daisyvar.awk

```
BEGIN {                                                    1
   # Initialize main variables                             2
   area_black    = 0.01;                                   3
   area_gray     = 0.01;                                   4
   area_white    = 0.01;                                   5
   area_suit     = 1.0;                                    6
   solar_const   = 917;                                    7
   Stefan        = 5.67E-08;                               8
   albedo_soil   = 0.5;                                    9
   albedo_black  = 0.25;                                  10
   albedo_gray   = 0.5;                                   11
   albedo_white  = 0.75;                                  12
   death_rate    = 0.3;                                   13
   q_factor      = 20;                                    14
   threshold     = 0.01;                                  15
}                                                         16
```

separate function library is not really necessary in the case of the Daisyworld model, which is a relatively short program and where the user-defined functions are highly specific to this particular piece of code, the approach is nevertheless adopted here to illustrate how it is implemented.

Thus, Program 9.4, daisy4.awk, can be split into three separate files, daisyvar.awk (Program 9.5), daisy5.awk (Program 9.6) and daisyfns.awk (Program 9.7). The first of these contains the code initializing the main variables used in the model, the second contains the bulk of the code, including the calls to the user-defined functions, and the third contains the definitions of the user-defined functions. Note that two of these files consist of BEGIN blocks, namely daisyvar.awk (Program 9.5) and daisy5.awk (Program 9.6). gawk will happily parse two or more BEGIN blocks (and, indeed, two or more main and END blocks), although the order in which they are listed on the command line is significant, as is shown below.

Re-Evaluation

Program 9.6 and Program 9.7 can be run together via the command line as follows:

```
gawk -f daisyvar.awk -f daisy5.awk -f daisyfns.awk > ↺      7
     ↺daisy5.dat
```

This should produce the same result as Program 9.4, daisy4.awk. Note that the name of each of the three program files is preceded by a separate command-line switch, -f, which indicates that the corresponding file contains code to be interpreted, rather than data to be processed. As noted above, the order in which the program files are given on the command line is significant: the file containing the variable initialization statements (daisyvar.awk) must, in this instance, be placed before the code in which the variables are first used (daisy5.awk); otherwise, the variables will be uninitialized

Program 9.6: daisy5.awk

```
BEGIN {                                                                        1
  for(luminosity=0.5;luminosity<=1.7;luminosity+=0.025){                       2
    # Run model until it reaches steady state for                              3
    # current solar luminosity                                                 4
    do{                                                                        5
      prev_temp_global=temp_global;                                            6
                                                                               7
      area_avail=area_suit- \                                                  8
        (area_black+area_white+area_gray);                                     9
      albedo_global=(area_avail*albedo_soil)+ \                               10
        (area_black*albedo_black)+ \                                          11
        (area_white*albedo_white) + \                                         12
        (area_gray*albedo_gray);                                              13
      temp_global=(((solar_const*luminosity* \                               14
        (1-albedo_global))/Stefan)**0.25)-273;                               15
                                                                              16
      temp_black=fn_temp(albedo_black);                                       17
      temp_gray=fn_temp(albedo_gray);                                         18
      temp_white=fn_temp(albedo_white);                                       19
                                                                              20
      growth_black=fn_growth(temp_black);                                     21
      growth_gray=fn_growth(temp_gray);                                       22
      growth_white=fn_growth(temp_white);                                     23
                                                                              24
      area_black+=fn_area(area_black,growth_black);                           25
      area_gray+=fn_area(area_gray,growth_gray);                              26
      area_white+=fn_area(area_white,growth_white);                          27
                                                                              28
      # Do not allow daisies to become extinct                               29
      if(area_black<0.01){area_black=0.01};                                  30
      if(area_gray<0.01){area_gray=0.01};                                    31
      if(area_white<0.01){area_white=0.01};                                  32
                                                                              33
      # Check whether global average temperature for current                 34
      # and previous model run differ by more than threshold                 35
      temp_difference=temp_global-prev_temp_global                           36
      if(temp_difference<0){temp_difference*=-1}                             37
    } while(temp_difference>threshold)                                       38
                                                                              39
    printf("%5.3f %7.4f %6.4f %6.4f %6.4f\n", \                             40
      luminosity, temp_global, area_black, \                                 41
      area_gray, area_white);                                                42
                                                                              43
  }                                                                          44
}
```

Program 9.7: daisyfns.awk

```
# Function to calculate local temperatures              1
function fn_temp(albedo_daisy)                          2
{                                                       3
  return ((q_factor*(albedo_global-albedo_daisy))+ \    4
    temp_global);                                        5
}                                                       6
                                                        7
# Function to calculate daisy growth rates              8
function fn_growth(temp_daisy)                           9
{                                                      10
  return (1-(0.003265*((22.5-temp_daisy)**2)));        11
}                                                      12
                                                       13
# Function to calculate change in daisy area           14
function fn_area(area_daisy,growth_daisy)              15
{                                                      16
  return (area_daisy*((area_avail*growth_daisy)- \     17
    death_rate));                                      18
}                                                      19
```

(i.e., they will be assumed to have the value zero or contain the null string) when the code is first run, which is not what is intended here.

Exercise 9.5: Which of the following command lines produces an error, and why?

```
gawk -f daisy5.awk -f daisyfns.awk -f daisyvar.awk
gawk -f daisyvar.awk -f daisyfns.awk -f daisy5.awk
gawk -f daisyfns.awk -f daisyvar.awk -f daisy5.awk
```

What is wrong with the following command line?

```
gawk -f daisyvar.awk daisyfns.awk daisy5.awk
```

9.5 SENSITIVITY ANALYSIS AND UNCERTAINTY ANALYSIS

It is important to investigate the extent to which the results generated by a model are sensitive to the parameter values with which it is initialized. If a small change in the value of an input parameter produces a large change in the values output from the model, the model is said to be sensitive to the parameter and the parameter is said to have a high influence on the model (Ford 1999). Conversely, if a large change in the value of an input parameter produces a small change in the values output from the model, the model is said to be insensitive to the parameter and the parameter is said to have a low influence on the model. The process of establishing the sensitivity of a model to its input parameters is known as sensitivity analysis (Saltelli *et al.* 2000, Saltelli *et al.* 2004). Knowledge gained by performing a sensitivity analysis of a model can help to elucidate the way in which the modeled system functions and to identify those parameters of the model whose values need to be specified most accurately. Ultimately, the aim of sensitivity analysis is to focus attention on the critical parts of the model and the environmental system that it purports to represent.

In practice, sensitivity analysis is usually performed by perturbing the values of the model parameters in known amounts and measuring the effects that these variations have on the model outputs. The simplest method is to vary the value of one parameter at a time, while the values of the other parameters are held constant. This approach, known as one-at-a-time (OAT) or univariate sensitivity analysis, is used to quantify the influence that each parameter exerts on the model and is the method adopted here. It has a significant limitation, however, because it cannot describe the sensitivity of the model to simultaneous changes in two or more parameters, each of which may have a limited influence when considered in isolation, but which combine to produce major changes in the model output. This problem requires the use of more sophisticated multivariate sensitivity analysis techniques, such as Monte Carlo simulation with simple random or Latin Hypercube sampling, which are beyond the scope of this book (Saltelli *et al.* 2000, Saltelli *et al.* 2004).

Figure 9.12 and Figure 9.13 show the results of two separate OAT sensitivity analyses performed on the three-species implementation of the Daisyworld model, `daisy4.awk`. In the first of these, the model is run a number of times, each time using a different value for the albedo of the black daisies ($0.1 \leq A_{black} \leq 0.45$), while the values of the other parameters are held constant (Figure 9.12); in the second, it is the albedo of the white daisies that is varied ($0.55 \leq A_{white} \leq 0.90$), while the other parameters are fixed (Figure 9.13). Figure 9.12 suggests that the value used to represent the albedo of the black daisies has a small but significant impact on the form of the relationship between solar luminosity and the globally averaged temperature of Daisyworld when the solar luminosity is between 0.7 and 0.9; in general terms, the planet warms up faster as the albedo of the black daisies decreases. This parameter has little impact on the temperature of Daisyworld, however, when the solar luminosity exceeds 0.9. By contrast, the albedo of the white daisies has a significant influence on the temperature of Daisyworld when the solar luminosity exceeds 1.0, but very little effect below this value.

The sensitivity analyses outlined above can also be used to explore the relationship between the albedo of the daisy species and the fractional area of the planet surface that they occupy. Thus, Figure 9.14 and Figure 9.15 show that the albedo of the black daisies has a significant influence on the fractional areas covered by both the black daisies and the white daisies at different solar luminosities.

Uncertainty analysis is the corollary of sensitivity analysis. It examines the propagation of uncertainty in the values of the model parameters (and sometimes in the structure and formulation of the model) through to uncertainty in the model outputs. For instance, the exact value of a parameter may not be known. Instead, it may be

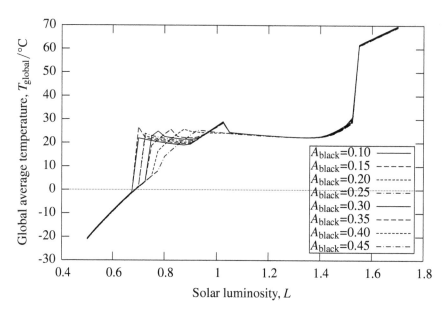

Figure 9.12: Sensitivity analysis of the three-species Daisyworld model, `daisy4.awk`, showing variation in the globally averaged temperature of the planet as a function of the albedo of the black daisies ($0.10 \leq A_{black} \leq 0.45$).

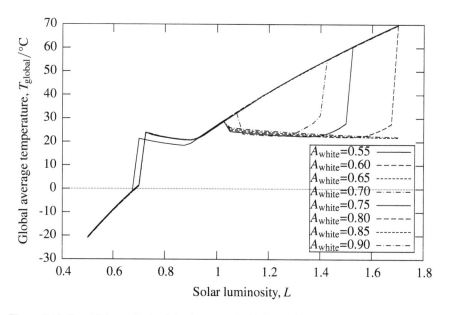

Figure 9.13: Sensitivity analysis of the three-species Daisyworld model, `daisy4.awk`, showing variation in the globally averaged temperature of the planet as a function of the albedo of the white daisies ($0.55 \leq A_{white} \leq 0.90$).

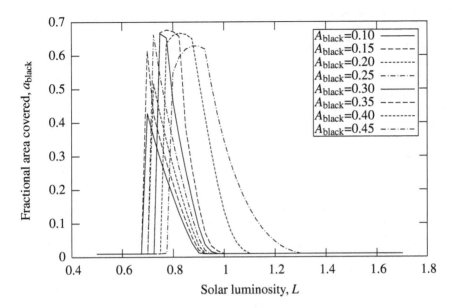

Figure 9.14: Sensitivity analysis of the three-species Daisyworld model, `daisy4.awk`, showing variation in the fractional area of the planet covered by black daisies as a function of the albedo of the black daisies ($0.10 \leq A_{\text{black}} \leq 0.45$).

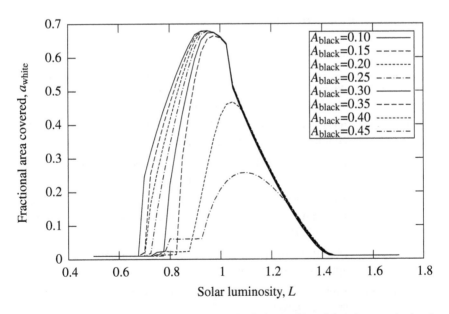

Figure 9.15: Sensitivity analysis of the three-species Daisyworld model, `daisy4.awk`, showing variation in the fractional area of the planet covered by white daisies as a function of the albedo of the black daisies ($0.10 \leq A_{\text{black}} \leq 0.45$).

thought to lie somewhere between fixed upper and lower limits, or perhaps to vary about a mean value according to a given standard deviation. The input value for this parameter is therefore constrained, but not prescribed. The uncertainty in the value that should be used for this parameter results in variation (uncertainty) in the model outputs, which can be quantified. In crude terms, therefore, the modeler wishes to know whether a 10% difference in the value of the parameter causes a 1%, 10% or 100% change in the model outputs. This type of uncertainty analysis is not performed here, but is left as an exercise for the interested reader.

9.6 SUMMARY

This chapter examines a model of biospheric feedback between the biota (different species of daisy) on an imaginary planet, known as Daisyworld, and their abiotic environment (Watson and Lovelock 1983). The model combines many of the topics covered in the preceding chapters, including constrained population growth and inter-specific competition (Chapter 8), and the interaction of solar radiation with various surface materials (Chapter 5 through Chapter 7). Several implementations of the Daisyworld model are presented, including one in which the solar luminosity is held constant and others in which it gradually increases over time. The sensitivity of the model to the number of daisy species (biodiversity) and to variations in the input parameter values is also explored. The computational model is also modularized with user-defined functions.

SUPPORTING RESOURCES

The following resources are provided in the `chapter9` sub-directory of the CD-ROM:

`daisy1.awk`	**gawk** implementation of the two-species Daisyworld model for conditions of constant solar luminosity.
`daisy2.awk`	**gawk** implementation of the two-species Daisyworld model for conditions of increasing solar luminosity.
`daisy3.awk`	**gawk** implementation of the three-species Daisyworld model for conditions of increasing solar luminosity.
`daisy4.awk`	**gawk** implementation of the two-species Daisyworld model for conditions of increasing solar luminosity, employing user-defined functions.
`daisyvar.awk`	Modular **gawk** program to initialize the variables for the three-species Daisyworld model.
`daisyfns.awk`	Library of user-defined **gawk** functions required by the three-species Daisyworld model.

`daisy5.awk`	Modular **gawk** program implementing the three-species Daisy-world model under conditions of increasing solar luminosity, employing user-defined functions. Also requires `daisyvar.awk` and `daisyfns.awk`.
`daisy.bat`	Command line instructions used in this chapter.
`daisy.plt`	**gnuplot** commands used to generate the figures presented in this chapter.
`daisy.gp`	**gnuplot** commands used to generate figures presented in this chapter as EPS files.

Chapter 10

Modeling Incident Solar Radiation and Hydrological Networks over Natural Terrain

Topics

- Calculating the gradient and aspect of terrain slopes from a DEM

- Estimating the direct, diffuse and total solar irradiance on sloping terrain

- Modeling hydrological networks (local drainage direction) using a DEM

Methods and techniques

- Processing 2D arrays in gawk

- Plotting vectors and arrows in 2D and 3D using gnuplot

10.1 INTRODUCTION

The Daisyworld model, which is explored in the preceding chapter, considers a highly idealized planet, the surface of which is assumed to be flat and level. This assumption simplifies the model in various ways, the most important of which is that radiation from Daisyworld's sun is incident on every part of the planet surface at the same angle (i.e., normal [perpendicular] to the surface). The same assumption cannot be applied generically to models of solar radiation interaction with Earth's land surface, which is rarely flat or level over significant distances, but slopes at different angles (gradients) and in different directions (aspects) according to the local terrain. As a result, those parts of the land surface that face toward the sun tend to receive more solar radiation than those that face away. More generally, the amount of solar radiation received at Earth's surface varies according to the slope of the terrain and the angle of the sun. The latter is, of course, a function of the latitude of the site, the time of day and the day of year, as is demonstrated in Chapter 5. The analysis begun in Chapter 5 is extended here to produce a model that simulates the amount of solar radiation received by the undulating terrain around Llyn Efyrnwy.

Terrain slope also exerts an effect on other environmental processes, perhaps the most obvious example of which is rainfall run-off (Beven 2001). Thus, the route that precipitation takes from the point at which it falls on the ground, via a sequence of streams and rivers, to the sea or to a lake is primarily controlled by the slope of the local terrain. It is relatively easy to derive comprehensive spatial data on terrain slope (i.e., gradient and aspect) from a DEM. This information can be used, in turn, to model the hydrological network (i.e., the local drainage direction) of an area (Moore 1996). From this, it is possible to simulate the hydrological response of the study area, as well as other properties related to soil erosion and sediment transport (Wainwright and Mulligan 2004). The second part of this chapter therefore introduces a model to predict the hydrological network for the Llyn Efyrnwy area, based on the DEM introduced in Chapter 2. The structure of the network is compared to the "blue line" features (i.e., streams and rivers) identified in the corresponding 1:50,000 scale OS topographic map.

10.2   VISUALIZING DIGITAL ELEVATION DATA AS AN ARRAY

Recall from Chapter 2 that the DEM of Llyn Efyrnwy, eyfynwy.dem, consists of a 51×51 element grid of elevation values sampled at regular spatial intervals (100 m) on the ground. In addition to the 3D visualizations presented in Chapter 2 (Figure 2.18, page 50), these data can be viewed as a form of planimetric map by issuing the following commands in gnuplot:

```
reset                                                              1
unset key                                                          2
set xlabel "Easting, m"                                            3
set ylabel "Northing, m"                                           4
set xtics 295000, 1000, 300000                                     5
set ytics 315000, 1000, 320000                                     6
set dgrid3d 51,51,16                                               7
set size square                                                    8
set tics out                                                       9
set style line 1 lt -1 lw 0.01                                     10
set palette gray gamma 1                                           11
set pm3d at s hidden3d 1 map                                       12
set colorbox vertical user origin 0.8,0.185 size 0.04,0.655        13
set cblabel "Elevation, m"                                         14
splot 'efyrnwy.dem' u 1:2:3 w pm3d                                 15
```

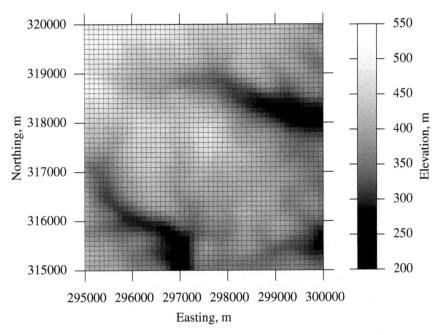

Figure 10.1: Planimetric visualization of the terrain elevation at Llyn Efyrnwy.

Most of these instructions have been introduced in the preceding chapters and so should be familiar by now. They produce the plot shown in Figure 10.1. Lines 1 to 15 instruct **gnuplot** to create a solid surface model rendered as a grayscale (line 11) planimetric map (line 12). The gray tones in the plot represent the elevation values, with black representing the lowest points in the DEM and white representing the highest ones. Line 9 instructs **gnuplot** to place the tic-marks facing outward and line 10 specifies the particular line type used to mark the data grid. Lines 13 and 14 specify the position and label, respectively, of the continuous grayscale key, which appears on the right-hand side of the plot. Some experimentation with the parameter values used to control the position of the key is normally required to obtain an aesthetically pleasing result. The significance of the resulting plot (Figure 10.1) is that it highlights the fact that the DEM is a regular rectangular grid of elevation values. It is important to note, however, that the elevation values relate to the intersections on the grid, rather than the grayscale rendered cells that **gnuplot** produces in this visualization.

10.3   HANDLING MULTI-DIMENSIONAL ARRAYS IN gawk

Although **gnuplot** can be used to display data organized in the form of a grid or array, **gawk** is needed to transform the data into the required format (e.g., to calculate the gradient and aspect of the terrain from the original elevation values). The methods used to handle arrays in **gawk** are introduced in Chapter 7 (page 184). The arrays used in that chapter are, however, one-dimensional (i.e., they store a list of values). By con-

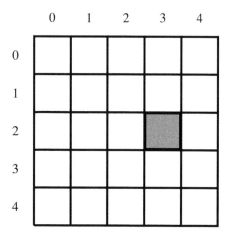

Figure 10.2: Graphical representation of a 2D array, `data`. The shaded cell is `data[2,3]`.

trast, the data examined in this chapter are inherently multi-dimensional. The dimensions are defined by the Easting, Northing and elevation of the data points. All three dimensions are measured in meters relative to ODN, using the OS coordinate system. gawk handles multi-dimensional arrays such as this by using a separate index for each dimension. Thus, while each element in a 1D array is referenced by a single unique index (`data[1]`, `data[2]`, `data[3]`, ..., `data[x]`), the elements of a 2D array are identified by two indexes separated by a comma (e.g., `data[1,1]`, `data[1,2]`, `data[1,3]`, ..., `data[2,1]`, ..., `data[y,x]`). The way in which gawk stores and processes the indexes is different from that of other programming languages (Robbins 2001), but the details need not concern us here.

Figure 10.2 is a graphical representation of a 2D array, `data`. The array contains 25 elements, which are arranged in a 5 × 5 element grid or matrix. It is usual for the first index of an array, such as this, to denote the row number, referenced from the top downward, and for the second index to indicate the column number, referenced from left to right. In some computer programming languages, such as C, the numbers used for the indexes start at zero (0); in others, such as FORTRAN, they begin at one (1). These two systems are commonly referred to as zero-based and one-based arrays, respectively. gawk is essentially agnostic in this respect, so that either system can be used. Hence, `data[2,3]` refers to the shaded cell in Figure 10.2.

Observant readers will notice that the array indexing system described above is different from the grid referencing system used by the OS, which is normally given in the following order: Easting (equivalent to column number), measured east to west, and Northing (equivalent to row number), measured south to north (i.e., referenced from the bottom up; see, for example, Figure 10.1). gawk is capable of coping with either of these array indexing systems. The onus is placed on the person who implements the computational model to make clear which system is being used and to be consistent in this respect, otherwise all manner of unintended consequences, errors and misunderstandings may arise.

Program 10.1: readarr.awk

```
# Reads digital elevation data stored in the form     1
# of triplets (Easting, Northing, elevation) into     2
# a 2D GAWK array.                                     3
#                                                      4
# Usage: gawk -f readarr.awk input_file               5
#                                                      6
# Variables:                                           7
# ---------                                            8
# E         Easting                                    9
# N         Northing                                  10
# elevation Array in which DEM data are stored        11
                                                       12
{                                                      13
  E=$1;                 # Read in Easting             14
  N=$2;                 # Read in Northing            15
                                                       16
  elevation[E,N]=$3;    # Read elevation data into array  17
}                                                      18
```

Program 10.1, readarr.awk, illustrates how digital elevation data held in the file efyrnwy.dem (Figure 2.12, page 44) can be read into a 2D gawk array. Apart from the comments, the program is contained entirely within the main pattern-action block (lines 13 to 18). The program takes the Easting and Northing values from the first and second fields of the current data record and stores these in the variables E and N, respectively (lines 14 and 15). These variables are then employed as the indexes to a 2D array, elevation, which is used to hold the corresponding elevation value, read from the third field of each record (line 17). Although the program does nothing with the data in the array, it illustrates how an array can be populated from a data file. This simple program forms the basis of a more extensive program presented in the next section, which calculates the gradient and aspect of every cell in a DEM array.

10.4 DETERMINING TERRAIN GRADIENT AND ASPECT

10.4.1 Formulation

Various methods have been developed to calculate gradient and aspect values from digital elevation data stored in a 2D array (Burrough and McDonnell 1998). The method used here is that of Zevenbergen and Thorne (1987), which represents a good compromise between accuracy and computational simplicity (Skidmore 1989). This method calculates the gradient and aspect of each cell in an array (elevation[E,N]) by measuring the difference in elevation between the cells immediately to the north and south of it (elevation[E,N+Δ] and elevation[E,N-Δ], respectively; where Δ is the cell size in meters) and between the cells immediately to the east and west of it (elevation[E-Δ,N] and elevation[E+Δ,N], respectively; Figure 10.3). These values

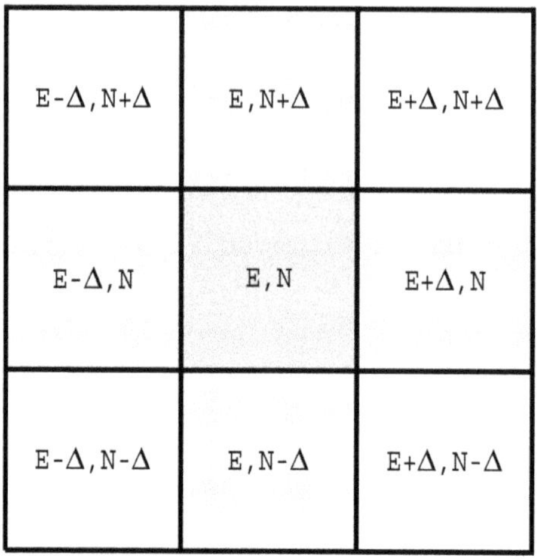

E-Δ,N+Δ	E,N+Δ	E+Δ,N+Δ
E-Δ,N	E,N	E+Δ,N
E-Δ,N-Δ	E,N-Δ	E+Δ,N-Δ

Figure 10.3: Relative indexing used to calculate the local gradient and aspect of a given cell, elevation[E,N], in an array containing digital elevation data, where E and N are the Easting and Northing, respectively, of the cell and where Δ is the cell size.

are used to calculate the gradient of the central cell in the north-south (NS) and east-west (EW) directions. The EW gradient can be expressed more formally as follows:

$$\mathrm{gradEW} = \frac{\mathrm{elevation}[E-\Delta,N] - \mathrm{elevation}[E+\Delta,N]}{2\Delta} \tag{10.1}$$

Similarly, the NS gradient is given by Equation 10.2.

$$\mathrm{gradNS} = \frac{\mathrm{elevation}[E,N+\Delta] - \mathrm{elevation}[E,N-\Delta]}{2\Delta} \tag{10.2}$$

The overall gradient of the cell, measured in arc degrees (°), can be computed from these two values using Equation 10.3.

$$\theta = \tan^{-1}\left(\sqrt{\mathrm{gradEW}^2 + \mathrm{gradNS}^2}\right) \tag{10.3}$$

Finally, the aspect of the cell relative to true north is given by Equation 10.4.

$$\phi = 180 + \tan^{-1}\left(-\frac{\mathrm{gradEW}}{\mathrm{gradNS}}\right) \tag{10.4}$$

10.4.2   Implementation

Equation 10.1 through Equation 10.4 are implemented in Program 10.2, gradasp.awk. The code used in this program can be interpreted as follows.

Lines 1 to 24 contain a series of comments that explain the purpose of the program and how it should be run from the command line, and that list the main variables used in the code. The BEGIN block (lines 26 to 29) initializes two variables, PI and rad2deg:

Program 10.2: gradasp.awk (lines 1 to 45; comments, BEGIN block and main pattern-action block).

```
# Takes a 2D DEM and for each cell determines the local      1
# gradient and aspect using the method of Zevenbergen        2
# and Thorne (1987). Prints out two files, each              3
# containing the coordinates of the cell concerned           4
# and either (i) its gradient or (ii) its aspect.            5
#                                                            6
# Usage: gawk -f gradasp.awk -v size=value input_file        7
#              [> output_file]                                8
#                                                            9
# Variables:                                                 10
# ----------                                                 11
# min_E       eastern-most Easting in DEM                     12
# max_E       western-most Easting in DEM                     13
# min_N       southern-most Northing in DEM                   14
# min_N       northern-most Northing in DEM                   15
# this_E      Easting of current cell                         16
# this_N      Northing of current cell                        17
# size        DEM cell size (metres)                          18
# gradNS      Gradient of current cell measured North-South   19
# gradEW      Gradient of current cell measured East-West     20
# gradient    Gradient of current cell                        21
# aspect      Aspect of current cell                          22
# elevation   2D array storing the DEM                         23
# rad2deg     Conversion factor - radians to degrees           24
                                                             25
BEGIN{                                                       26
  PI=3.14159263359;                                          27
  rad2deg=360.0/(2*PI);                                      28
}                                                            29
                                                             30
{                                                            31
  E=$1;                   # Reading in Easting               32
  N=$2;                   # Reading in Northing              33
                                                             34
  # Initialize minimum and maximum Easting and Northing       35
  # of the DEM                                               36
  if(NR==1){min_E=E; max_E=E; min_N=N; max_N=N}              37
                                                             38
  if(E<min_E){min_E=E}    # Determine minimum and            39
  if(E>max_E){max_E=E}    # maximum Eastings and             40
  if(N<min_N){min_N=N}    # Northings (i.e., limits)         41
  if(N>max_N){max_N=N}    # of DEM                           42
                                                             43
  elevation[E,N]=$3;      # Read elevation data into array   44
}                                                            45
```

Program 10.1 (continued): gradasp.awk (lines 46 to 77; END block).

```
                                                                    46
END {                                                               47
  # Pass across DEM by Easting                                      48
  for(this_E=min_E+size; this_E<max_E; this_E+=size){               49
                                                                    50
    # Pass up through DEM by Northing                               51
    for(this_N=min_N+size; this_N<max_N; this_N+=size){             52
                                                                    53
      # Calculate Zevenbergen and Thorne's (1987) parameters       54
      gradEW=(elevation[this_E-size,this_N]- \                      55
            elevation[this_E+size,this_N])/(2*size);                56
      gradNS=(elevation[this_E,this_N+size]- \                      57
            elevation[this_E,this_N-size])/(2*size);                58
                                                                    59
      # Calculate the gradient (degrees)                            60
      gradient[this_E,this_N]=rad2deg*\                             61
            (gradNS*gradNS + gradEW*gradEW)**0.5;                   62
      # Print this value out to a file                              63
      printf("%6i %6i %5.2f\n", this_E, this_N,                     64
        gradient[this_E,this_N]) > "gradient.dem";                  65
                                                                    66
      # Calculate the aspect (degrees)                              67
      aspect=180+rad2deg*atan2(-gradEW,gradNS);                     68
      # Print this value out to a file                              69
      printf("%6i %6i %3i\n", this_E, this_N,                       70
          aspect) > "aspect.dem";                                   71
                                                                    72
    }                                                               73
                                                                    74
  }                                                                 75
                                                                    76
}                                                                   77
```

the former should need no explanation; the latter is a factor used to convert angular values from radians to degrees.

The main pattern-action block (lines 31 to 45) reads the digital elevation data from the input file into a 2D array, `elevation`. The Easting and Northing values (fields 1 and 2 of the data file) are used as the indexes to the array (lines 32, 33 and 44), and the corresponding elevation values are read into the array from the third field of the file (line 44). The code on lines 37 and 39 to 42 determines the minimum and maximum Eastings and Northings of the data set. This information is used later on in the program, in the `END` block.

The gradient and aspect of each cell in the array, `elevation`, is calculated in the `END` block (lines 47 to 77). This is achieved using two `for` loops to navigate a cell at a time through the array, east to west and south to north: the outer loop (lines 49 to 75) controls the Easting; the inner loop (lines 52 to 73) controls the Northing. The variables `this_E` and `this_N` are used to store the location of the current cell for which gradient and aspect values are calculated. Note that the `for` loops start one cell in from the eastern-most edge (`min_E+size`; line 49) and one cell up from the southern-most edge (`min_N+size`; line 52) of the array. This is because the gradient and aspect values of a given cell are calculated with reference to the cells immediately above and below it, as well as to either side of it. Clearly, this is not possible for the cells around the edge of the array. For the same reason, the `for` loops stop processing just before they reach the northern-most and western-most edges of the array (`this_E<max_E`, line 49, and `this_N<max_N`, line 52). Consequently, the data files for gradient and aspect that are output from this program are two columns and two rows smaller than the input DEM (i.e., they are both 49×49 element grids). This feature is illustrated in Figure 10.4.

Inside the `for` loops, lines 55 and 56 calculate the EW gradient of the current cell (`gradEW`; Equation 10.1), lines 57 and 58 calculate its NS gradient (`gradNS`; Equation 10.2) and lines 61 and 62 calculate its overall gradient (`gradient`; Equation 10.3). The latter value is printed out on lines 64 and 65, along with the Easting and the Northing of the cell to which it relates. Note that the output is written to a file called `gradient.dem` in the working directory, rather than being sent to the computer screen. This outcome is achieved using the redirection symbol and the appropriate file name enclosed in double quotation marks (`> "gradient.dem"`; line 65). The output file is opened once only while the program is running, such that the Easting, Northing and gradient values for subsequent cells are appended to the end of the named file (Robbins 2001).

Finally, the aspect (or azimuth) of the cell is calculated on line 68. Note that the built-in function used by **gawk** to calculate the inverse tangent of an angle, $\tan^{-1}$, is `atan2`; there is no `atan` function (Robbins 2001). The output from this function is given in radians, and the result is converted into arc degrees through multiplication by the conversion factor `rad2deg`. To convert this angle into arc degrees relative to due north, 180° is added to the result. Lines 70 and 71 write the Easting, Northing and aspect of the current cell to a file called `aspect.dem`, which is created in the working directory, using the redirection method described above.

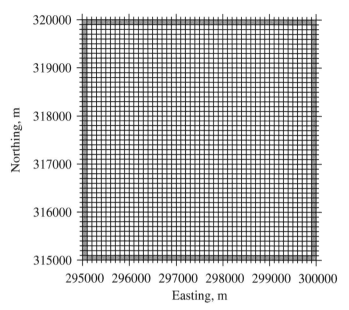

Figure 10.4: Relationship between the size of the data array given as input to Program 10.2, gradasp.awk, (gray and white cells combined) and the output data arrays that it produces (white cells only).

10.4.3 Evaluation

Program 10.2 can be run from the command line as follows:

```
gawk -f gradasp.awk -v size=100 efyrnwy.dem                          1
```

Note that the grid cell size, size, measured in meters, is specified on the command line. As noted previously, the program creates two output files, gradient.dem and aspect.dem, in the working directory. The former contains the gradient (measured in arc degrees; °) of each cell in the input DEM; the latter contains the corresponding aspect values (measured in arc degrees relative to true north). Note that the names of the output data files do not have to provided on the command line because these are specified in the program itself (lines 65 and 73 of Program 10.2, gradasp.awk).

10.4.4 Visualization

The gradient data produced by Program 10.2 can be visualized in **gnuplot** as follows:

```
set xrange [295000:300000]                                           16
set yrange [320000:315000]                                           17
set cblabel "Gradient, arc degrees"                                  18
splot 'gradient.dem' u 1:2:3 w pm3d                                  19
```

These instructions generate the plot shown in Figure 10.5. The aspect data can be similarly visualized by entering the following **gnuplot** commands:

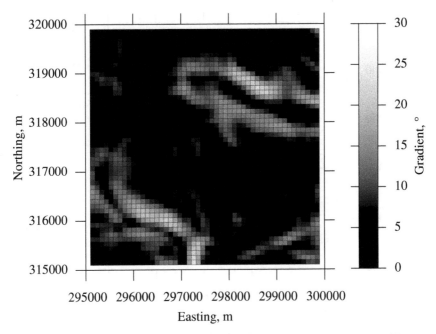

Figure 10.5: Planimetric visualization of the terrain gradient (arc degrees) at Llyn Efyrnwy.

```
set cbrange [0:360]                                      20
set cbtics 0,30,360                                      21
set cblabel "Aspect, arc degrees"                        22
set palette gray rgbformulae -13,-13,-13                 23
splot 'aspect.dem' u 1:2:3 w pm3d                        24
```

These commands produce the plot shown in Figure 10.6. Note that the instruction on line 20 sets the range of values to be plotted on the continuous grayscale key, limiting this to 0 to 360. Line 21 controls the tic-marks placed beside the key. Line 22 specifies the text used to label the key. Line 23 instructs gnuplot to use a grayscale palette that runs from black (0°) to white (180°) and back to black again (360°), rather than the normal linear grayscale palette (white to black); this command ensures that north-facing slopes are shaded dark gray through to black, while the south-facing ones are shaded light gray through to white.

Figure 10.5 and Figure 10.6 assist in the visual interpretation of the terrain around Llyn Efyrnwy, especially Figure 10.6, which is similar in appearance to a shaded relief map. The value of the data used to create these figures is exploited more fully in the following section, in which they are used to calculate the direct, diffuse and total (global) solar irradiance incident on every cell in the DEM. Information of this type is important to various types of study concerned with the surface radiation regime, including those attempting to model the surface energy budget (Monteith and Unsworth 1995, Campbell and Norman 1998) or the growth of crops (Russell *et al.* 1989, Bouman 1992, Goudriaan and van Laar 1994).

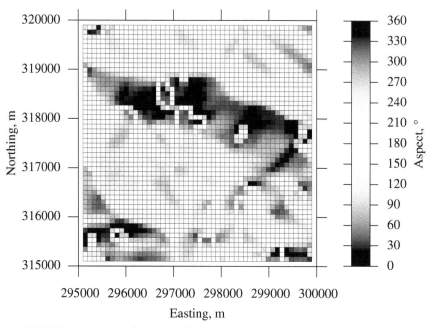

Figure 10.6: Planimetric visualization of the terrain aspect (arc degrees relative to true north) at Llyn Efyrnwy.

10.5   SOLAR IRRADIANCE ON SLOPING TERRAIN

Chapter 5 presents an analysis of the temporal variation in solar radiation incident at Llyn Efyrnwy based on a computational model and measurements made *in situ*. The computational model employed in that context calculates both the direct and the diffuse solar radiation incident on a horizontal element of Earth's surface using the mathematical formulations given by Campbell and Norman (1998), which are adapted from those of Liu and Jordan (1960). In this section, the analysis is extended to simulate the amount of direct and diffuse solar radiation incident on the undulating terrain around Llyn Efyrnwy, taking into account the zenith and azimuth angles of the sun and the slope (gradient and aspect) of the terrain.

10.5.1   Formulation

To calculate the solar radiation incident on an inclined surface, the angle between the sun and the surface normal must be known. This angle is given by Equation 10.5.

$$\zeta = \cos^{-1}\left(\cos\theta\cos\Psi + \sin\theta\sin\Psi\cos\left(\phi - \phi_s\right)\right) \qquad (10.5)$$

where ζ is the angle between the normal to the inclined surface and the direction of the sun, θ is the gradient of the surface slope, Ψ is the solar zenith angle, ϕ is the aspect (azimuth angle) of the surface slope, and ϕ_s is the azimuth angle of the sun (Iqbal 1983, Campbell and Norman 1998, Twidell and Weir 2006). The equations used to calculate the gradient and aspect (azimuth) of a sloping terrain facet are presented in the previous section, while those used to calculate the solar zenith angle are presented in Chapter 5. The solar azimuth angle, ϕ_s, measured clockwise with respect to due north, is given by Equation 10.6 for the morning (for hour angles ≤ 0) and by Equation 10.7 for the afternoon (for hour angles > 0).

$$\phi_s = 180 - \cos^{-1}\left(\frac{-(\sin\delta - \cos\Psi\sin\phi_L)}{\cos\phi_L\sin\Psi}\right) \tag{10.6}$$

$$\phi_s = 180 + \cos^{-1}\left(\frac{-(\sin\delta - \cos\Psi\sin\phi_L)}{\cos\phi_L\sin\Psi}\right) \tag{10.7}$$

where δ is the solar declination angle and θ_L is the latitude of the site (Campbell and Norman 1998). Note that the equation used to calculate the solar declination angle is presented in Chapter 5 (Equation 5.13).

The direct solar irradiance on an inclined surface can be calculated by modifying Equation 5.8 as follows:

$$I_{\text{direct}} = E_0\tau^m\cos\zeta \tag{10.8}$$

where E_0 is the total exo-atmospheric solar irradiance (Equation 5.7; assumed to be 1380 W·m^{-2} here), τ is the atmospheric transmittance (assumed to be 0.7 here), and m is the atmospheric air mass (Equation 5.11). Likewise, the diffuse solar irradiance on an inclined surface can be approximated by modifying Equation 5.9 as follows:

$$I_{\text{diffuse}} = \left(\frac{1 + \cos\theta}{2}\right)0.3\left(1 - \tau^m\right)E_0 \tag{10.9}$$

The term $(1 + \cos\theta)/2$, often referred to as the sky view factor, describes the fraction of the sky hemisphere that is visible from a point on the surface. As the gradient of the terrain, θ, increases, the fraction of the sky hemisphere visible from that point decreases and the amount of diffuse solar radiation it receives reduces concomitantly. Note that this is a relatively simple model of diffuse irradiance on a sloping surface. More sophisticated models also take into account the solar radiation reflected from adjacent terrain, which contributes to the diffuse radiation received at a given point. This phenomenon is sometimes known as the adjacency effect (Vermote *et al.* 1997), but its inclusion is beyond the scope of the discussion presented here.

10.5.2 Implementation

Program 10.3, demirrad.awk, implements Equation 10.5 through Equation 10.9 in a computational model. This program produces three output files, each of which relates to a different component of the incident solar radiation (i.e., direct, diffuse and total). The program is quite long, but it can be broken down into a number of discrete sections to help understand how it functions.

Lines 1 to 45 are comments, which describe what the program does, explain how it is operated from the command line, and list the primary variables employed in the

Program 10.3: demirrad.awk (lines 1 to 45; comments).

```
# Takes a 2D DEM in the form Easting, Northing, elevation    1
# and determines the local gradient and aspect of           2
# each cell using the method of Zevenbergen and             3
# Thorne (1987). It then uses this information to           4
# calculate the direct, diffuse and total solar irradiance  5
# on each cell given prescribed values of total exo-        6
# atmospheric solar irradiance, atmospheric transmittance   7
# and atmospheric pressue (at the altitude of the site and  8
# at sea level), and user-defined values of the latitude    9
# of the site, the day of year and the solar hour angle.    10
#                                                           11
# Usage: gawk -f dsirrad.awk -v size=value -v latitude=value 12
#              -v DOY=value -v hour_angle=value input_file   13
#              [> output_file]                               14
#                                                           15
# Variables:                                                16
# ----------                                                17
# E_0        Exo-atmospheric solar irradiance (W.m^{-2})    18
# tau        Atmospheric transmittance                      19
# p_alt      Atmospheric pressure at average altitude (mbar) 20
# p_sea      Atmospheric pressure at sea level (mbar)        21
# rad2deg    Conversion factor - radians to degrees          22
# deg2rad    Conversion factor - degrees to radians          23
# latitude   Latitude of site (initially in degrees)         24
# hour_angle Solar hour angle at site (initially degrees)    25
# zenith     Solar zenith angle (radians)                    26
# azimuth    Solar azimuth angle (radians)                   27
# declinatn  Solar declination angle (degrees)               28
# air_mass   Atmospheric air mass                            29
# zeta       Angle between sun and surface normal (radians)   30
# I_direct   Direct solar irradiance on surface (W.m^{-2})   31
# I_diffuse  Diffuse solar irradiance on surface (W.m^{-2})  32
# I_total    Total solar irradiance on surface (W.m^{-2})    33
# sky_view   Sky view factor                                 34
# min_E      eastern-most Easting in DEM (metres)             35
# max_E      western-most Easting in DEM (metres)             36
# min_N      southern-most Northing in DEM (metres)           37
# min_N      northern-most Northing in DEM (metres)           38
# this_E     Easting of current cell (metres)                 39
# this_N     Northing of current cell (metres)                40
# size       DEM cell size (metres)                           41
# grad_NS    N-Sg radient of current cell (degrees)           42
# grad_EW    E-W gradient of current cell (degrees)           43
# gradient   Gradient of current cell (degrees)               44
# aspect     Aspect of current cell (degrees)                 45
```

Program 10.2 (continued): demirrad.awk (lines 46 to 88; BEGIN block).

```
                                                                46
BEGIN{                                                          47
  # Initialize key variables                                   48
  E_0=1380.0;                                                   49
  tau=0.7;                                                      50
  p_alt=1000.0;                                                 51
  p_sea=1013.0;                                                 52
  PI=3.14159263359;                                             53
  rad2deg=360/(2*3.1415927);                                    54
  deg2rad=1/rad2deg;                                            55
                                                                56
  # Convert latitude and hour angle values provided            57
  # via command line from degrees to radians                   58
  latitude*=deg2rad;                                            59
  hour_angle*=deg2rad;                                          60
                                                                61
  # Calculate solar declination angle (radians)                62
  declinatn=(-23.4*deg2rad)* \                                  63
    cos(deg2rad*(360*(DOY+10)/365));                            64
                                                                65
  # Calculate solar zenith angle (radians)                     66
  if(hour_angle<(deg2rad*90)   hour_angle>(-90*deg2rad)){       67
    zenith=fn_acos(sin(latitude)*sin(declinatn)+\               68
      cos(latitude)*cos(declinatn)*cos(hour_angle));            69
  } else {                                                      70
    zenith=fn_acos(sin(latitude)*sin(declinatn)-\               71
      cos(latitude)*cos(declinatn)*cos(hour_angle));            72
  }                                                             73
                                                                74
  # Calculate solar azimuth angle (radians)                    75
  if(hour_angle>0){                                             76
    azimuth=(deg2rad*180)+fn_acos(-1*(sin(declinatn)-\          77
      cos(zenith)*sin(latitude))/\                              78
(cos(latitude)*sin(zenith)));                                   79
  } else {                                                      80
    azimuth=(deg2rad*180)-fn_acos(-1*(sin(declinatn)-\          81
      cos(zenith)*sin(latitude))/\                              82
(cos(latitude)*sin(zenith)));                                   83
  }                                                             84
                                                                85
  # Calculate atmospheric air mass                             86
  air_mass=(p_alt/p_sea)/cos(zenith);                           87
}                                                               88
```

Program 10.2 (continued): demirrad.awk (lines 89 to 131; main pattern-action and END blocks).

```
                                                                              89
{                                                                             90
    # Read in Easting and Northing of current cell from DEM                   91
    E=$1;                                                                     92
    N=$2;                                                                     93
                                                                              94
    # Initialize min. and max. Easting and Northing of DEM                    95
    if(NR==1){min_E=E; max_E=E; min_N=N; max_N=N}                             96
                                                                              97
    # Determine min. and max. Eastings and Northings of DEM                   98
    if(E<min_E){min_E=E}                                                      99
    if(E>max_E){max_E=E}                                                      100
    if(N<min_N){min_N=N}                                                      101
    if(N>max_N){max_N=N}                                                      102
                                                                              103
    # Read elevation data into array (dem)                                    104
    dem[E,N]=$3;                                                              105
}                                                                             106
                                                                              107
END{                                                                          108
    # Pass across DEM by Easting                                              109
    for(this_E=min_E+size;this_E<max_E;this_E+=size){                         110
        # Pass up through DEM by Northing                                     111
        for(this_N=min_N+size;this_N<max_N;this_N+=size){                     112
                                                                              113
            # Calculate Zevenbergen and Thorne's parameters                   114
            grad_EW=(dem[this_E-size,this_N]- \                               115
                    dem[this_E+size,this_N])/(2*size);                        116
            grad_NS=(-dem[this_E,this_N-size]+ \                              117
                    dem[this_E,this_N+size])/(2*size);                        118
                                                                              119
            # Calculate the gradient (radians)                                120
            gradient=deg2rad*(360/(2*PI))*(grad_NS*grad_NS +\                 121
                grad_EW*grad_EW)**0.5;                                        122
                                                                              123
            # Calculate the aspect (radians)                                  124
            aspect=PI+(atan2(-grad_EW,grad_NS));                              125
                                                                              126
            # Calculate angle of incidence of solar radiation                 127
            # for the inclined surface                                        128
            cos_zeta=(cos(gradient)*\                                         129
                cos(zenith))+(sin(gradient)*\                                 130
                sin(zenith)*cos(azimuth-aspect));                             131
```

Program 10.2 (continued): demirrad.awk (lines 132 to 176; END block (continued)).

```
132
                 # Calculate the direct solar irradiance            133
                 I_direct=(E_0)*(tau**air_mass)*cos_zeta;           134
                 if(I_direct<0){I_direct=0}                         135
                                                                    136
                 # Print out the Easting, Northing and direct solar 137
                 # irradiance for this cell                         138
                 printf("%6d %6d %6.1f\n", this_E, this_N,          139
                    I_direct) > "efyrnwy.dir";                      140
                                                                    141
                 # Calculate the diffuse solar irradiance           142
                 sky_view=0.5*(1+cos(gradient));                    143
                 I_diffuse=sky_view*0.3*(1-(tau**air_mass))*\       144
                    (E_0)*cos(zenith);                              145
                 if(I_diffuse<0){I_diffuse=0}                       146
                                                                    147
                 # Print out the Easting, Northing and diffuse solar 148
                 # irradiance for this cell                         149
                 printf("%6d %6d %6.1f\n", this_E, this_N,          150
                    I_diffuse) > "efyrnwy.dif";                     151
                                                                    152
                 # Calculate global solar irradiance on surface     153
                 I_global=I_direct+I_diffuse;                       154
                 if(I_global<0){I_global=0}                         155
                                                                    156
                 # Print out the Easting, Northing and diffuse solar 157
                 # irradiance for this cell                         158
                 printf("%6d %6d %6.1f\n", this_E, this_N,          159
                    I_global) > "efyrnwy.tot";                      160
                                                                    161
           }                                                        162
       }                                                            163
   }                                                                164
                                                                    165
function fn_abs(x){                                                 166
   if(x<0){return -1*x} else {return x}                             167
}                                                                   168
                                                                    169
function fn_acos(x){                                                170
   if(x*x<=1){                                                      171
       return 1.570796 - atan2(x/sqrt(1-x*x),1);                    172
   } else {                                                         173
       return 0;                                                    174
   }                                                                175
}                                                                   176
```

code. Lines 47 to 88 contain code placed in the BEGIN block. These lines initialize several of the major variables (lines 49 to 55), convert some of the variables from arc degrees into radians (lines 59 and 60), and calculate the solar declination angle (lines 63 and 64), solar zenith angle (lines 67 to 73), solar azimuth angle (lines 76 to 84) and the atmospheric air mass (line 87). Much of this code is taken directly from Program 5.2, solarrad.awk, which is presented in Chapter 5. Lines 76 to 84 are new, however, and they implement Equation 10.6 and Equation 10.7, respectively.

Lines 90 to 106 contain code placed in the main pattern-action block. This code is identical to the main block of Program 10.2, gradasp.awk, given earlier in this chapter, and it performs the same purpose, namely to read the digital elevation data from the input data file into the 2D array, dem.

Lines 108 to 164 contain code placed in the END block. These lines of code represent an extension to the END block presented in Program 10.2, gradasp.awk. Thus, they not only calculate the gradient and aspect (azimuth) of each cell in the array (lines 110 to 125), they also determine the angle between the sun and the surface normal for that cell (Equation 10.5; lines 129 to 131) and the amount of direct solar radiation (Equation 10.9; lines 134 and 135), diffuse solar radiation (Equation 10.8; lines 143 to 146) and total solar radiation (Equation 10.9; lines 154 to 155) incident upon it. The latter three sets of values are written to separate files on lines 139 and 140, lines 150 and 151, and lines 159 and 160, respectively.

A number of the calculations performed by the code outlined above make use of two user-defined functions, which are also included in Program 10.3. The first of these determines the absolute value of a given input (fn_abs; lines 166 to 168); this function is used to convert negative values into positive ones. The second calculates the inverse cosine of an angle (fn_acos; lines 170 to 176), because **gawk** does not provide a built-in function that serves this purpose.

10.5.3   Evaluation

Program 10.3 is a fairly sophisticated computational model, which brings together most of the programming techniques that have been introduced throughout this book. To run this program from the command line, the user must provide a number of input values, including the cell size of the DEM, the latitude of the study site, the day of year and the time of day, in addition to the name of the file that contains the digital elevation data on which the program operates. For example, to determine the direct, diffuse and total solar radiation incident on the terrain around Llyn Efyrnwy (size=100 and latitude=52.756) at solar noon (hour_angle=0) on June 21 (the northern summer solstice; DoY=172), the appropriate command line is as follows:

```
gawk -f demirrad.awk -v size=100 -v latitude=52.756 -v DOY↩   2
    ↪=172 -v hour_angle=0 efyrnwy.dem
```

Remember that the names of the three output data files do not have to be specified on the command line because these details are given explicitly in the program (lines 139 and 140, 150 and 151, and 159 and 160).

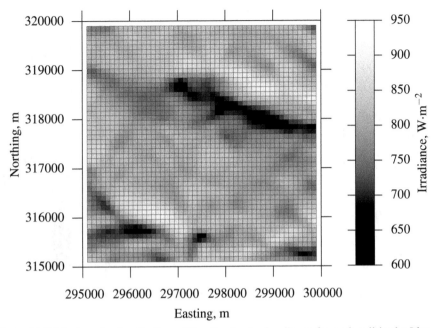

Figure 10.7: Planimetric visualization of the direct solar irradiance for each cell in the Llyn Efyrnwy DEM at solar noon on DoY = 172 (June 21).

10.5.4 Visualization

The direct solar irradiance data simulated using Program 10.3 can be visualized in gnuplot with the following commands:

```
set palette gray                          25
set zrange [*:*]                          26
set cbrange [*:*]                         27
set cbtics autofreq                       28
set cblabel "Irradiance W.m^{-2}"         29
splot 'efyrnwy.dir' u 1:2:3 w pm3d        30
```

These instructions produce the plot shown in Figure 10.7. Note that the instruction on line 25, above, resets the standard grayscale palette (black through to white); lines 26 and 27 reset the range of values that are plotted on the z-axis and on the grayscale key, respectively, so that they are determined automatically by gnuplot; line 28 instructs gnuplot to determine automatically the number of tic-marks plotted on the grayscale key; line 29 specifies the text used to label the key; and line 30 plots the direct solar irradiance data. It is evident from Figure 10.7 that the predicted values of direct solar irradiance range from about $600\,\mathrm{W{\cdot}m^{-2}}$ on some north-facing slopes to just under $950\,\mathrm{W{\cdot}m^{-2}}$ on some south-facing ones. This difference can be seen most clearly in the relatively steep valley running roughly WNW–ESE between Northings 318000 and 319000 on the eastern side of the study area. This valley contains the Afon Conwy, a large stream that drains into the Afon Efyrnwy below Llyn Efyrnwy.

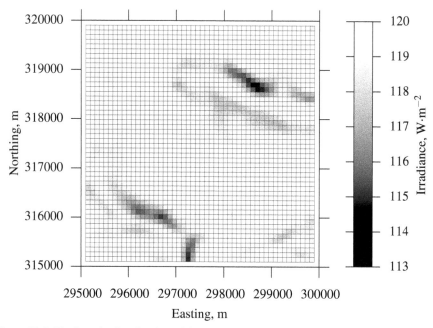

Figure 10.8: Planimetric visualization of the diffuse solar irradiance for each cell in the Llyn Efyrnwy DEM at solar noon on DoY = 172 (June 21).

Similar plots can also be constructed for the diffuse solar irradiance (Figure 10.8) and the total solar irradiance (Figure 10.9), respectively, using the following gnuplot commands:

```
splot 'efyrnwy.dif' u 1:2:3 w pm3d                              31
```

and

```
splot 'efyrnwy.tot' u 1:2:3 w pm3d                              32
```

Note that the estimated values of diffuse solar irradiance are smaller in absolute terms and more limited in their range ($113\,\mathrm{W \cdot m^{-2}}$ to $120\,\mathrm{W \cdot m^{-2}}$) than those of the direct solar irradiance. Also, some of the lowest values are recorded on the south-facing slopes of the Afon Conwy valley because of their steep gradient and, hence, restricted sky view factor.

Exercise 10.1: Using Program 10.3, demirrad.awk, simulate the direct, diffuse and total solar irradiance on the terrain at Llyn Efyrnwy at 8 am on June 21 and at solar noon on January 1. Use gnuplot to visualize the results obtained from each of these experiments.

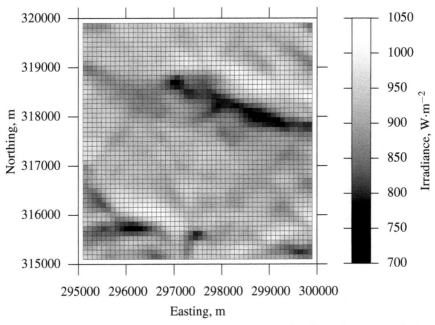

Figure 10.9: Planimetric visualization of the total (global) solar irradiance for each cell in the Llyn Efyrnwy DEM at solar noon on DoY = 172 (June 21).

10.6 MODELING HYDROLOGICAL NETWORKS

Information on drainage basins, such as the catchment boundaries (or watersheds) and the spatial location and topological connectivity of stream and river networks, is important to many studies in hydrology and geomorphology (Burrough and Mc-Donnell 1998, Beven 2001, Mulligan 2004). Some of the information required by these studies can be obtained directly from automated analyses of DEMs including, most notably, the flow paths that precipitation falling on the terrain subsequently trace across the land surface (Burrough and McDonnell 1998, Mulligan 2004). Knowledge of these pathways can be used, in turn, to model the movement and accumulation of water, soil and environmental contaminants over the terrain.

Various methods have been developed that can be used to model flow paths across a DEM, the simplest of which is known as the D8 or "eight-point pour" algorithm (Moore 1996, Burrough and McDonnell 1998, Mulligan 2004). This algorithm makes the assumption that water flows from each cell in the DEM to one, and only one, of its eight immediate neighbors, whichever defines the direction of the steepest downhill gradient (Figure 10.10). The D8 algorithm has a number of limitations for hydrological modeling, the most notable of which is that flow is restricted to the eight cardinal (north, south, east and west) and ordinal (north-east, south-east, south-west and north-west) directions. A corollary of this is that flow dispersion is not taken into account. Nevertheless, the D8 algorithm provides a useful starting point for modeling hydrological networks and is the method employed here.

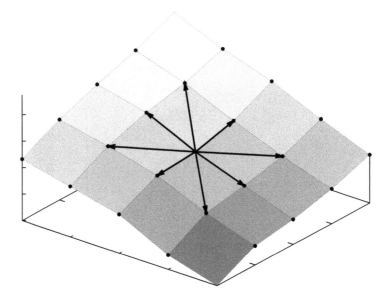

Figure 10.10: Potential directions in which water may flow across a DEM in the D8 or "eight-point pour" algorithm.

Figure 10.11 illustrates the application of the D8 algorithm to an example DEM. The number in each cell of the array represents its elevation in meters. The arrows show the direction of steepest downhill gradient from each cell to one of its eight immediate neighbors. Collectively, the arrows define the flow paths across the DEM, which are also known as local drainage direction (LDD) vectors.

10.6.1   Implementation

Program 10.4, d81dd.awk, presents an implementation in **gawk** of the D8 method for determining the local drainage directions across a DEM. This program shares many common elements with Program 10.2 and Program 10.4, including a set of explanatory comments (lines 1 to 27), the main pattern-action block (lines 29 to 45), and the two for loops in the END block, which are used to traverse east to west and south to north across the DEM array (lines 49 and 51). d81dd.awk differs from the preceding two programs, however, in terms of how it processes the data stored in the array, dem. Lines 56, 58 and 59 define two further nested for loops, which are used to examine the elevation values in a 3 × 3 cell window centered on the current cell in the array. The elevation of the central cell in the window is compared with those of its eight neighbors to establish which of these results in the steepest downhill gradient (lines 62 to 74). The Easting and Northing of that cell, and its gradient with respect to the central cell, are recorded (lines 70 to 72). The program prints out the Easting and Northing of the current cell (i.e., the central cell in the 3 × 3 window, from which flow occurs), the relative coordinates of its steepest downhill neighbor (i.e., the cell into which flow occurs) and their respective elevation values (lines 82 to 85).

Program 10.4: d8ldd.awk (lines 1 to 45; comments and main pattern-action block).

```
# Implementation of the D8 ("eight point pour") algorithm.        1
# Derives local drainage direction (LDD) from a DEM (see          2
# [Burrough and McDonnell 1998]). Prints out coordinates of       3
# each cell, the relative coordinates to the neighboring          4
# cell in the direction of steepest downhill descent, plus        5
# the elevations of these two cells.                              6
#                                                                 7
# Usage: gawk -f d8ldd.awk -v size=value input [> output]         8
#                                                                 9
# Variables:                                                      10
# ----------                                                      11
# size        DEM cell size (meters)                              12
# min_E       western-most Easting in DEM (meters)                13
# max_E       eastern-most Easting in DEM (meters)                14
# min_N       northern-most Northing in DEM (meters)              15
# min_N       southern-most Northing in DEM (meters)              16
# this_E      Easting of current cell (meters)                    17
# this_N      Northing of current cell (meters)                   18
# offset_E    Offset (Easting) from current cell (meters)         19
# offset_N    Offset (Northing) from current cell (meters)        20
# steep_E     Offset (Easting) to neighboring cell with           21
#             steepest gradient (meters)                          22
# steep_N     Offset (Northing) to neighboring cell with          23
#             steepest gradient (meters)                          24
# gradient    Gradient between the current cell and one of        25
#             its eight immediate neighbors (degrees)             26
# steepest    Steepest gradient in neighborhood (degrees)         27
#                                                                 28
{                                                                 29
    # Read in Easting and Northing of current cell from DEM       30
    E=$1;                                                         31
    N=$2;                                                         32
                                                                 33
    # Initialize min. and max. Easting and Northing of DEM        34
    if(NR==1){min_E=E; max_E=E; min_N=N; max_N=N}                 35
                                                                 36
    # Determine min. and max. Eastings and Northings of DEM       37
    if(E<min_E){min_E=E}                                          38
    if(E>max_E){max_E=E}                                          39
    if(N<min_N){min_N=N}                                          40
    if(N>max_N){max_N=N}                                          41
                                                                 42
    # Read elevation data into array (DEM)                        43
    dem[E,N]=$3;                                                  44
}                                                                 45
```

Program 10.2 (continued): d8ldd.awk (lines 46 to 89; END block).

```
                                                                              46
END{                                                                          47
    # Pass across DEM by Easting                                             48
    for(this_E=min_E+size;this_E<max_E;this_E+=size){                        49
        # Pass up through DEM by Northing                                    50
        for(this_N=min_N+size;this_N<max_N;this_N+=size){                    51
                                                                              52
            # Initialize steep downhill descent                             53
            steepest=-999;                                                   54
            # Loop through eight neighboring cells, by Easting              55
            for(offset_E= -size;offset_E<=size;offset_E+=size){             56
                # ...and now loop by Northing                               57
                for(offset_N= -size;offset_N<=size;                         58
                    offset_N+=size){                                        59
                                                                              60
                    # Ignore center cell, from which flow occurs            61
                    if(!((offset_E==0) && (offset_N==0))){                   62
                        gradient=dem[this_E,this_N]-\                        63
                    dem[this_E+offset_E,this_N+offset_N];                    64
                        gradient/=sqrt((offset_E*offset_E)+\                 65
                    (offset_N*offset_N));                                    66
                        if(gradient>steepest){                              67
                            # Steepest downhill gradient detected.           68
                            # Record gradient and direction.                69
                            steepest=gradient;                               70
                            steep_E=offset_E;                                71
                            steep_N=offset_N;                                72
                        }           # End of inner 'if' block                73
                    }               # End of outer 'if' block                74
                                                                              75
                }                   # End of inner 'for' loop                76
            }                       # End of outer 'for' loop                77
                                                                              78
            # Print coordinates of cell from which flow                     79
            # occurs and relative coordinates of cell into                  80
            # which flow occurs, plus their elevations.                     81
            printf("%6i %6i %6i %6i %6.1f %6.1f\n",\                         82
                this_E, this_N, steep_E, steep_N, \                          83
                dem[this_E,this_N], \                                        84
                dem[this_E+steep_E,this_N+steep_N]);                        85
                                                                              86
        }                           # End Northing 'for' loop               87
    }                               # End Easting 'for loop                  88
}                                   # End 'END' block                        89
```

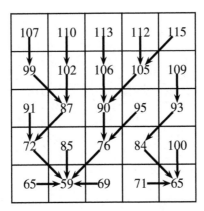

Figure 10.11: Example of a small DEM array showing the LDD vectors derived using the D8 ("eight-point pour") algorithm.

10.6.2 Evaluation

Program 10.4 can be run from the command line as follows:

```
gawk -f d8ldd.awk -v size=100 efyrnwy.dem > efyrnwy.ld8
```
3

Note that the DEM cell size (100 m) is given on the command line (`size=100`) and that the results are redirected to the file `efyrnwy.ld8` located in the working directory.

10.6.3 Visualization

The data generated using Program 10.4 can be visualized using the following gnuplot command:

```
plot 'efyrnwy.dd8' u 1:2:3:4 w vector nohead
```
33

This instruction produces the plot shown in Figure 10.12. The `vector` data style used here draws vectors between, in this case, E, N and $E + \Delta E$, $N + \Delta N$; it requires four columns of data, namely E, N, ΔE and ΔN, which are provided in the first four fields of the input data file `efyrnwy.ld8`. The `nohead` directive instructs gnuplot not to draw an arrowhead at the end of the vector, which simplifies the resulting figure in this instance. Note that it is also possible to superimpose the "blue line" features (i.e., the stream and river networks derived from the 1:50,000 scale OS topographic map), which are contained in the file `rivers.dat`, on the LDD vectors as follows:

```
plot 'efyrnwy.dd8' u 1:2:3:4 w vector nohead,\
     'rivers.dat' w l lt 3 lw 2
```
34
35

These instructions produce the plot shown in Figure 10.13, which shows a high degree of correspondence between the LDD vectors derived from the Llyn Efyrnwy DEM and the "blue line" features extracted from the OS topographic map of this area. The blue line features are clearly fewer in number and more limited in extent than the LDD vectors, owing to the selection, abstraction and generalization processes that

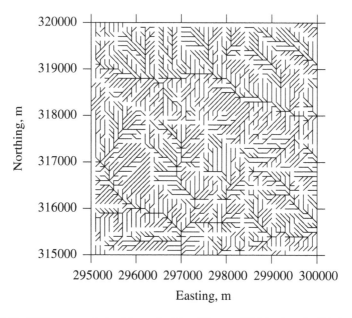

Figure 10.12: LDD vectors derived from the Llyn Efyrnwy DEM using the D8 ("eight-point pour") algorithm.

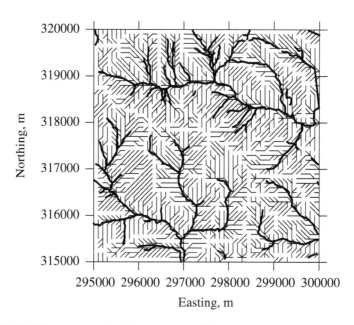

Figure 10.13: LDD vectors derived from the Llyn Efyrnwy DEM using the D8 ("eight-point pour") algorithm with OS stream network superimposed.

are inherent in map production (Buttenfield and McMaster 1991), but all of the major hydrological routes are successfully identified.

10.6.4 Modified Implementation

It is possible to make a few small modifications to the END block of Program 10.4 so that the LDD vectors it creates can be visualized in 3D using gnuplot's facility for drawing arrows with arbitrary start and end points in terms of the x-, y- and z-axes. The revised code is given in Program 10.5, d8arrows.awk, of which only the modified END block is presented here. Program 10.5 makes use of a further variable, arrow_number, which is initialized on line 49, to number the arrows that represent the LDD vectors. The number of each arrow and its start and end points in terms of Easting, Northing and elevation are printed out on lines 81 to 86. The arrow number is incremented on line 87, before the main for loops move on to consider the next cell in the DEM array.

10.6.5 Evaluation of the Modified Implementation

Program 10.5 can be run from the command line as follows:

```
gawk -f d8arrows.awk -v size=100 efyrnwy.dem > efyrnwy.arr    4
```

Once again, the DEM cell size (100 m) is given on the command line (size=100) and the results are redirected to a file, efyrnwy.arr, in the working directory. An extract from the output file, efyrnwy.arr, is presented in Figure 10.14.

10.6.6 Visualization of the LDD Vectors on a Solid Surface Model

The output from Program 10.5 can be visualized in gnuplot by issuing the following commands, most of which have been used elsewhere in this chapter:

```
load 'efyrnwy.arr'                                                      36
unset key                                                              37
set style line 1 lt -1 lw 0.01                                        38
set dgrid3d 51,51,16                                                  39
set pm3d at s hidden3d 1                                              40
set palette gray                                                      41
set colorbox vertical user origin 0.8,0.185 size 0.04,0.655          42
set cblabel "Elevation, m"                                            43
set view 45,45,1,1                                                    44
set surface                                                            45
splot 'efyrnwy.dem' w pm3d                                            46
```

These commands produce the plot shown in Figure 10.15, which gives a powerful impression of the flow paths that water takes across the terrain around Llyn Efyrnwy. Line 36 loads the set arrow instructions, which are contained in the file efyrnwy.arr, into gnuplot. The syntax of these instructions is set arrow tag from x1,y1,z1 to x2,y2,z2 nohead front, where tag is an integer that identifies a particular arrow, x1,y1,z1 denotes the start point of that arrow, x2,y2,z2 indicates its end point, nohead

Program 10.5: d8arrows.awk (lines 47 to 90 only; END block).

```
END{                                                                           47
    # Number of arrow used in 3D visualization                                 48
    arrow_number=1;                                                             49
    # Pass across DEM by Easting                                               50
    for(this_E=min_E+size;this_E<max_E;this_E+=size){                          51
        # Pass up through DEM by Northing                                      52
        for(this_N=min_N+size;this_N<max_N;this_N+=size){                      53
            # Initialize steep downhill descent                                54
            steepest=-999;                                                     55
            # Loop through eight neighboring cells, by Easting                 56
            for(offset_E= -size;offset_E<=size;offset_E+=size){                57
                # ...and now loop by Northing                                  58
                for(offset_N= -size;offset_N<=size;                            59
                    offset_N+=size){                                           60
                    # Ignore center cell, from which flow occurs               61
                    if(!((offset_E==0) && (offset_N==0))){                     62
                        gradient=dem[this_E,this_N]-\                          63
                    dem[this_E+offset_E,this_N+offset_N];                      64
                        gradient/=sqrt((offset_E*offset_E)+\                   65
                    (offset_N*offset_N));                                      66
                        if(gradient>steepest){                                 67
                            # Steepest downhill gradient detected.             68
                            # Record gradient and direction.                   69
                            steepest=gradient;                                 70
                            steep_E=offset_E;                                  71
                            steep_N=offset_N;                                  72
                        }               # End of inner "if" block              73
                    }                   # End of outer "if" block              74
                }                       # End of inner "for" loop              75
            }                           # End of outer "for" loop              76
                                                                               77
            # Print coordinates of cell from which flow                       78
            # occurs and relative coordinates of cell into                    79
            # which flow occurs, plus their elevations.                       80
            printf("set arrow %i from %6i,%6i,%6.1f to \                      81
                %6i,%6i,%6.1f", arrow_number, \                               82
                this_E, this_N, dem[this_E,this_N], \                         83
                this_E+steep_E, this_N+steep_N, \                             84
                dem[this_E+steep_E,this_N+steep_N]);                          85
            printf(" nohead front lt 1 lw 2.0\n");                            86
            ++arrow_number;    # Increment arrow number                       87
        }                      # End Northing "for" loop                      88
    }                          # End Easting "for" loop                        89
}                              # End "END" block                              90
```

```
set arrow  1 from 295100,315100, 408.0 to 295100,315000, 401.0 nohead front lt 1 lw 2.0
set arrow  2 from 295100,315200, 416.0 to 295100,315100, 408.0 nohead front lt 1 lw 2.0
set arrow  3 from 295100,315300, 421.0 to 295200,315400, 412.0 nohead front lt 1 lw 2.0
set arrow  4 from 295100,315400, 419.0 to 295200,315500, 403.0 nohead front lt 1 lw 2.0
set arrow  5 from 295100,315500, 408.0 to 295100,315600, 395.0 nohead front lt 1 lw 2.0
set arrow  6 from 295100,315600, 395.0 to 295100,315700, 382.0 nohead front lt 1 lw 2.0
set arrow  7 from 295100,315700, 382.0 to 295100,315800, 361.0 nohead front lt 1 lw 2.0
set arrow  8 from 295100,315800, 361.0 to 295200,315800, 357.0 nohead front lt 1 lw 2.0
set arrow  9 from 295100,315900, 385.0 to 295100,315800, 361.0 nohead front lt 1 lw 2.0
set arrow 10 from 295100,316000, 396.0 to 295200,315900, 376.0 nohead front lt 1 lw 2.0
set arrow 11 from 295100,316100, 400.0 to 295200,316200, 383.0 nohead front lt 1 lw 2.0
set arrow 12 from 295100,316200, 399.0 to 295200,316300, 374.0 nohead front lt 1 lw 2.0
set arrow 13 from 295100,316300, 398.0 to 295200,316300, 374.0 nohead front lt 1 lw 2.0
set arrow 14 from 295100,316400, 394.0 to 295200,316400, 365.0 nohead front lt 1 lw 2.0
set arrow 15 from 295100,316500, 386.0 to 295200,316600, 349.0 nohead front lt 1 lw 2.0
set arrow 16 from 295100,316600, 382.0 to 295200,316600, 349.0 nohead front lt 1 lw 2.0
set arrow 17 from 295100,316700, 369.0 to 295200,316600, 349.0 nohead front lt 1 lw 2.0
set arrow 18 from 295100,316800, 370.0 to 295200,316800, 353.0 nohead front lt 1 lw 2.0
set arrow 19 from 295100,316900, 368.0 to 295200,317000, 336.0 nohead front lt 1 lw 2.0
set arrow 20 from 295100,317000, 359.0 to 295200,317000, 336.0 nohead front lt 1 lw 2.0
set arrow 21 from 295100,317100, 347.0 to 295200,317000, 336.0 nohead front lt 1 lw 2.0
set arrow 22 from 295100,317200, 374.0 to 295100,317100, 347.0 nohead front lt 1 lw 2.0
set arrow 23 from 295100,317300, 388.0 to 295100,317200, 374.0 nohead front lt 1 lw 2.0
set arrow 24 from 295100,317400, 401.0 to 295100,317300, 388.0 nohead front lt 1 lw 2.0
set arrow 25 from 295100,317500, 417.0 to 295100,317400, 401.0 nohead front lt 1 lw 2.0
```

Figure 10.14: Extract from the data file efyrnwy.arr produced by Program 10.5, d8arrows.awk (first 25 records).

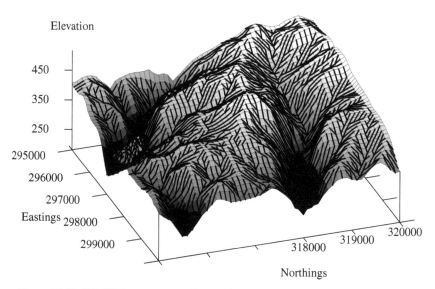

Figure 10.15: D8 LDD vectors superimposed onto a 3D visualization of the Llyn Efyrnwy DEM.

specifies that the arrow should be drawn without an arrowhead (i.e., as a straight line segment) and front instructs gnuplot to plot the arrow on top of the 3D surface. The resulting arrows are superimposed on the solid surface plot of the Llyn Efyrnwy DEM, which is produced by the commands on lines 37 to 46.

10.7   SUMMARY AND FURTHER DIRECTIONS

This chapter provides an introduction to some of the methods that can be employed to model the amount of direct, diffuse and total solar radiation received on sloping terrain, and to simulate hydrological networks (i.e., local drainage directions), using digital elevation data stored in a 2D array. Many of the numerical and computational techniques introduced in the preceding chapters are combined here to address these challenges. In one sense, therefore, this chapter is the culmination of the various topics covered throughout the book. Viewed from a wider perspective, however, it is only just the beginning, an entry point to the field of environmental modeling. There are many routes that the reader might wish to explore from here, according to his or her areas of interest. These routes could involve computational implementations of environmental models expressed in mathematical form, which are published in the scientific literature, or perhaps the development of new mathematical models *ab initio*. The skills developed throughout this book should enable the reader to engage in either of these activities.

Beyond this point, the reader should consult more widely the scientific literature concerned with different aspects of environmental modeling. Excellent overviews are presented by Jakeman *et al.* (1993), Ford (1999), Deaton and Winebrake (2000) and

Wainwright and Mulligan (2004). Two texts by Harte (1988, 2001) are also highly recommended; these books cover a wide range of environmental problems and they adopt a didactic approach which enhances the reader's skills in developing conceptual models and in formulating these models in mathematical terms.

Issues concerning the computational implementation of a range of ecological and evolutionary models are presented in Wilson (2000) and Donovan and Welden (2001, 2002). The first of these three books uses the C programming language; the second and third employ standard spreadsheet packages. All three books present material which is complementary to that introduced here.

Further coverage of hydrological modeling can be found in Abbott and Refsgaard (1996) and Beven (1998, 2001). These books cover several hydrological models and their applications, and provide sufficient detail to help the reader construct computational implementations of the models concerned.

Over the last 40 years, the computational models designed to represent various aspects of Earth's climate system have become ever more sophisticated and, hence, increasingly complex. An excellent introduction to this evolving field is provided by McGuffie and Henderson-Sellers (2005), who cover topics ranging from basic energy balance models and Daisyworld, through Earth system models of intermediate complexity (EMIC), to coupled climate system models.

Increasingly, given the explicit spatial nature of many environmental problems, many types of environmental models are being implemented in the framework of a GIS. In this context, instructive overviews of the various applications of GIS to problems in environmental modeling are given in a number of texts, including those by Goodchild *et al.* (1996), Clarke *et al.* (2002) and Brimicombe (2003).

The field of environmental modeling is also considered in a number of academic journals, which present research papers on various aspects of environmental modeling and software development. The main journals include *Geographical and Environmental Modelling* (Taylor & Francis), *Environmental Modelling and Software* (Elsevier) and *Environmental Modeling and Assessment* (Springer). The reader may wish to consult these and other related journals for research and review articles on specific topics of interest, and to be informed of the latest developments in the field.

Finally, whichever of these or other directions the reader subsequently decides to take, the skills acquired from this book should make the way ahead more accessible and hopefully more enjoyable. Good luck! *Pob lwc*!

SUPPORTING RESOURCES

The following resources are provided in the `chapter10` sub-directory of the CD-ROM:

`readarr.awk`	Program to read digital elevation data triplets (Easting, Northing, elevation) into a 2D **gawk** array.
`gradasp.awk`	**gawk** program to calculate terrain slope (gradient and aspect) from digital elevation data.

`demirrad.awk`	**gawk** program to calculate the direct, diffuse and total solar irradiance on each cell in a DEM.
`d81dd.awk`	**gawk** program to determine the local drainage direction for each cell in a DEM.
`rivers.dat`	Data file containing E, N coordinates of the "blue line" features (i.e., streams and rivers) extracted from the 1:50,000 scale OS topographic map covering Llyn Efyrnwy.
`dem.bat`	Command line instructions used in this chapter.
`dem.plt`	**gnuplot** commands used to generate the figures presented in this chapter.
`dem.gp`	**gnuplot** commands used to generate figures presented in this chapter as EPS files.

Appendix A

Installing and Running the Software

A.1 Introduction

Both of the software packages used in this book — gawk and gnuplot — are "free"; that is, they are covered by the GNU's Not Unix (GNU) General Public License (Appendix B) and by the gnuplot license (Appendix C), respectively. These licenses allow the programs to be copied and distributed freely to others, which means that they can be placed on any number of computers (provided, of course, that one has right of access to do so). You are strongly encouraged to read the terms of the licenses, which are given in full in Appendix B and Appendix C.

This appendix provides instructions on how to install the gawk and gnuplot software on a personal computer running either GNU/Linux or Microsoft Windows. This is preceded by a brief introduction to some basic computing concepts.

A.2 Some Basic Computing Concepts

A.2.1 Operating System

An operating system (OS) is a software program or collection of software programs that controls the overall operation of a computer (British Computer Society 1998). The OS performs a number of essential tasks, including the management of the computer's hardware (its memory, disk drives, keyboard, mouse and VDU) and the provision of basic services to other software programs and to the computer user (access to files, network connections, scheduling and switching rapidly between different tasks [multi-tasking], and graphical and audio operations). The part of the OS that handles access to files and programs stored on the computer's disk drives is sometimes known as the disk operating system (DOS).

A computer user normally interacts with the OS in one of three ways: (i) by typing instructions on the keyboard, which are entered and interpreted via the command-

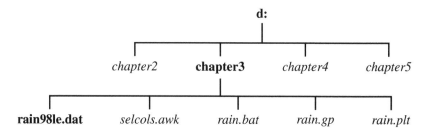

Figure A.1: Graphical representation of some files and directories on the CD-ROM, illustrating the full path to the file `rain98le.dat` under Microsoft Windows (`d:/chapter3/rain98le.dat`).

line interface (CLI); (ii) by point-and-click methods associated with a graphical user interface (GUI); or (iii) by voice control. Each of these approaches has its strengths and weaknesses. For the most part, the exercises in this book make use of the CLI approach.

A.2.2   Files, Directories, Paths and File Systems

In computing terms, a file is a collection of related data, which is treated as a single entity and assigned a name with which it is referenced. A file can contain textual, numerical, graphical or audio data, a software program, or various other types of information. A file can be stored either temporarily in the computer's memory, while the computer is switched on, or more permanently on one of its disk drives.

Depending on which OS is installed on the computer, the name of a file may be limited to just a few letters (a ... z, A ... Z) and digits (0 ... 9) or it may be possible to use a long sequence of letters, digits, blank spaces and other characters (e.g., `;$%()-@^_{}`) to describe the contents and purpose of the file. On many OSs, file names are prohibited from containing certain characters (`/\?*:|"<>`). One of the most basic, and therefore ubiquitous, file naming conventions is the so-called 8.3 system. In the 8.3 system, the filename consists of a base name of up to eight characters (a ... z, 0 ... 9, - and _) and an optional extension composed of a period (.) and up to three further characters; the filename extension is frequently used to indicate the nature and contents of the file (`.dat`, `.txt`, `.doc`). The 8.3 system is case insensitive, so that `MyFile.dat` and `myfile.dat` refer to the same file. This system is employed by Microsoft DOS, versions of Microsoft Windows prior to Windows 95, and the ISO 9660:1999 file system used on many CD-ROMs. It is also the file naming convention employed throughout this book.

A file system is used to organize the files stored on a computer disk drive so that they are uniquely referenced and can be accessed rapidly. Most file systems are hierarchical, with files organized into named directories (also known as folders). Each directory can store a collection of files, as well as other directories, which are known as sub-directories (Figure A.1 and Figure A.2). In the 8.3 file system, directory and sub-directory names can consist of up to eight alphanumeric characters (a ... z, 0 ... 9, - and _). Sub-directories can contain further sub-directories. Thus, the whole file

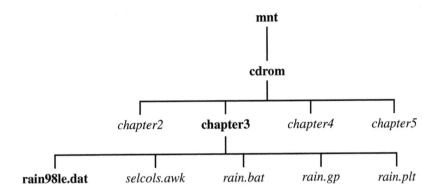

Figure A.2: Graphical representation of some files and directories on the CD-ROM, illustrating the full path to the file `rain98le.dat` under GNU/Linux (`/mnt/cdrom/chapter3 /rain98le.dat`).

system extends in a branching tree structure from the master or root directory. The collection of all directories and sub-directories on a given disk drive is often referred to as a directory tree, and a part of the tree is sometimes known as a sub-tree (British Computer Society 1998).

The location of a given file or directory in the file system is described by its path. A path can be either absolute or relative. The absolute (or full) pathname of a file or directory typically describes its location relative to the root directory of the disk drive on which it is stored. For instance, the full pathname of the file `rain98le.dat` on the CD-ROM may be `d:\chapter3\rain98le.dat` under Microsoft Windows (Figure A.1) and `/mnt/cdrom/chapter3rain98le.dat` under GNU/Linux (Figure A.2). The relative pathname usually describes the location of a file or directory with reference to the current working directory. Thus, assuming that the user is currently working in the `chapter4` directory on the CD-ROM, the relative pathname of the file `rain98le.dat` is `..\chapter3\rain98le.dat` under Microsoft Windows (Figure A.1) and `../chapter3rain98le.dat` under GNU/Linux (Figure A.2), where the symbol `..` means "move up one level in the directory tree".

A.3 Installing and Running gnuplot

gnuplot is an interactive data and function plotting utility designed for use on personal computers running GNU/Linux, UNIX, MacOS X or Microsoft Windows, in addition to a number of other operating systems. It supports many types of 2D and 3D plots (using lines, points, boxes, contours, vector fields and surfaces), as well as various other specialized plot types. gnuplot can direct its output to the computer screen, pen plotters, printers, and various forms of graphics file (EMF, LATEX, JPEG, PNG, PS, PDF and SVG).

The current official version of gnuplot, used throughout this book, is version 4.0, which was released on April 15, 2004. Copies of the software intended for use on computers running either GNU/Linux or Microsoft Windows are included on the

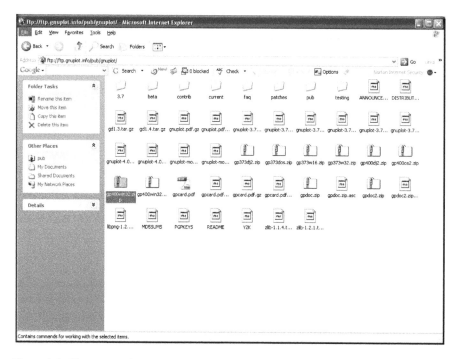

Figure A.3: The gnuplot download page viewed in a standard web browser (ftp://ftp.gnuplot. info/pub/gnuplot/).

CD-ROM; copies for these and other operating systems can also be obtained from the gnuplot homepage (http://www.gnuplot.info). Note that a new version of gnuplot (v4.1) is currently under development. This version can also be downloaded from the gnuplot homepage.

The remainder of this section describes how to download, install and run gnuplot on a personal computer running either GNU/Linux or a recent version of Microsoft Windows.

A.3.1 Instructions for Microsoft Windows

1. a) Installing from the gnuplot web site

It is recommended that you download the latest version of the software from the gnuplot web site, assuming that you have access to a reasonably fast network connection. Using a web browser, go to ftp://ftp.gnuplot.info/pub/gnuplot/ and locate the file marked gp400win32.zip (Figure A.3). This file is a compressed (zipped) archive, approximately 2.9 MB in size, which contains the gnuplot software, documentation, examples and associated files. Save this file on your hard disk, either by dragging and dropping it onto the Microsoft Windows desktop or by right-clicking the relevant icon and then selecting the Save as... option.

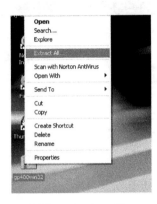

(a) Extracting the archive

(b) File extraction wizard.

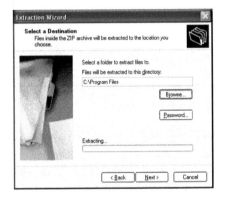

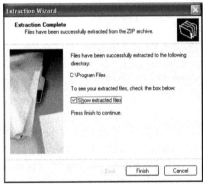

(c) Selecting where to extract the archive.

(d) Completing the extraction procedure.

Figure A.4: Installing gnuplot on Microsoft Windows XP. See text for details.

b) Installing from the CD-ROM

The zip archive `gp400win32.zip` is also available on the CD-ROM, and can be found in the `software` directory (`d:\software\gp400win32.zip`).

2. To unpack the archive, right-click on the `gp400win32.zip` icon and select the `Extract All...` option (Figure A.4a), which will start the file extraction wizard (Figure A.4b). Continue by clicking the `Next >` button.

3. You should now instruct the computer where to install the software. Using the `Browse...` button, select the `C:\Program Files` folder (Figure A.4c) if you have permission to do so; otherwise, accept the default location suggested by the installation wizard or choose another folder to which you have access. Now click `Next >` (Figure A.4c) and then `Finish` (Figure A.4d).

4. At this point, you should be presented with a file explorer, open at the folder

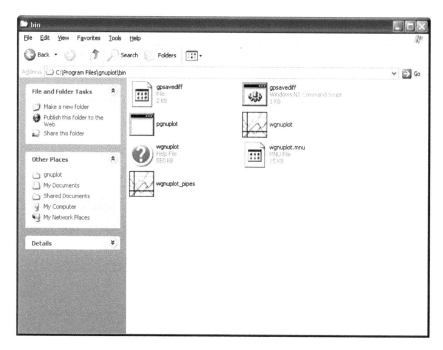

Figure A.5: Folder containing the gnuplot executable.

in which you unpacked the **gnuplot** archive (`C:\Program Files`). Using the file explorer, click open the `gnuplot` folder and, from there, the `bin` folder. The latter contains the **gnuplot** executable (i.e., program; Figure A.5). Locate the `wgnuplot` program icon, then right-click on this to create a shortcut. Drag the shortcut icon onto the Microsoft Windows desktop.

5. To start **gnuplot**, double-click the shortcut icon on the Microsoft Windows desktop to open the interface shown in Figure A.6. **gnuplot** is now ready to run.

A.3.2   Instructions for GNU/Linux

1. A version of **gnuplot** is included with most major distributions of GNU/Linux, in which case it is possible to start **gnuplot** by opening a GNU/Linux console, or an xterm, and then typing `gnuplot`. This should present you with a command-line interface similar to that shown in Figure A.7.

2. If **gnuplot** is already installed on your system, check which version is running. Figure A.7, for instance, indicates that the system concerned is using version 4.0 of the software. If your system reports an older version, especially if it precedes version 3.7, you should consider installing a newer one.

3. a) Installing from an RPM

   The procedure used to install **gnuplot** on GNU/Linux systems differs slightly from one distribution to another. Many distributions make use of the RPM

Figure A.6: GUI version of gnuplot for Microsoft Windows.

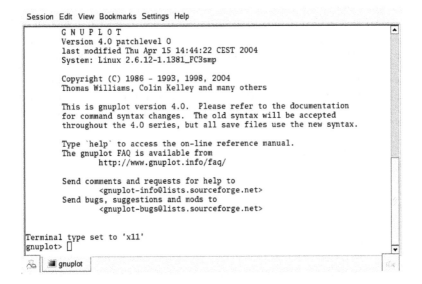

Figure A.7: Command-line interface for gnuplot running under GNU/Linux.

Package Manager (RPM) installation system, however, which is a convenient solution in most cases. To use this facility, search for the most recent version of gnuplot for your system on one of the RPM repository web sites (e.g., http://rpmfind.net/linux/RPM/) and download this file onto your computer. Log onto your computer as super-user (if you don't know what this is, then you probably don't have the necessary permissions to install new software on your system; instead, consult the system administrator) and type

```
rpm -i gnuplot-4.x.x-x.xxxx.rpm
```

in a GNU/Linux console or xterm, replacing x.x-x.xxx with the appropriate values for the version you have downloaded, for example,

```
rpm -i gnuplot-4.0.0-5.i586.rpm
```

Log out as super-user. gnuplot is now available on your system.

b) Installing from source

gnuplot can also be installed on a system by downloading and compiling the relevant source code. Although this is not a difficult procedure, it is generally not recommended for novice users. For more experienced users, however, the instructions are as follows:

- Download the latest version of the source code from the gnuplot web site (ftp.gnuplot.info/pub/gnuplot/gnuplot-4.0.0.tar.gz).

- Open a GNU/Linux console or xterm and extract the gnuplot archive by typing tar zxvf gnuplot-4.0.0.tar.gz (or something similar, if you have downloaded a slightly different version of the software).

- Change to the directory into which you extracted the gnuplot archive (cd gnuplot-4.0.0). Read the installation instructions contained in the file called INSTALL, checking any special requirements appropriate to your system.

- Compile and install the software by typing the following commands:

```
./configure                                              1
make                                                     2
make install                                             3
```

Note that numerous different options can be passed to the compiler via the ./configure command (see the INSTALL file), and that you will almost certainly need super-user access to the system to install gnuplot in either /usr/bin or /usr/local/bin.

- gnuplot should now be available on your system.

A.4 Installing and Running gawk

gawk is a high-level programming language that, among many other things, can be used to read, manipulate, reformat and output data contained in ASCII text files in a quick, convenient and relatively easy manner. There are several versions of awk: the one used here is the GNU Project (http://www.gnu.org/) implementation, known as gawk (Dougherty 1996, Robbins 2001).

Version 3.1.3 of gawk, which was released in June 2004, is used throughout this book. Copies of this software, designed for use with either GNU/Linux or Microsoft Windows, are included on the CD-ROM; copies for these and other operating systems can also be obtained from the gawk homepage (http://www.gnu.org/software/gawk/ gawk.html).

The remainder of this section describes how to download, install and run gawk on a personal computer running either the GNU/Linux operating system or a recent version of Microsoft Windows.

A.4.1 Instructions for Microsoft Windows

1. a) Installing from the GnuWin32 web site

It is recommended that you download gawk from the GnuWin32 web site (http: //gnuwin32.sourceforge.net/), if you have access to a reasonably fast network connection. To do so, point your web browser at http://gnuwin32.sourceforge. net/packages.html. This web page lists a number of software tools released under GPL and similar "open source" licenses, which are commonly available on GNU/Linux and UNIX systems, that have been ported for use on 32-bit computers running Microsoft Windows.

Locate the entry for gawk (marked "Gawk 3.1.3 pattern scanning and process-ing") and click on the link marked Setup on this line. This will direct you to a list of servers from which the software can be obtained. Choose the server closest to you geographically. You should then be presented with a pop-up window indicating the name of the file that you are about to download (e.g., gawk-3.1.3-2.exe), the size of the file (e.g., 6.35 MB), and three buttons that determine how or whether to proceed. Click on the button marked Run to start installing the software on your computer. You may be asked to confirm, via another pop-up window, that the installation procedure should indeed now run, in which case you should once again click the button marked Run.

You should now follow the instructions presented by the installation wizard (Figure A.8 and Figure A.9). In most cases, all that is required is to accept the default suggestions, and to click the Next button to proceed to the subsequent step. There are, however, two important exceptions to this general rule. The first is that you will be asked to confirm whether or not you agree to the terms and conditions of the gnuplot software license (given in full in Appendix C). To proceed you must click on the I accept the agreement option (Figure A.8b). The second is that you will be asked to identify where on your system the

(a) Starting the installing wizard.

(b) Accepting the license agreement.

(c) Deciding where to install.

(d) Selecting the components to install.

(e) Additional installation tasks.

(f) Starting the installation procedure.

Figure A.8: Some of the steps involved in installing **gawk** on Microsoft Windows.

Figure A.9: Further steps involved in installing **gawk** on Microsoft Windows.

software should be installed: it is recommended that you click on the `Browse` button and select the `C:\Program Files\GnuWin32` folder (Figure A.8c) if you have permission to do so; otherwise, accept the default location suggested by the installation wizard or choose another folder to which you have access.

b) Installing from the CD-ROM

The program used to set up **gawk** on your computer (`gawk-3.1.3-2.exe`) is also provided in the `software` directory on the CD-ROM (`d:\software\gawk -3.1.3-2.exe`). Use a file explorer to locate this file, double-click on its icon, and follow the instructions presented by the installation wizard, described above.

2. **gawk** is a command-line tool (i.e., it does not have a GUI that makes use of point-and-click techniques). To use **gawk**, therefore, one must run the CLI. In Microsoft Windows XP, the CLI (known in the past as the DOS prompt) can be found via `Start→All Programs→Accessories→Command Prompt`.

3. Before using **gawk**, however, Microsoft Windows must be told where to find the **gawk** utility (`gawk.exe`). The most convenient way of doing this is to add the folder in which `gawk.exe` is installed to the environment variable known as the `Path`. This can be achieved by right-clicking on the `My Computer` icon in the `Start` menu (Figure A.10a), and from there selecting the `Properties` option. Select the `Advanced` tab in the `Properties` window, and then click on the `Environment Variables` button. A new window should appear (Figure A.10b). Search in the lower pane of this window, marked `System variables`, for the `Path` entry. Highlight this line in the textbox and then click on the `Edit` button. This process should produce a further pop-up window entitled `Edit System Variable` (Figure A.10c). In the text entry box that is marked `Variable value:`, add the following text to the end of the current path `;C:\ Program Files\GnuWin32\bin` (or another pathname, if you have installed **gawk** elsewhere on your system) and then click the `OK` button.

4. The **gawk** utility should now be available on your system. To test this, open a command-line window (see above) and type the command `gawk`. You should be presented with a response similar to that shown in Figure A.11.

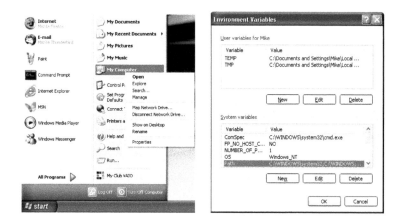

(a) Accessing the Windows environ-   (b) The `Path` environment variable.
ment variables.

(c) Modifying the `Path`.

Figure A.10: Instructing Microsoft Windows where to find **gawk** on the system.

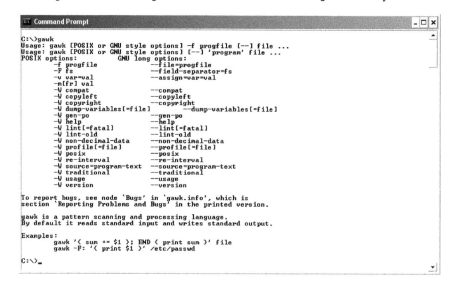

Figure A.11: Running **gawk** under Microsoft Windows.

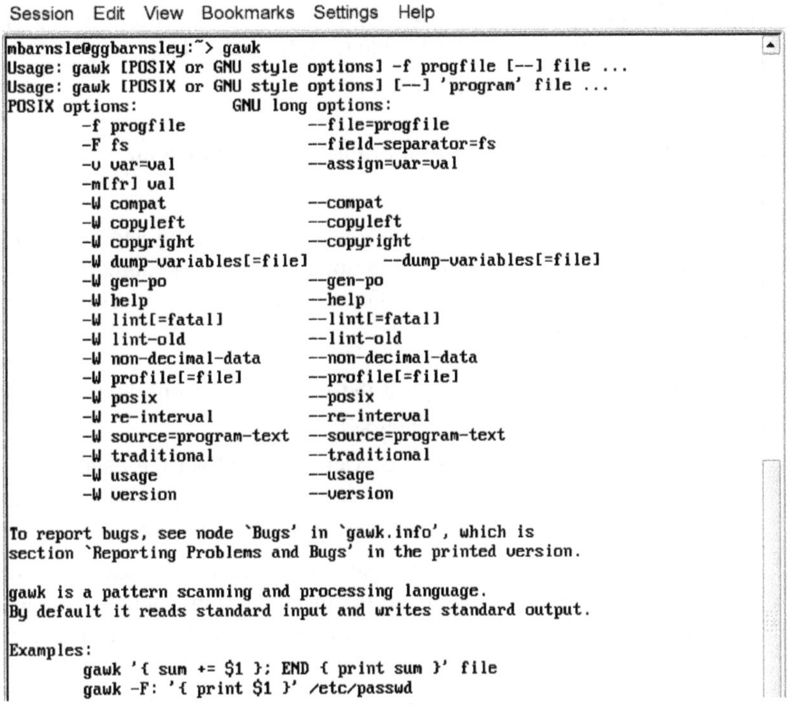

```
 Session  Edit  View  Bookmarks  Settings  Help

mbarnsle@ggbarnsley:~> gawk
Usage: gawk [POSIX or GNU style options] -f progfile [--] file ...
Usage: gawk [POSIX or GNU style options] [--] 'program' file ...
POSIX options:            GNU long options:
        -f progfile               --file=progfile
        -F fs                     --field-separator=fs
        -v var=val                --assign=var=val
        -m[fr] val
        -W compat                 --compat
        -W copyleft               --copyleft
        -W copyright              --copyright
        -W dump-variables[=file]         --dump-variables[=file]
        -W gen-po                 --gen-po
        -W help                   --help
        -W lint[=fatal]           --lint[=fatal]
        -W lint-old               --lint-old
        -W non-decimal-data       --non-decimal-data
        -W profile[=file]         --profile[=file]
        -W posix                  --posix
        -W re-interval            --re-interval
        -W source=program-text    --source=program-text
        -W traditional            --traditional
        -W usage                  --usage
        -W version                --version

To report bugs, see node `Bugs' in `gawk.info', which is
section `Reporting Problems and Bugs' in the printed version.

gawk is a pattern scanning and processing language.
By default it reads standard input and writes standard output.

Examples:
        gawk '{ sum += $1 }; END { print sum }' file
        gawk -F: '{ print $1 }' /etc/passwd
```

Figure A.12: gawk running in a GNU/Linux console.

A.4.2 Instructions for GNU/Linux

1. A version of gawk is included with most major distributions of GNU/Linux, in which case it is possible to use the gawk utility by typing gawk in a GNU/Linux console or xterm. Running gawk in this way without further instructions causes it to print out basic information on how it should be used (Figure A.12). The version of gawk available on the system can be checked by typing gawk --version on the command line.

2. If gawk is not already available on your system, you can install it in one of two ways described below.

   a) Installing from an RPM

   The procedure used to install gawk from an RPM on GNU/Linux systems is much the same as that described for gnuplot in Section A.3.2. Search for the most recent version of gawk for your system on one of the RPM repository web sites (e.g., http://rpmfind.net/linux/RPM/) and download this file onto your computer. Log onto your computer as super-user (if you don't know what this is, then you probably don't have the necessary permissions to install new software on your system; instead, consult the system administrator) and type

   ```
   rpm -i gawk-3.1.x.xxxx.rpm
   ```

in a GNU/Linux console or xterm, replacing x.x.x.xxx with the appropriate values for the version you have downloaded, for example,

```
rpm -i gawk-3.1.5.i586.rpm
```

Log out as super-user. **gawk** is now available on your system.

b) Installing from source

gawk can also be installed on a system by downloading and compiling the relevant source code. Although this is not a difficult procedure, it is generally not recommended for novice users. For more experienced users, however, the instructions are as follows:

- Download the latest version of the source code from the **gawk** web site (ftp.gnu.org/pub/gawk-3.1.5.tar.gz).

- Open a GNU/Linux console or xterm and extract the **gawk** archive by typing tar zxvf gawk-3.1.5.tar.gz (or something similar, if you have downloaded a slightly different version of the software).

- Change to the directory into which you extracted the **gawk** archive; for instance, cd gawk-3.1.5. Read the installation instructions contained in the file called INSTALL, checking any special requirements appropriate to your system.

- Compile and install the software by typing the following commands:

```
./configure                                                        1
make                                                               2
make install                                                       3
```

- Note that numerous different options can be passed to the compiler via the ./configure command (see the INSTALL file), and that you will almost certainly need super-user access to the system to install **gawk** in either /usr/bin or /usr/local/bin.

- **gawk** should now be available on your system.

Appendix B

GNU General Public License

Version 2, June 1991

Copyright (©) 1989, 1991 Free Software Foundation, Inc.

59 Temple Place – Suite 330, Boston, MA 02111-1307, USA

B.1 PREAMBLE

The licenses for most software are designed to take away your freedom to share and change it. By contrast, the GNU General Public License is intended to guarantee your freedom to share and change free software—to make sure the software is free for all its users. This General Public License applies to most of the Free Software Foundation's software and to any other program whose authors commit to using it. (Some other Free Software Foundation software is covered by the GNU Library General Public License instead.) You can apply it to your programs, too.

When we speak of free software, we are referring to freedom, not price. Our General Public Licenses are designed to make sure that you have the freedom to distribute copies of free software (and charge for this service if you wish), that you receive source code or can get it if you want it, that you can change the software or use pieces of it in new free programs; and that you know you can do these things.

To protect your rights, we need to make restrictions that forbid anyone to deny you these rights or to ask you to surrender the rights. These restrictions translate to certain responsibilities for you if you distribute copies of the software, or if you modify it.

For example, if you distribute copies of such a program, whether gratis or for a fee, you must give the recipients all the rights that you have. You must make sure that they, too, receive or can get the source code. And you must show them these terms so they know their rights.

We protect your rights with two steps: (1) copyright the software, and (2) offer you this license which gives you legal permission to copy, distribute and/or modify the software.

Also, for each author's protection and ours, we want to make certain that everyone understands that there is no warranty for this free software. If the software is modified by someone else and passed on, we want its recipients to know that what they have is not the original, so that any problems introduced by others will not reflect on the original authors' reputations.

Finally, any free program is threatened constantly by software patents. We wish to avoid the danger that redistributors of a free program will individually obtain patent licenses, in effect making the program proprietary. To prevent this, we have made it clear that any patent must be licensed for everyone's free use or not licensed at all.

The precise terms and conditions for copying, distribution and modification follow.

B.2   TERMS AND CONDITIONS FOR COPYING, DISTRIBUTION AND MODIFICATION

1. This License applies to any program or other work which contains a notice placed by the copyright holder saying it may be distributed under the terms of this General Public License. The "Program", below, refers to any such program or work, and a "work based on the Program" means either the Program or any derivative work under copyright law; that is to say, a work containing the Program or a portion of it, either verbatim or with modifications and/or translated into another language. (Hereinafter, translation is included without limitation in the term "modification".) Each licensee is addressed as "you".

   Activities other than copying, distribution and modification are not covered by this License; they are outside its scope. The act of running the Program is not restricted, and the output from the Program is covered only if its contents constitute a work based on the Program (independent of having been made by running the Program). Whether that is true depends on what the Program does.

2. You may copy and distribute verbatim copies of the Program's source code as you receive it, in any medium, provided that you conspicuously and appropriately publish on each copy an appropriate copyright notice and disclaimer of warranty; keep intact all the notices that refer to this License and to the absence of any warranty; and give any other recipients of the Program a copy of this License along with the Program.

   You may charge a fee for the physical act of transferring a copy, and you may at your option offer warranty protection in exchange for a fee.

3. You may modify your copy or copies of the Program or any portion of it, thus forming a work based on the Program, and copy and distribute such modifica-

tions or work under the terms of Section 1 above, provided that you also meet all of these conditions:

(a) You must cause the modified files to carry prominent notices stating that you changed the files and the date of any change.

(b) You must cause any work that you distribute or publish, that in whole or in part contains or is derived from the Program or any part thereof, to be licensed as a whole at no charge to all third parties under the terms of this License.

If the modified program normally reads commands interactively when run, you must cause it, when started running for such interactive use in the most ordinary way, to print or display an announcement including an appropriate copyright notice and a notice that there is no warranty (or else, saying that you provide a warranty) and that users may redistribute the program under these conditions, and telling the user how to view a copy of this License. (Exception: if the Program itself is interactive but does not normally print such an announcement, your work based on the Program is not required to print an announcement.)

These requirements apply to the modified work as a whole. If identifiable sections of that work are not derived from the Program, and can be reasonably considered independent and separate works in themselves, then this License, and its terms, do not apply to those sections when you distribute them as separate works. But when you distribute the same sections as part of a whole which is a work based on the Program, the distribution of the whole must be on the terms of this License, whose permissions for other licensees extend to the entire whole, and thus to each and every part regardless of who wrote it.

Thus, it is not the intent of this section to claim rights or contest your rights to work written entirely by you; rather, the intent is to exercise the right to control the distribution of derivative or collective works based on the Program.

In addition, mere aggregation of another work not based on the Program with the Program (or with a work based on the Program) on a volume of a storage or distribution medium does not bring the other work under the scope of this License.

4. You may copy and distribute the Program (or a work based on it, under Section 2) in object code or executable form under the terms of Sections 1 and 2 above provided that you also do one of the following:

(a) Accompany it with the complete corresponding machine-readable source code, which must be distributed under the terms of Sections 1 and 2 above on a medium customarily used for software interchange; or,

(b) Accompany it with a written offer, valid for at least three years, to give any third party, for a charge no more than your cost of physically performing source distribution, a complete machine-readable copy of the corresponding source code, to be distributed under the terms of Sections 1 and 2 above on a medium customarily used for software interchange; or,

(c) Accompany it with the information you received as to the offer to distribute corresponding source code. (This alternative is allowed only for non-commercial distribution and only if you received the program in object code or executable form with such an offer, in accord with Subsection b above.)

The source code for a work means the preferred form of the work for making modifications to it. For an executable work, complete source code means all the source code for all modules it contains, plus any associated interface definition files, plus the scripts used to control compilation and installation of the executable. However, as a special exception, the source code distributed need not include anything that is normally distributed (in either source or binary form) with the major components (compiler, kernel, and so on) of the operating system on which the executable runs, unless that component itself accompanies the executable.

If distribution of executable or object code is made by offering access to copy from a designated place, then offering equivalent access to copy the source code from the same place counts as distribution of the source code, even though third parties are not compelled to copy the source along with the object code.

5. You may not copy, modify, sublicense, or distribute the Program except as expressly provided under this License. Any attempt otherwise to copy, modify, sublicense or distribute the Program is void, and will automatically terminate your rights under this License. However, parties who have received copies, or rights, from you under this License will not have their licenses terminated so long as such parties remain in full compliance.

6. You are not required to accept this License, since you have not signed it. However, nothing else grants you permission to modify or distribute the Program or its derivative works. These actions are prohibited by law if you do not accept this License. Therefore, by modifying or distributing the Program (or any work based on the Program), you indicate your acceptance of this License to do so, and all its terms and conditions for copying, distributing or modifying the Program or works based on it.

7. Each time you redistribute the Program (or any work based on the Program), the recipient automatically receives a license from the original licensor to copy, distribute or modify the Program subject to these terms and conditions. You may not impose any further restrictions on the recipients' exercise of the rights granted herein. You are not responsible for enforcing compliance by third parties to this License.

8. If, as a consequence of a court judgment or allegation of patent infringement or for any other reason (not limited to patent issues), conditions are imposed on you (whether by court order, agreement or otherwise) that contradict the conditions of this License, they do not excuse you from the conditions of this License. If you cannot distribute so as to satisfy simultaneously your obligations

under this License and any other pertinent obligations, then as a consequence you may not distribute the Program at all. For example, if a patent license would not permit royalty-free redistribution of the Program by all those who receive copies directly or indirectly through you, then the only way you could satisfy both it and this License would be to refrain entirely from distribution of the Program.

If any portion of this section is held invalid or unenforceable under any particular circumstance, the balance of the section is intended to apply and the section as a whole is intended to apply in other circumstances.

It is not the purpose of this section to induce you to infringe any patents or other property right claims or to contest validity of any such claims; this section has the sole purpose of protecting the integrity of the free software distribution system, which is implemented by public license practices. Many people have made generous contributions to the wide range of software distributed through that system in reliance on consistent application of that system; it is up to the author/donor to decide if he or she is willing to distribute software through any other system and a licensee cannot impose that choice.

This section is intended to make thoroughly clear what is believed to be a consequence of the rest of this License.

9. If the distribution and/or use of the Program is restricted in certain countries either by patents or by copyrighted interfaces, the original copyright holder who places the Program under this License may add an explicit geographical distribution limitation excluding those countries, so that distribution is permitted only in or among countries not thus excluded. In such case, this License incorporates the limitation as if written in the body of this License.

10. The Free Software Foundation may publish revised and/or new versions of the General Public License from time to time. Such new versions will be similar in spirit to the present version, but may differ in detail to address new problems or concerns.

    Each version is given a distinguishing version number. If the Program specifies a version number of this License which applies to it and "any later version", you have the option of following the terms and conditions either of that version or of any later version published by the Free Software Foundation. If the Program does not specify a version number of this License, you may choose any version ever published by the Free Software Foundation.

11. If you wish to incorporate parts of the Program into other free programs whose distribution conditions are different, write to the author to ask for permission. For software which is copyrighted by the Free Software Foundation, write to the Free Software Foundation; we sometimes make exceptions for this. Our decision will be guided by the two goals of preserving the free status of all derivatives of our free software and of promoting the sharing and reuse of software generally.

B.3   NO WARRANTY

1. Because the program is licensed free of charge, there is no warranty for the program, to the extent permitted by applicable law. Except when otherwise stated in writing the copyright holders and/or other parties provide the program "as is" without warranty of any kind, either expressed or implied, including, but not limited to, the implied warranties of merchantability and fitness for a particular purpose. The entire risk as to the quality and performance of the program is with you. Should the program prove defective, you assume the cost of all necessary servicing, repair or correction.

2. In no event unless required by applicable law or agreed to in writing will any copyright holder, or any other party who may modify and/or redistribute the program as permitted above, be liable to you for damages, including any general, special, incidental or consequential damages arising out of the use or inability to use the program (including but not limited to loss of data or data being rendered inaccurate or losses sustained by you or third parties or a failure of the program to operate with any other programs), even if such holder or other party has been advised of the possibility of such damages.

END OF TERMS AND CONDITIONS

Appendix C

Gnuplot License

Copyright 1986–1993, 1998, 2004 Thomas Williams, Colin Kelley

Permission to use, copy, and distribute this software and its documentation for any purpose with or without fee is hereby granted, provided that the above copyright notice appear in all copies and that both that copyright notice and this permission notice appear in supporting documentation.

Permission to modify the software is granted, but not the right to distribute the complete modified source code. Modifications are to be distributed as patches to the released version. Permission to distribute binaries produced by compiling modified sources is granted, provided you

1. distribute the corresponding source modifications from the released version in the form of a patch file along with the binaries,

2. add special version identification to distinguish your version in addition to the base release version number,

3. provide your name and address as the primary contact for the support of your modified version, and

4. retain our contact information in regard to use of the base software.

Permission to distribute the released version of the source code along with corresponding source modifications in the form of a patch file is granted with the same provisions 2 through 4 for binary distributions.

This software is provided "as is" without express or implied warranty to the extent permitted by applicable law.

Appendix D

Standards

D.1   International Standard Date and Time Notation

The International Organization for Standardization has published ISO 8610:2004, an international standard for the numerical representation of information on dates and times (http://www.iso.org/iso/en/prods-services/popstds/datesandtime.html). The aim of ISO 8610:2004 is to remove the potential for ambiguity and confusion that arises when different national standards are employed. Extracts from ISO 8610:2004 are given in Table D.1, where YYYY denotes the four-digit calendar year, MM is the two-digit ordinal number of the calendar month (01 refers to January and 12 refers to December), DD is the two-digit ordinal number of the day of the month (01 to 31), hh is the two-digit hour (00 to 23), mm is the two-digit minute (00 to 59), ss is the two-digit second (00 to 59) and .s is one or more digits denoting a decimal fraction of a second. Note that an uppercase T, known as the time designator, is used to separate the date and time components. Unless a specific time zone designator is employed, times are expressed in Coordinated Universal Time (UTC).

Table D.1: ISO 8601:2004 date and time notations.

Format	Notation	Example
Basic	YYYYMM	200512
	YYYYMMDD	20051224
	YYYYMMDDThhmm	20051224T2359
	YYYYMMDDThhmmss.s	20051224T2359.9
Extended	YYYY-MM	2005-12
	YYYY-MM-DD	2005-12-24
	YYYY-MM-DDThh:mm	2005-12-24T23:59
	YYYY-MM-DDThh:mm:ss.s	2005-12-24T23:59:59.9

Table D.2: SI Base Units

Base Quantity	SI Base Unit	
	Name	Symbol
length	meter	m
mass	kilogram	km
time	second	s
electric current	ampere	A
thermodynamic temperature	kelvin	K
amount of substance	mole	mol
luminous intensity	candela	cd

Two versions of the standard notation are recognized by ISO 8610:2004 (Table D.1). The first is known as the "basic" format, where the components of the date and time information are simply concatenated together (e.g., `20061224T235959`, meaning one second before midnight on December 24, 2006). The second is referred to as the "extended" format, where hyphens (`-`) are used to separate the year, month and day components of the date, and colons (`:`) are used to separate the hour, minute and second of the time (e.g., `2006-12-24T23:59:59`). The former offers the advantage of compactness; the latter is arguably easier for humans to read.

D.2   SI Units (Systéme International d'Unités)

The International System of Units is an international system for measurement that is used in the fields of science and commerce (Taylor 1995). Universally referred to as SI units (from the French *Systéme International d'Unités*), the system was first established in 1960. SI units are divided into three broad categories, namely (i) base units, (ii) derived units and (iii) supplementary units. The SI base units are listed in Table D.2, while examples of the derived and supplementary units are given in Table D.3. The International System of Units also defines SI prefixes that describe common multiples of the SI units (Table D.4).

There are rules and conventions for the proper usage of SI units in text documents and scientific figures (Taylor 1995). In particular, statements of the form "10 m × 5 m" are preferred to "10 × 5 m" because it makes it clear that the value 10 is measured in units of meters. Similarly, when referring to a range of values (in this example, temperature measured in degree Celsius), the form "20 °C to 30 °C" is preferred to "20 °C–30 °C" or "20 to 30 °C"; the form "(20 to 30) °C" is also acceptable. The SI documentation also specifies the formats that should be used to represent unit symbols and unit names. For example, the following forms are all acceptable: "kg/m^3", "$kg \cdot m^{-3}$" or "kilogram per cubic meter". Further information on these and other aspects of the SI units and nomenclature can be found in Taylor (1995).

Table D.3: SI derived and supplementary units

Derived Quantity	Name	Symbol	Expressed in Base SI Units	Expressed in SI Base units
area	square meters		m^2	m^2
volume	cubic meters		m^3	m^3
speed, velocity	meters per second		$m \cdot s^{-1}$	$m \cdot s^{-1}$
plane angle	radian	rad	$m \cdot m^{-1}$	$m \cdot m^{-1}$
solid angle	steradian	sr	$m^2 \cdot m^{-2}$	$m^2 \cdot m^{-2}$
frequency	hertz	Hz	s^{-1}	s^{-1}
force	newton	N	$m \cdot kg$	$m \cdot kg$
pressure	pascal	Pa	$N \cdot m^{-2}$	$m^{-1} \cdot kg \cdot s^{-2}$
energy	joule	J	$N \cdot m$	$m^2 \cdot kg \cdot s^{-2}$
power, radiant flux	watt	W	$J \cdot s^{-1}$	$m^2 \cdot kg \cdot s^{-3}$
Celsius temperature	degrees Celsius	°C	—	$s \cdot A$
heat flux density, irradiance	watt per square meter		$W \cdot m^{-2}$	—
radiant intensity	watt per steradian		$W \cdot sr^{-1}$	—
radiance	watt per square meter per steradian		$W \cdot m^{-2} \cdot sr^{-1}$	—

Table D.4: SI prefixes

Factor	Name	Symbol
10^{24}	yotta	Y
10^{21}	zetta	Z
10^{18}	exa	E
10^{15}	peta	P
10^{12}	tera	T
10^{9}	giga	G
10^{6}	mega	M
10^{3}	kilo	k
10^{2}	hecto	h
10^{1}	deka	da
10^{-1}	deci	d
10^{-2}	centi	c
10^{-3}	milli	m
10^{-6}	micro	μ
10^{-9}	nano	n
10^{-12}	pico	p
10^{-15}	femto	f
10^{-18}	atto	a
10^{-21}	zepto	z
10^{-24}	yocto	y

Appendix E

Solutions to Exercises

Exercise 2.1

```
plot 'le98temp.dat' using 3:1                                        1
```

Exercise 2.2

```
reset                                                                1
unset key                                                            2
set style data points                                               3
set xlabel "Easting, m"                                              4
set ylabel "Northing, m"                                             5
set zlabel "Elevation, m"                                            6
set xtics 295000, 1000, 300000                                       7
set ytics 315000, 1000, 320000                                       8
set ztics 250, 100, 550                                              9
splot 'efyrnwy.dem' u 1:2:3                                         10
pause -1 "Press the return key to continue"                        11
                                                                    12
set view 45, 45, 1, 1                                              13
replot                                                              14
pause -1 "Press the return key to continue"                        15
                                                                    16
set view 45, 60, 1, 1                                              17
replot                                                              18
pause -1 "Press the return key to continue"                        19
                                                                    20
set view 60, 60, 1, 1                                              21
replot                                                              22
pause -1  "Press the return key to continue"                       23
```

Exercise 2.3

```
reset                                                         1
unset key                                                     2
set style data lines                                          3
set xlabel "Easting, m"                                       4
set ylabel "Northing, m"                                      5
set zlabel "Elevation, m"                                     6
set xtics 295000, 1000, 300000                                7
set ytics 315000, 1000, 320000                                8
set ztics 250, 100, 550                                       9
set dgrid3d 51,51,16                                          10
set style line 9                                              11
set pm3d at s hidden3d 9                                      12
splot 'efyrnwy.dem' u 1:2:3 with pm3d                         13
pause -1 "Press the return key to continue"                   14
                                                              15
set pm3d at b                                                 16
splot 'efyrnwy.dem' u 1:2:3 with pm3d                         17
pause -1 "Press the return key to continue"                   18
                                                              19
set pm3d at t                                                 20
replot                                                        21
pause -1 "Press the return key to continue"                   22
                                                              23
set pm3d at bst                                               24
replot                                                        25
pause -1 "Press the return key to continue"                   26
```

Exercise 2.4

```
reset                                                         1
unset key                                                     2
set xlabel "Easting, m"                                       3
set ylabel "Northing, m"                                      4
set zlabel "Elevation, m"                                     5
set xtics 295000, 1000, 300000                                6
set ytics 315000, 1000, 320000                                7
set style data lines                                          8
set dgrid3d 51,51,16                                          9
unset surface                                                 10
set contour base                                              11
set cntrparam levels discrete 300, 325, 350, 375, 400, \     12
    425, 450, 475, 500                                        13
set terminal push                                             14
set terminal table                                            15
set output 'contours.dat'                                     16
splot 'efyrnwy.dem' using 1:2:3                               17
set output                                                    18
```

```
set terminal pop                                          19
plot 'contours.dat' index 0 u 1:2, \                      20
     'contours.dat' index 1 u 1:2, \                       21
     'contours.dat' index 2 u 1:2, \                       22
     'contours.dat' index 3 u 1:2, \                       23
     'contours.dat' index 4 u 1:2, \                       24
     'contours.dat' index 5 u 1:2, \                       25
     'contours.dat' index 6 u 1:2, \                       26
     'contours.dat' index 7 u 1:2, \                       27
     'contours.dat' index 8 u 1:2                          28
```

Exercise 2.5

```
reset                                                     1
set key top right box                                     2
set xdata time                                            3
set timefmt "%Y%m%d%H%M"                                  4
set format x "%b"                                         5
set xlabel "Month, 1998"                                  6
set xrange ["199801010000":"199901010000"]               7
set ylabel "Temperature, degrees Celsius"                8
set yrange [-5:30]                                        9
set y2label "Precipitation, mm"                           10
set ytics nomirror                                        11
set y2tics nomirror                                       12
set term png                                              13
set output "myplot2.png"                                  14
plot 'le98temp.dat' u 2:3 t "Maximum temperature" w lp, \  15
     'le98rain.dat' u 2:3 axes x1y2 t "Precipitation" w i  16
set output                                                17
```

Exercise 3.1

```
gawk '{print $4, $5, $6, $7, $9/10}' rain98le.dat          1
```

```
YEAR MONTH DAY END_HOUR 0                                  1
1998 1 1 900 0                                             2
1998 1 1 2100 14.6                                         3
1998 1 2 900 5.2                                           4
1998 1 2 2100 1.6                                          5

1998 12 29 2100 0.2                                       713
1998 12 30 900 2.4                                        714
1998 12 30 2100 0                                         715
1998 12 31 900 0                                          716
1998 12 31 2100 0.8                                       717
```

```
gawk '{print $4 $5 $6 $7 $9/10}' rain98le.dat              2
```

```
YEARMONTHDAYEND_HOUR0                                          1
1998119000                                                     2
199811210014.6                                                 3
1998129005.2                                                   4
19981221001.6                                                  5

1998122921000.2                                                713
199812309002.4                                                 714
1998123021000                                                  715
199812319000                                                   716
1998123121000.8                                                717
```

Exercise 3.2

```
gawk -f selcols3.awk rain98le.dat                              1
```

```
199811900  0.000000                                            1
1998112100  14.600000                                          2
199812900  5.200000                                            3
1998122100  1.600000                                            4
199813900  35.800000                                           5

19981230900  2.400000                                          713
199812302100  0.000000                                         714
19981231900  0.000000                                          715
199812312100  0.800000                                         716
```

Program E.1: ex3-2.awk

```
(NR>1){printf("%i%i%i%i %f", $4, $5, $6, $7, $9/10.0)}        1
```

```
199811900  0.0000001998112100  14.600000199812900  5.200000..
```

Exercise 3.3

Program E.2: seltemp.awk

```
(NR>1){                                                        1
  printf("%3i %04i%02i%02iT%04i %7.2f %7.2f\n", \              2
    NR, $4, $5, $6, $7, $9/10.0, $10/10.0);                    3
}                                                              4
```

```
gawk -f seltemp.awk temp98le.dat > temp98le.out                    1
```

```
    2 19980101T0900    1.80    -0.30                                1
    3 19980101T2100    7.70     1.60                                2
    4 19980102T0900    6.80     3.90                                3
    5 19980102T2100    6.50     3.80                                4
    6 19980103T0900    9.70     4.80                                5

  714 19981230T0900    9.00     6.10                              713
  715 19981230T2100    7.40     4.10                              714
  716 19981231T0900    5.40     3.70                              715
  717 19981231T2100    7.00     4.80                              716
```

```
reset                                                              1
set xdata time                                                     2
set timefmt "%Y%m%dT%H%M"                                          3
set format x "%d/%m"                                               4
set xlabel "Day/Month, 1998"                                       5
set ylabel "Temperature, t/degree Celsius"                         6
set xrange ["19980401T0000":"19980701T0000"]                       7
set yrange [-5:30]                                                 8
set key top right box                                              9
set data style linespoints                                        10
plot 'le98temp.out' u 2:3 t "Maximum temperature", \              11
     'le98temp.out' u 2:4 t "Minimum temperature"                 12
```

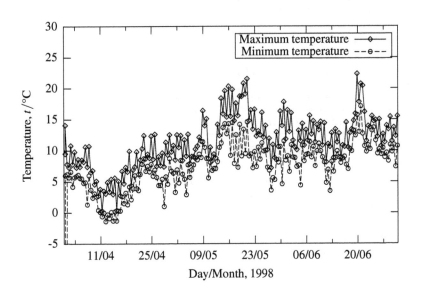

Figure E.1: Output from Exercise 3.3

Exercise 4.1

Program E.3: winddir.awk

```
(NR>1 && $9!=-999){                                              1
  if($9>345   $9<=15){++N}                                       2
  if($9>15 && $9<=45){++NNE}                                     3
  if($9>45 && $9<=75){++ENE}                                     4
  if($9>75 && $9<=105){++E}                                      5
  if($9>105 && $9<=135){++ESE}                                   6
  if($9>135 && $9<=165){++SSE}                                   7
  if($9>165 && $9<=195){++S}                                     8
  if($9>195 && $9<=225){++SSW}                                   9
  if($9>225 && $9<=255){++WSW}                                   10
  if($9>255 && $9<=285){++W}                                     11
  if($9>285 && $9<=315){++WNW}                                   12
  if($9>315 && $9<=345){++NNW}                                   13
  ++num_obs;                                                     14
}                                                                15
END{                                                             16
  printf("%3i %5.2f\n", 0, N/num_obs);                           17
  printf("%3i %5.2f\n", 30, NNE/num_obs);                        18
  printf("%3i %5.2f\n", 60, ENE/num_obs);                        19
  printf("%3i %5.2f\n", 90, E/num_obs);                          20
  printf("%3i %5.2f\n", 120, ESE/num_obs);                       21
  printf("%3i %5.2f\n", 150, SSE/num_obs);                       22
  printf("%3i %5.2f\n", 180, S/num_obs);                         23
  printf("%3i %5.2f\n", 210, SSW/num_obs);                       24
  printf("%3i %5.2f\n", 240, WSW/num_obs);                       25
  printf("%3i %5.2f\n", 270, W/num_obs);                         26
  printf("%3i %5.2f\n", 300, WNW/num_obs);                       27
  printf("%3i %5.2f\n", 330, NNW/num_obs);                       28
}                                                                29
```

```
gawk -f winddir.awk wind98le.dat > wind98le.dir          1
```

```
# Create a standard x-y plot                                       1
unset key                                                          2
set xlabel "Wind direction (degrees relative to north)"            3
set ylabel "Relative frequency"                                    4
plot 'wind98le.dir' w i lw 2                                       5
pause -1 "Press the return key to continue"                        6
# Create a polar plot                                              7
set angles degrees       # Angles measured in degrees              8
set polar                # Produce a polar plot                    9
set grid polar 15.0      # Mark grid at 15 degree intervals       10
unset border             # Turn off normal borders                11
set xtics axis           # Plot the xtics along the x-axis        12
set ytics axis           # Plot the ytics along the y=axis        13
```

```
set xrange [-0.30:0.30]                                    14
set yrange [-0.30:0.30]                                    15
set size square          # Make the output square          16
plot 'wind98le.dir' u (90-$1):($2) w i lw 2                17
```

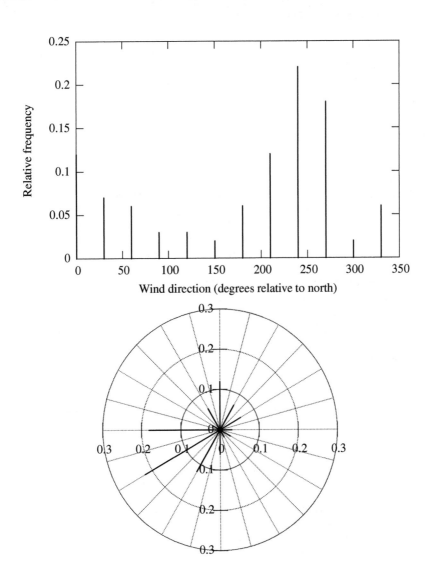

Figure E.2: Output from Exercise 4.1

Exercise 4.2

```
phi_prime(u_prime,c,k)=exp((-1.0*(u_prime/c)**k))          1
freq_cut_in=phi_prime(3.0,5.32883,1.55037)                 2
freq_cut_out=phi_prime(20.0,5.32883,1.55037)               3
print freq_cut_in-freq_cut_out                             4
```

```
0.662989225124349
```

Exercise 4.3

```
unset key                                                            1
set xtics auto                                                       2
set xrange [0:20]                                                    3
rho=1.225                  # Air density kg/m^3 at sea level          4
radius=3.0                 # Rotor radius in meters                   5
area=pi*(radius**2)        # Area swept by turbine rotors             6
Cp=0.4                     # Power coefficient                        7
Ng=0.75                    # Generator efficiency                     8
Nb=0.9                     # Mechanical efficiency                    9
set dummy u                                                          10
Power(u)=area*(rho*(u**3)/2)/1000                                    11
set xlabel "Wind speed, u_0/m.s^{-1}"                               12
set ylabel "Wind power density, P/kW.m^{-2}"                        13
plot Power(u)*Cp*Ng*Nb/area lw 2                                    14
```

Exercise 4.4

Program E.4: meanpwr.awk

```
{                                                                    1
  sum_power+=$2;                                                     2
  ++sum_hours;                                                       3
}                                                                    4
                                                                     5
END{                                                                 6
  printf("Mean output while WECS active: %6.4f kW.h\n", \            7
    sum_power/sum_hours);                                            8
  printf("Mean output throughout year:   %6.4f kW.h\n", \            9
    sum_power/(365*24));                                            10
}                                                                   11
```

```
gawk -f meanpwr.awk wind98le.pwr > meanpwr.out
```

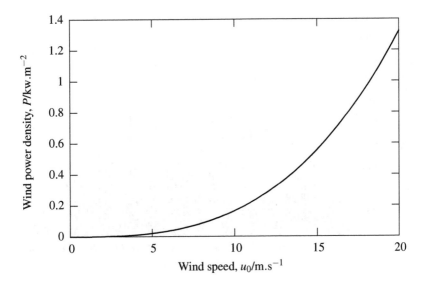

Figure E.3: Output from Exercise 4.3

```
Mean output while WECS active: 1.9322 kW.h
Mean output throughout year:   1.1856 kW.h
```

Exercise 5.1

Program E.5: ex5-1.awk

```
(NR>1 && $7==1200 && $9>=0){                                    1
  printf("%4i%02i%02iT%04i %4i\n", $4, $5, $6, $7, $9);         2
}                                                               3
```

```
gawk -f ex5-1.awk radt98le.dat > ex5-1.dat
```

```
reset                                                           1
unset key                                                       2
set xdata time                                                  3
set timefmt "%Y%m%dT%H%M"                                       4
set format x "%b"                                               5
set xlabel "Month, 1998"                                        6
set ylabel "Solar irradiance, W.h.m^-2"                         7
set xrange ["19980101T1200":"19990101T1200"]                    8
plot 'ex5-1.dat' u 1:2 w i                                      9
```

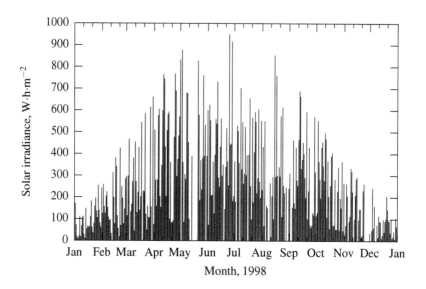

Figure E.4: Output from Exercise 5.1

Exercise 5.2

Program E.6: ex5-2.awk

```
BEGIN{                                                                    1
    latitude=52.756;              # Latitude (degrees)                    2
    E_0=1380.0;                   # Exo-atmos. solar irradiance           3
    tau=0.7;                      # Atmos. transmittance                  4
    p_alt=1000;                   # Atmos. pressure (altitude)            5
    p_sea=1013;                   # Atmos. pressure (sea level)           6
    DOY=172;                      # Day of year (June 21)                 7
                                                                          8
    deg2rad=(2*3.1415927)/360;    # Degrees to radians                    9
                                                                          10
    latitude*=deg2rad;            # Latitude in radians                   11
                                                                          12
    # Solar declination angle                                            13
    declination = (-23.4*deg2rad)* \                                     14
        cos(deg2rad*(360*(DOY+10)/365));                                 15
                                                                          16
    # for-loop to cycle through 24 hours in a day                        17
    for(hour_angle=-180;hour_angle<=180;hour_angle+=15){                 18
        # Cosine of the solar zenith angle                               19
        cos_zenith=sin(latitude)*sin(declination)+ \                     20
            cos(latitude)*cos(declination)*\                             21
    cos(hour_angle*deg2rad);                                             22
                                                                          23
```

```
    # Atmospheric air mass                                          24
    air_mass=(p_alt/p_sea)/cos_zenith;                              25
                                                                    26
    # Direct, diffuse and global solar irradiance                  27
    I_direct=(E_0)*(tau**air_mass)*cos_zenith;                      28
    I_diffuse=0.3*(1-(tau**air_mass))*(E_0)*cos_zenith;            29
    I_global=I_direct+I_diffuse;                                    30
                                                                    31
    hour=12+(hour_angle/15.0);                                      32
                                                                    33
    # Output results                                                34
    if(I_global>=0){                                                35
      printf("%4i %7.3f %7.3f %7.3f\n", \                          36
        hour, I_global, I_direct, I_diffuse);                       37
    }                                                               38
  }                                                                 39
}                                                                   40
```

```
gawk -f ex5-2.awk > ex5-2.out
```

```
set xdata time                                                     1
set timefmt "%Y%m%dT%H%M"                                          2
set xrange ["19980621T0000":"19980622T0000"]                      3
set x2range [0:24]                                                4
set key top left Left box                                         5
set xlabel "Time"                                                 6
set xtics nomirror                                               7
set format x "%H:%M"                                             8
set ylabel "Global solar irradiance, W.m^2"                     9
plot 'radt98le.out' u 1:2 t "Measured" w lp lw 2 pt 7, \        10
     'ex5-2.out' axes x2y1 t "Modeled" w lp lw 2 pt 7           11
```

Exercise 6.1

Program E.7: ex6-1.awk

```
# Simple model of solar radiation interaction with a mixed   1
# soil, vegetation and snow surface. The program calculates  2
# spectral reflectance of a surface covered by specified     3
# areal fractions of soil, vegetation (leaves) and snow      4
#                                                            5
# Usage: gawk -f mixture.awk -v rho_leaf=value \            6
#               -v rho_soil=value -v rho_snow \             7
#               [ > output_file ]                           8
#                                                            9
```

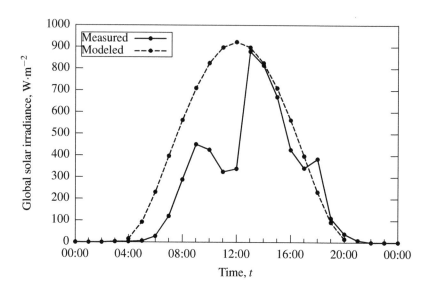

Figure E.5: Output from Exercise 5.2

```
# Variables:                                                                10
# area_leaf    Fractional area covered by leaves                            11
# rho_leaf     Leaf spectral reflectance                                    12
# rho_soil     Soil spectral reflectance                                    13
# rho_surface  Average spectral reflectance of surface                      14
                                                                            15
BEGIN{                                                                      16
   for(area_leaf=0;area_leaf<=1;area_leaf+=0.1){                            17
      for(area_snow=0;area_snow<=1-area_leaf;area_snow+=0.1){               18
         area_soil=1-(area_leaf+area_snow);                                 19
         rho_surface=(area_leaf*rho_leaf)+ \                                20
            area_snow*rho_snow + \                                          21
            (area_soil*rho_soil);                                           22
            printf("%3.1f %3.1f %3.1f %4.2f\n", \                           23
         area_leaf, area_snow, area_soil, rho_surface);                     24
      }                                                                     25
   }                                                                        26
}                                                                           27
```

```
gawk -f ex6-1.awk -v rho_leaf=0.475 -v rho_snow=0.75 -v ↺            1
   ↺rho_soil=0.125 > ex6-1.out
```

```
0.0 0.0 1.0 0.12                                                    1
0.0 0.1 0.9 0.19                                                    2
0.0 0.2 0.8 0.25                                                    3
0.0 0.3 0.7 0.31                                                    4
0.0 0.4 0.6 0.38                                                    5
```

```
0.8  0.0  0.2  0.40                                          61
0.8  0.1  0.1  0.47                                          62
0.8  0.2  0.0  0.53                                          63
0.9  0.0  0.1  0.44                                          64
0.9  0.1  0.0  0.50                                          65
1.0  0.0  0.0  0.47                                          66
```

Exercise 6.2

```
gawk -f 3layers.awk -v rho_leaf=0.15 -v tau_leaf=0.07 -v ↩   1
    ↪rho_soil=0.05 > ex6-2a.dat
gawk -f 3layers.awk -v rho_leaf=0.07 -v tau_leaf=0.15 -v ↩   2
    ↪rho_soil=0.05 > ex6-2b.dat
gawk -f 3layers.awk -v rho_leaf=0.15 -v tau_leaf=0.07 -v ↩   3
    ↪rho_soil=0.20 > ex6-2c.dat
gawk -f 3layers.awk -v rho_leaf=0.07 -v tau_leaf=0.15 -v ↩   4
    ↪rho_soil=0.20 > ex6-2d.dat
gawk -f 3layers.awk -v rho_leaf=0.0 -v tau_leaf=0.475 -v ↩   5
    ↪rho_soil=0.125 > ex6-2e.dat
gawk -f 3layers.awk -v rho_leaf=0.475 -v tau_leaf=0.0 -v ↩   6
    ↪rho_soil=0.125 > ex6-2f.dat
```

```
reset                                                                        1
set xlabel "Leaf Area Index, LAI"                                            2
set ylabel "Spectral reflectance, rho_canopy"                                3
set style data linespoints                                                   4
set key bottom right                                                         5
set yrange [0:*]                                                             6
plot 'ex6-2a.dat' \                                                          7
    t "rho_leaf=0.15, tau_leaf=0.07, rho_soil=0.05", \                       8
    'ex6-2b.dat' \                                                           9
    t "rho_leaf=0.07, tau_leaf=0.15, rho_soil=0.05", \                      10
    'ex6-2c.dat' \                                                          11
    t "rho_leaf=0.15, tau_leaf=0.07, rho_soil=0.20", \                      12
    'ex6-2c.dat' \                                                          13
    t "rho_leaf=0.07, tau_leaf=0.15, rho_soil=0.20"                         14
pause -1                                                                    15
                                                                           16
set key top right                                                          17
plot 'ex6-2a.dat' \                                                        18
    t "rho_leaf=0.0, tau_leaf=0.475, rho_soil=0.125", \                    19
    'ex6-2d.dat' \                                                         20
    t "rho_leaf=0.475, tau_leaf=0.0, rho_soil=0.125", \                    21
    'ex6-2e.dat' \                                                         22
    t "rho_leaf=0.0, tau_leaf=0.0, rho_soil=0.125"                         23
pause -1                                                                    24
```

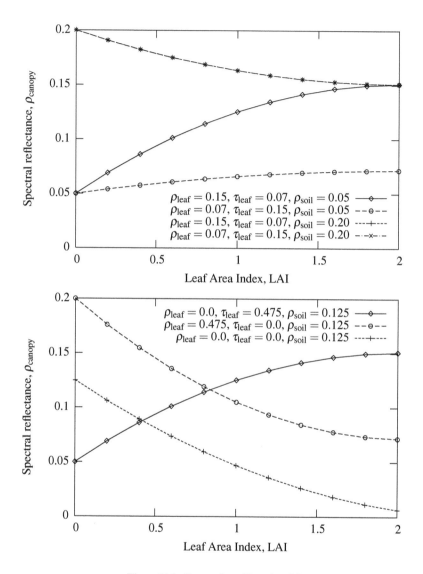

Figure E.6: Output from Exercise 6.2

Exercise 7.1

```
gawk -f analytic.awk -v R_Leaf=0.07 -v T_Leaf=0.07 -v ↪     1
    ↪R_Soil=0.06
gawk -f analytic.awk -v R_Leaf=0.18 -v T_Leaf=0.18 -v ↪     2
    ↪R_Soil=0.05
```

```
0.0702952
0.181635
```

The spectral reflectance of the vegetation canopy at blue wavelengths predicted by the analytical model is approximately 0.703, while at green wavelengths it is roughly 0.182.

Exercise 7.2

```
gawk -f iterate3.awk -v R_Leaf=0.475 -v T_Leaf=0.475 -v ↪    1
    ↪R_Soil=0.75 -v layers=10 -v threshold=0.001
```

```
0.0010   16  0.7192
```

The numerical model, iterate3.awk, takes 16 iterations to reach a stable solution given input values of $R_L = 0.475$, $T_L = R_L$, $R_S = 0.75$, 10 leaf-layers and a threshold value of 0.001.

Exercise 8.1

```
gawk -f discrete.awk -v pop_init=100 -v lambda=0.95 -v ↪    1
    ↪period=50 > ex8-1a.dat
gawk -f discrete.awk -v pop_init=100 -v lambda=0.90 -v ↪    2
    ↪period=50 > ex8-1b.dat
gawk -f discrete.awk -v pop_init=100 -v lambda=0.85 -v ↪    3
    ↪period=50 > ex8-1c.dat
```

```
reset                                                        1
set xlabel "Time, t"                                         2
set ylabel "Population size, N"                              3
set style data points                                        4
set key top right box                                        5
                                                             6
plot 'ex8-1a.dat' t "lambda=0.95", \                         7
     'ex8-1b.dat' t "lambda=0.90", \                         8
     'ex8-1c.dat' t "lambda=0.85"                            9
pause -1                                                      10
                                                             11
set key bottom left box                                      12
set logscale y                                               13
replot                                                       14
pause -1                                                     15
```

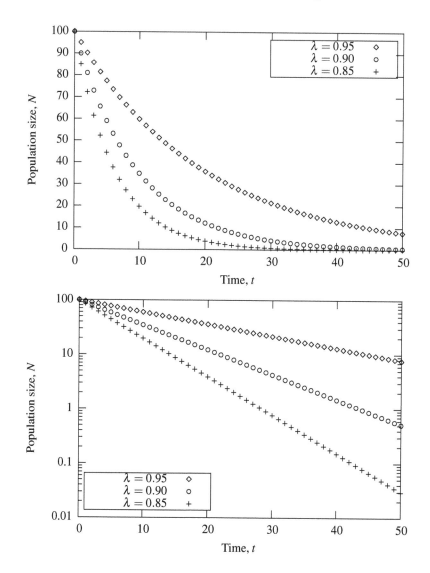

Figure E.7: Ouptut from Exercise 8.1

Exercise 8.2

```
gawk -f continue.awk -v pop_init=100 -v growth_rate=-0.05↩   1
     ↪ -v period=50 > ex8-2a.dat
gawk -f continue.awk -v pop_init=100 -v growth_rate=-0.10↩   2
     ↪ -v period=50 > ex8-2b.dat
gawk -f continue.awk -v pop_init=100 -v growth_rate=-0.15↩   3
     ↪ -v period=50 > ex8-2c.dat
```

```
reset                                                          1
set xlabel "Time, t"                                           2
set ylabel "Population size, N"                                3
set style data lines                                           4
set key top right box                                          5
                                                               6
plot 'ex8-2a.dat' t "r=-0.05", \                               7
     'ex8-2b.dat' t "r=-0.10", \                               8
     'ex8-2c.dat' t "r=-0.15"                                  9
pause -1                                                       10
                                                              11
set key bottom left box                                       12
set logscale y                                                13
replot                                                        14
pause -1                                                      15
```

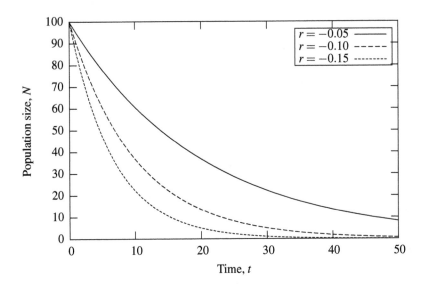

Figure E.8: First plot from Exercise 8.2

Exercise 8.3

```
gawk -f cntlogst.awk -v pop_init=1000 -v growth_rate=0.07↺  1
     ↺ -v carry_cap=100 -v period=50 > ex8-3.dat
```

```
reset                                                          1
set xlabel "Time, t"                                           2
set ylabel "Population size, N"                                3
set style data lines                                           4
```

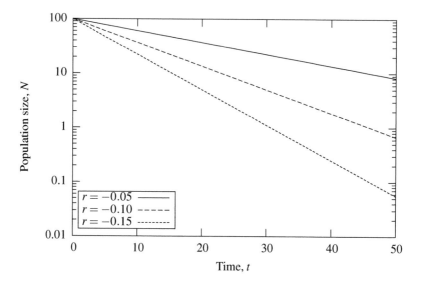

Figure E.9: Second plot from Exercise 8.2

```
set key top right box                                            5
plot 'ex8-3.dat' t "r=0.07, K=100"                              6
pause -1                                                         7
```

Exercise 8.4

This exercise is left to the discretion of the reader to experiment freely with the predator-prey model and its outputs.

Exercise 9.1

```
gawk -f daisy1.awk -v luminosity=0.8 > ex9-1_08.dat           1
gawk -f daisy1.awk -v luminosity=0.9 > ex9-1_09.dat           2
gawk -f daisy1.awk -v luminosity=1.0 > ex9-1_10.dat           3
gawk -f daisy1.awk -v luminosity=1.1 > ex9-1_11.dat           4
gawk -f daisy1.awk -v luminosity=1.2 > ex9-1_12.dat           5
```

```
reset                                                           1
set yrange [0:0.7]                                              2
set style data lines                                            3
set key top right box                                           4
set xlabel "Time, t"                                            5
set ylabel "Fractional area covered by black daisies"          6
plot 'ex9-1_08.dat' u 1:3 t "L=0.8", \                         7
     'ex9-1_09.dat' u 1:3 t "L=0.9", \                         8
```

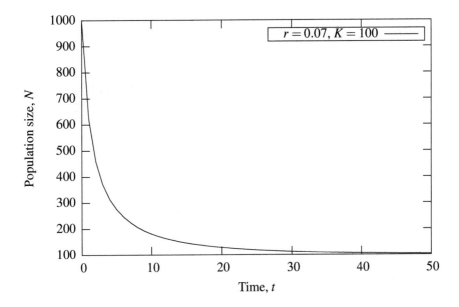

Figure E.10: Output from Exercise 8.3

```
      'ex9-1_10.dat' u 1:3 t "L=1.0", \          9
      'ex9-1_11.dat' u 1:3 t "L=1.1", \          10
      'ex9-1_12.dat' u 1:3 t "L=1.2"             11
pause -1                                         12
                                                 13
set ylabel "Fractional area covered by white daisies"  14
plot 'ex9-1_08.dat' u 1:4 t "L=0.8", \          15
      'ex9-1_09.dat' u 1:4 t "L=0.9", \          16
      'ex9-1_10.dat' u 1:4 t "L=1.0", \          17
      'ex9-1_11.dat' u 1:4 t "L=1.1", \          18
      'ex9-1_12.dat' u 1:4 t "L=1.2"             19
pause -1                                         20
```

Exercise 9.2

Program E.8: daisy2x.awk

```
BEGIN {                                          1
  # Initialize main variables                    2
  area_daisy   = 0.01;                            3
  area_suit    = 1.0;                             4
  solar_const  = 917;                             5
  Stefan_Boltz = 5.67E-08;                        6
  albedo_soil  = 0.5;                             7
  death_rate   = 0.3;                             8
```

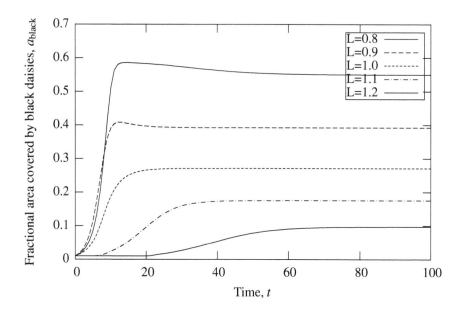

Figure E.11: First plot from Exercise 9.1

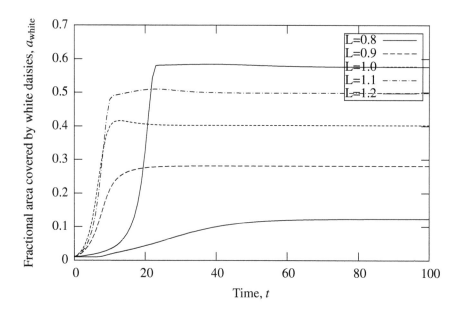

Figure E.12: Second plot from Exercise 9.1

```
q_factor     = 20;                                                    9
threshold    = 0.01;                                                 10
                                                                     11
for(luminosity=0.5;luminosity<=1.7;luminosity+=0.025){              12
  # Run model until it reaches steady state for                     13
  # current solar luminosity                                        14
  do{                                                               15
    prev_temp_global=temp_global;                                   16
    area_avail=area_suit-area_daisy;                                17
    albedo_global=(area_avail*albedo_soil)+ \                       18
      (area_daisy*albedo_daisy);                                    19
    temp_global=(((solar_const*luminosity* \                        20
      (1-albedo_global))/Stefan_Boltz)**0.25)-273;                  21
    temp_daisy=((q_factor*(albedo_global-albedo_daisy))+ \22
      temp_global);                                                 23
    growth_daisy=(1-(0.003265*((22.5-temp_daisy)**2)));            24
    area_daisy+=area_daisy*((area_avail*growth_daisy)- \           25
      death_rate);                                                  26
    # Do not allow daisies to become extinct                       27
    if(area_daisy<0.01){area_daisy=0.01};                          28
    # Check whether global average temperature for                 29
    # current and previous run differ by > threshold               30
    temp_difference=temp_global-prev_temp_global                    31
    if(temp_difference<=0){temp_difference*=-1}                     32
  } while(temp_difference>threshold)                                33
  printf("%5.3f %7.4f %6.4f\n", \                                   34
    luminosity, temp_global, area_daisy);                           35
  }                                                                 36
}                                                                   37
```

```
gawk -f daisy2x.awk -v albedo_daisy=0.25 > ex9-2blk.dat           1
gawk -f daisy2x.awk -v albedo_daisy=0.75 > ex9-2wht.dat           2
```

```
reset                                                              1
set key top left Left box                                          2
set xlabel "Solar luminosity, L"                                   3
set style data lines                                               4
                                                                   5
set ylabel "Globally averaged temperature, T/deg C"               6
plot 'ex9-2blk.dat' u 1:2 t "Black daisies only", \               7
    'ex9-2wht.dat' u 1:2 t "White daisies only", \                8
    'daisy2.dat' u 1:2 t "Black and white daisies"                9
pause -1                                                          10
                                                                  11
set yrange [0:0.8]                                                12
set ylabel "Fractional area covered, a"                           13
plot 'ex9-2blk.dat' u 1:3 t "Black daisies only", \              14
    'ex9-2wht.dat' u 1:3 t "White daisies only"                  15
pause -1                                                          16
```

Exercise 9.3

Program E.9: daisy3x.awk

```
BEGIN {                                                                    1
  # Initialize main variables                                             2
  area_black    = 0.01;                                                   3
  area_gray     = 0.01;                                                   4
  area_white    = 0.01;                                                   5
  area_suit     = 1.0;                                                    6
  solar_const   = 917;                                                    7
  Stefan_Boltz  = 5.67E-08;                                               8
  albedo_soil   = 0.5;                                                    9
  albedo_black  = 0.25;                                                   10
  # albedo_gray  = 0.5;                                                   11
  albedo_white  = 0.75;                                                   12
  death_rate    = 0.3;                                                    13
  q_factor      = 20;                                                     14
  threshold     = 0.01;                                                   15
                                                                           16
  for(luminosity=0.5;luminosity<=1.7;luminosity+=0.025){                  17
    # Run model until it reaches steady state for                        18
    # current solar luminosity                                            19
    do{                                                                    20
      prev_temp_global=temp_global;                                       21
      area_avail=area_suit- \                                             22
        (area_black+area_white+area_gray);                               23
      albedo_global= (area_avail*albedo_soil)+ \                         24
        (area_black*albedo_black)+ \                                      25
        (area_white*albedo_white) + \                                     26
        (area_gray*albedo_gray);                                          27
      temp_global=(((solar_const*luminosity* \                           28
        (1-albedo_global))/Stefan_Boltz)**0.25)-273;                     29
      temp_black=((q_factor*(albedo_global-albedo_black))+ \30
        temp_global);                                                     31
      temp_gray=((q_factor*(albedo_global-albedo_gray))+ \    32
        temp_global);                                                     33
      temp_white=((q_factor*(albedo_global-albedo_white))+ \34
        temp_global);                                                     35
      growth_black=(1-(0.003265*((22.5-temp_black)**2)));               36
      growth_gray=(1-(0.003265*((22.5-temp_gray)**2)));                 37
      growth_white=(1-(0.003265*((22.5-temp_white)**2)));               38
      area_black+=area_black*((area_avail*growth_black)- \              39
        death_rate);                                                      40
      area_gray+=area_gray*((area_avail*growth_gray)- \                 41
        death_rate);                                                      42
      area_white+=area_white*((area_avail*growth_white)- \              43
        death_rate);                                                      44
      # Do not allow daisies to become extinct                           45
```

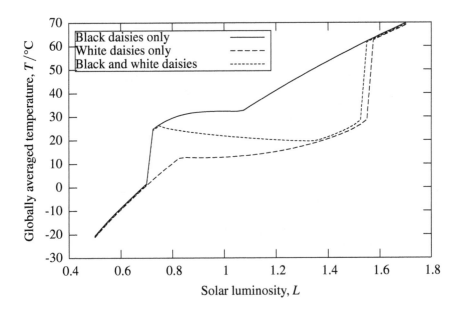

Figure E.13: First plot from Exercise 9.2

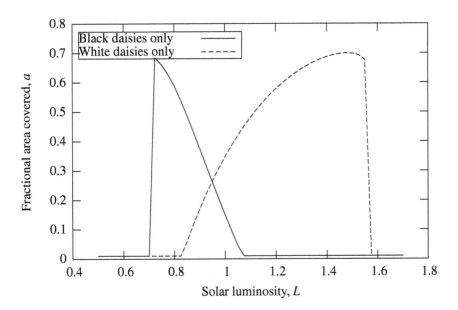

Figure E.14: Second plot from Exercise 9.2

```
      if(area_black<0.01){area_black=0.01};              46
      if(area_gray<0.01){area_gray=0.01};                47
      if(area_white<0.01){area_white=0.01};              48
      # Check whether global average temperature for     49
      # current and previous run differ by > threshold   50
      temp_difference=temp_global-prev_temp_global        51
      if(temp_difference<0){temp_difference*=-1}          52
    } while(temp_difference>threshold)                    53
    printf("%5.3f %7.4f %6.4f %6.4f %6.4f\n", \           54
      luminosity, temp_global, area_black, \              55
      area_gray, area_white);                             56
  }                                                       57
}                                                         58
```

```
gawk -f daisy3x.awk -v albedo_gray=0.3 > ex9-3_03.dat    1
gawk -f daisy3x.awk -v albedo_gray=0.4 > ex9-3_04.dat    2
gawk -f daisy3x.awk -v albedo_gray=0.5 > ex9-3_05.dat    3
gawk -f daisy3x.awk -v albedo_gray=0.6 > ex9-3_06.dat    4
gawk -f daisy3x.awk -v albedo_gray=0.7 > ex9-3_07.dat    5
```

```
reset                                                     1
set style data lines                                      2
set key top left box                                      3
set xlabel "Solar luminosity, L"                          4
set ylabel "Globally averaged temperature, T_global/degC" 5
plot 'ex9-3_03.dat' t "a_gray=0.3", \                     6
     'ex9-3_04.dat' t "a_gray=0.4", \                     7
     'ex9-3_05.dat' t "a_gray=0.5", \                     8
     'ex9-3_06.dat' t "a_gray=0.6", \                     9
     'ex9-3_07.dat' t "a_gray=0.7"                        10
pause -1                                                  11
                                                         12
set key top right box                                    13
set ylabel "Fractional area of black daisies, a_black"   14
plot 'ex9-3_03.dat' u 1:3 t "a_gray=0.3", \              15
     'ex9-3_04.dat' u 1:3 t "a_gray=0.4", \              16
     'ex9-3_05.dat' u 1:3 t "a_gray=0.5", \              17
     'ex9-3_06.dat' u 1:3 t "a_gray=0.6", \              18
     'ex9-3_07.dat' u 1:3 t "a_gray=0.7"                 19
pause -1                                                  20
                                                         21
set key top left box                                     22
set ylabel "Fractional area of white daisies, a_white"   23
plot 'ex9-3_03.dat' u 1:4 t "a_gray=0.3", \              24
     'ex9-3_04.dat' u 1:4 t "a_gray=0.4", \              25
     'ex9-3_05.dat' u 1:4 t "a_gray=0.5", \              26
     'ex9-3_06.dat' u 1:4 t "a_gray=0.6", \              27
     'ex9-3_07.dat' u 1:4 t "a_gray=0.7"                 28
pause -1                                                  29
```

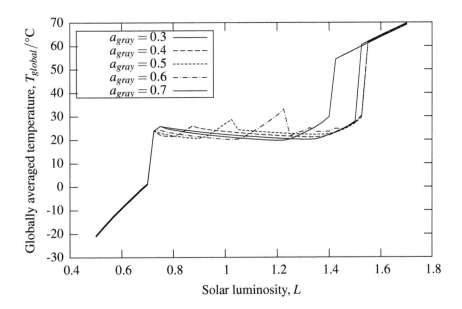

Figure E.15: First plot from Exercise 9.3

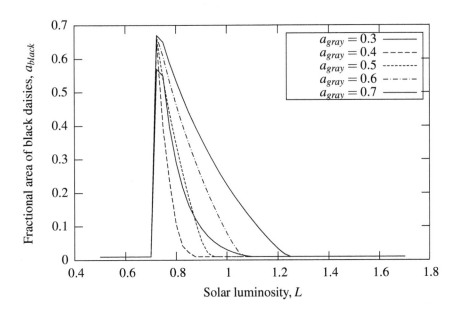

Figure E.16: Second plot from Exercise 9.3

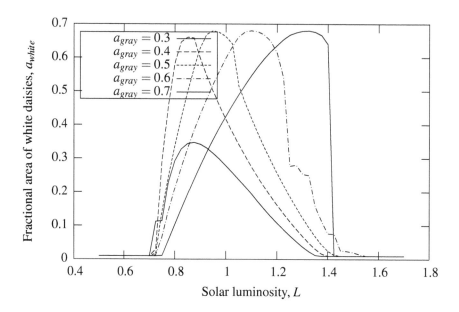

Figure E.17: Third plot from Exercise 9.3

Exercise 9.4

Program E.10: daisy4x.awk

```
BEGIN {                                                      1
                                                             2
    # Initialize main variables                              3
    area_black      = 0.01;                                  4
    area_dkgray     = 0.01;                                  5
    area_gray       = 0.01;                                  6
    area_palegray   = 0.01;                                  7
    area_white      = 0.01;                                  8
    area_suit       = 1.0;                                   9
    solar_const     = 917;                                   10
    Stefan          = 5.67E-08;                              11
    albedo_soil     = 0.5;                                   12
    albedo_black    = 0.22;                                  13
    albedo_dkgray   = 0.375;                                 14
    albedo_gray     = 0.5;                                   15
    albedo_palegray = 0.675;                                 16
    albedo_white    = 0.75;                                  17
    death_rate      = 0.3;                                   18
    q_factor        = 20;                                    19
    threshold       = 0.01;                                  20
                                                             21
```

```
for(luminosity=0.5;luminosity<=1.7;luminosity+=0.025){      22
  # Run model until it reaches steady state for              23
  # current solar luminosity                                 24
  do{                                                        25
    prev_temp_global=temp_global;                            26
                                                             27
    area_avail=area_suit- \                                  28
      (area_black+area_dkgray+area_gray+\                    29
      area_palegray+area_white);                             30
    albedo_global=(area_avail*albedo_soil)+ \                31
      (area_black*albedo_black)+ \                           32
      (area_dkgray*albedo_dkgray)+ \                         33
      (area_gray*albedo_gray)+ \                             34
      (area_palegray*albedo_palegray)+ \                     35
      (area_white*albedo_white);                             36
    temp_global=(((solar_const*luminosity* \                37
      (1-albedo_global))/Stefan)**0.25)-273;                 38
                                                             39
    temp_black=fn_temp(albedo_black);                        40
    temp_dkgray=fn_temp(albedo_dkgray);                      41
    temp_gray=fn_temp(albedo_gray);                          42
    temp_palegray=fn_temp(albedo_palegray);                  43
    temp_white=fn_temp(albedo_white);                        44
                                                             45
    growth_black=fn_growth(temp_black);                      46
    growth_dkgray=fn_growth(temp_dkgray);                    47
    growth_gray=fn_growth(temp_gray);                        48
    growth_palegray=fn_growth(temp_palegray);                49
    growth_white=fn_growth(temp_white);                      50
                                                             51
    area_black+=fn_area(area_black,growth_black);            52
    area_dkgray+=fn_area(area_dkgray,growth_dkgray);         53
    area_gray+=fn_area(area_gray,growth_gray);               54
    area_palegray+=fn_area(area_palegray,growth_palegray);   55
    area_white+=fn_area(area_white,growth_white);            56
                                                             57
    # Do not allow daisies to become extinct                58
    if(area_black<0.01){area_black=0.01};                    59
    if(area_dkgray<0.01){area_dkgray=0.01};                  60
    if(area_gray<0.01){area_gray=0.01};                      61
    if(area_palegray<0.01){area_palegray=0.01};              62
    if(area_white<0.01){area_white=0.01};                    63
                                                             64
    # Check whether global average temperature for           65
    # current and previous model run differ by more          66
    # than threshold                                         67
    temp_difference=temp_global-prev_temp_global             68
    if(temp_difference<0){temp_difference*=-1}               69
  } while(temp_difference>threshold)                         70
```

```
                                                                          71
    printf("%5.3f %7.4f %6.4f %6.4f %6.4f %6.4f %6.4f\n", \           72
       luminosity, temp_global, area_black, \                          73
       area_dkgray, area_gray, \                                       74
       area_palegray, area_white);                                     75
  }                                                                     76
}                                                                       77
                                                                        78
# Function to calculate local temperatures                             79
function fn_temp(albedo_daisy)                                          80
{                                                                       81
  return ((q_factor*(albedo_global-albedo_daisy))+ \                   82
    temp_global);                                                       83
}                                                                       84
                                                                        85
# Function to calculate daisy growth rates                             86
function fn_growth(temp_daisy)                                          87
{                                                                       88
  return (1-(0.003265*((22.5-temp_daisy)**2)));                        89
}                                                                       90
                                                                        91
# Function to calculate change in daisy area                           92
function fn_area(area_daisy,growth_daisy)                               93
{                                                                       94
  return (area_daisy*((area_avail*growth_daisy)- \                     95
    death_rate));                                                       96
}                                                                       97
```

```
gawk -f daisy4x.awk > daisy4x.dat                                        1
```

```
reset                                                                    1
set style data lines                                                     2
set key top left box                                                     3
set xlabel "Solar luminosity, L"                                         4
                                                                         5
set ylabel "Globally averaged temperature, T_global/degC"               6
plot 'daisy4x.dat' t "5 daisy species", \                               7
     'daisy3.dat' t "3 daisy species", \                                8
     'daisy2.dat' t "2 daisy species", \                                9
     'nolife.dat' t "0 daisy species"                                   10
pause -1                                                                11
```

Exercise 9.5

The command line `gawk -f daisy5.awk -f daisyfns.awk -f daisyvar.awk` results in a "division by zero" error, because the values of the main variables are initialized after the main pattern-action block.

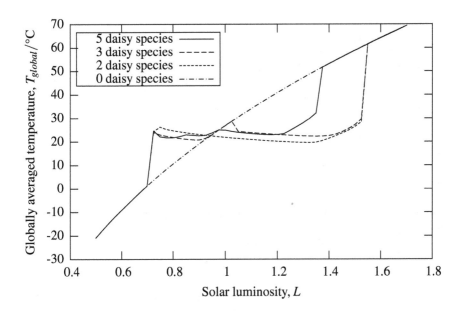

Figure E.18: Output from Exercise 9.4

The command line `gawk -f daisyvar.awk daisyfns.awk daisy5.awk` produces no output because it treats the files `daisyfns.awk` and `daisy5.awk` as input data files (cf. program files), since neither is preceded by a `-f` command-line switch. **gawk** therefore interprets the rules in `daisyvar.awk`, which initializes several variables. This program file contains no instructions on how to process the "data" contained in the files `daisyfns.awk` and `daisy5.awk`, however, so **gawk** produces no output.

Exercise 10.1

To simulate the direct, diffuse and total solar irradiance on the terrain around Llyn Efyrnwy at 8 am on June 21:

```
gawk -f demirrad.awk -v size=100 -v latitude=52.756 -v ↵     1
    ↵DOY=172 -v hour_angle=-60 efyrnwy.dem
```

To visualize the results:
```
reset                                                        1
unset key                                                    2
set xlabel "Easting, m"                                      3
set ylabel "Northing, m"                                     4
set xtics 295000, 1000, 300000                               5
set ytics 315000, 1000, 320000                               6
set xrange [295000:300000]                                   7
set yrange [320000:315000]                                   8
set dgrid3d 51,51,16                                         9
set size square                                             10
```

```
set tics out                                                        11
set style line 1 lt -1 lw 0.01                                      12
set palette gray gamma 1                                            13
set pm3d at s hidden3d 1 map                                        14
set colorbox vertical user origin 0.8,0.185 size 0.04,0.655         15
set cblabel "Irradiance, W.m^{-2}"                                  16
splot 'efyrnwy.tot' u 1:2:3 w pm3d                                  17
pause -1                                                            18
splot 'efyrnwy.dif' u 1:2:3 w pm3d                                  19
pause -1                                                            20
splot 'efyrnwy.tot' u 1:2:3 w pm3d                                  21
pause -1                                                            22
```

To simulate the direct, diffuse and total solar irradiance on the terrain around Llyn Efyrnwy at solar noon on January 1:

```
gawk -f demirrad.awk -v size=100 -v latitude=52.756 -v ↩    2
    ↪DOY=1 -v hour_angle=0 efyrnwy.dem
```

To visualize the results, repeat the **gnuplot** commands shown above. Figure E.19 shows the total solar irradiance estimated by the second of these two simulations.

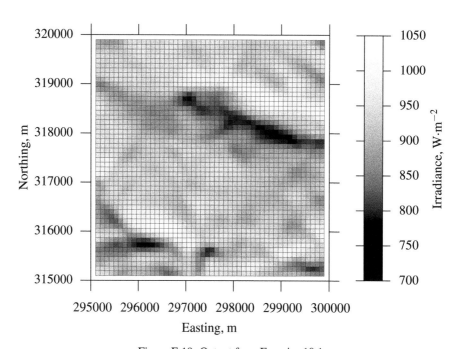

Figure E.19: Output from Exercise 10.1

Appendix F

Acronyms and Abbreviations

6S	Second Simulation of the Satellite Signal in the Solar Spectrum
ASCII	American Standard Code for Information Interchange
AWS	automatic weather station
awk	Aho, Weinberger and Kernighan
BADC	British Atmospheric Data Centre
CD-ROM	Compact Disc Read-Only Memory
CH$_4$	methane
CLI	command-line interface
CO$_2$	carbon dioxide
CSV	comma-separated value
DD	decimal degrees
DEM	digital elevation model
DGVM	dynamic global vegetation model
DMS	degrees, minutes and seconds
DOS	disk operating system
EMF	Enhanced Metafile Format
EMIC	Earth system models of intermediate complexity
EPS	Encapsulated PostScript
ERC	emission reduction credit

ESE	east-south-east
fAPAR	fraction of absorbed photosynthetically active radiation
FDE	finite difference equation
FORTRAN	FORmula TRANslation
gawk	GNU AWK
GIS	Geographic Information System
GMT	Greenwich Mean Time
GNU	GNU's Not Unix
GPL	General Public License
GUI	graphical user interface
HFCs	hydrofluorocarbons
IDW	inverse distance weighting
ISO	International Organization for Standardization
JPEG	Joint Photographic Experts Group
LAI	leaf area index
LDD	local drainage direction
MODTRAN	MODerate spectral resolution atmospheric TRANSsmittance algorithm and computer model
N$_2$O	nitrous oxide
NCM	National Climate Message
NERC	Natural Environmental Research Council
NIR	near infra-red
OAT	one-at-a-time
ODN	Ordnance Datum Newlyn
OS	Ordnance Survey
OS	operating system
PDF	probability density function
PDF	Portable Document Format

PFCs	perfluorocarbons
PNG	Portable Network Graphics
PS	PostScript
RMSE	root mean square error
RPM	RPM Package Manager
RSPB	Royal Society for the Protection of Birds
SF$_6$	sulfur hexafluoride
SVG	Scalable Vector Graphics
TIN	triangulated irregular network
TSI	total solar irradiance
UNFCCC	United Nations Framework Convention on Climate Change
UK	United Kingdom of Great Britain and Northern Ireland
URL	uniform resource locator
UTC	Coordinated Universal Time
VDU	visual display unit
WECS	wind energy conversion system
WNW	west-north-west
WYSIWYG	What You See Is What You Get

Appendix G

List of Symbols

Chapter 4 Wind Speed and Wind Power

A cross-sectional area of turbine rotors (m^2)

c Weibull distribution scale parameter

C_p Power coefficient

k Weibull distribution shape parameter

$\Phi(u)$ Weibull distribution

P wind power (W)

N_g generator efficiency

N_b mechanical efficiency

ρ air density ($kg \cdot m^{-3}$)

r radius (m)

u wind speed ($m \cdot s^{-1}$)

u_0 wind speed at ground level ($m \cdot s^{-1}$)

$\bar{u}$ mean wind speed ($m \cdot s^{-1}$)

u' threshold wind speed {$m \cdot s^{-1}$)

Chapter 5 Solar Radiation at Earth's Surface

c	speed of light (approximately 3×10^8 m·s^{-1})
d	distance between Earth and the sun (approximately 1.4×10^{11} m)
E_0	total solar radiation incident at the top of Earth's atmosphere (W·m^{-2})
$E_0(\lambda)$	spectral distribution of solar radiation incident at the top of Earth's atmosphere (W·m^{-2}·μm^{-1})
h	Planck's constant (6.626×10^{-34} J·s)
I_{diffuse}	diffuse solar irradiance (W·m^{-2})
I_{direct}	direct solar irradiance (W·m^{-2})
I_{global}	global (total) solar irradiance (W·m^{-2})
k	Boltzmann's constant (1.3807×10^{-23} J·K^{-1})
λ	wavelength (m, μm or nm)
λ_{max}	wavelength of maximum emission (μm)
m	air mass number
M	emittance (W·m^{-2})
$M(\lambda)$	spectral emittance (W·m^{-2}·μm)
M_{Earth}	spectral emittance of Earth (W·m^{-2}·μm)
M_{sun}	spectral emittance of the sun (W·m^{-2}·μm)
π	pi (3.14159265358979)
Ψ	solar zenith angle ($^\circ$)
p_{alt}	atmospheric pressure at a given altitude (mbar)
p_{sea}	atmospheric pressure at sea level (mbar)
r_{sun}	radius of the sun (approximately 6.96×10^8 m)
σ	Stefan-Boltzmann constant (5.67×10^{-8} W·m^{-2}·K^{-4})
τ	atmospheric transmittance
T	temperature (K or °C)

Chapter 6 Light Interaction with a Plant Canopy

α absorptance

$\alpha(\lambda)$ spectral absorptance

$\alpha_{\text{leaf}}(\lambda)$ leaf spectral absorptance

A fractional area of ground covered by each leaf layer

E irradiance $(\text{W} \cdot \text{m}^{-2})$

$E(\lambda)$ spectral irradiance $(\text{W} \cdot \text{m}^{-2} \cdot \mu\text{m}^{-1})$

λ wavelength (μm)

LAI leaf area index

M radiant exitance $(\text{W} \cdot \text{m}^{-2})$

$M(\lambda)$ spectral radiance exitance $(\text{W} \cdot \text{m}^{-2} \cdot \mu\text{m}^{-1})$

$M_{\text{leaf}}(\lambda)$ spectral radiance exitance of leaves $(\text{W} \cdot \text{m}^{-2} \cdot \mu\text{m}^{-1})$

$M_{\text{soil}}(\lambda)$ spectral radiance exitance of soil $(\text{W} \cdot \text{m}^{-2} \cdot \mu\text{m}^{-1})$

$M_{\text{surface}}(\lambda)$ spectral radiance exitance of a mixed leaf and soil surface $(\text{W} \cdot \text{m}^{-2} \cdot \mu\text{m}^{-1})$

N number of leaf layers

ρ reflectance

$\rho(\lambda)$ spectral reflectance

$\rho_{\text{canopy}}(\lambda)$ canopy spectral reflectance

$\rho_{\text{leaf}}(\lambda)$ leaf spectral reflectance

$\rho_{\text{soil}}(\lambda)$ soil spectral reflectance

$\rho_{\text{surface}}(\lambda)$ spectral reflectance of a mixed leaf and soil surface

τ transmittance

$\tau(\lambda)$ spectral transmittance

$\tau_{\text{leaf}}(\lambda)$ leaf spectral transmittance

Chapter 7 Analytical and Numerical Solutions

$dI^\uparrow$	infinitesimal change in the flux traveling upward through a plant canopy (as a function of distance)
$dI^\downarrow$	infinitesimal change in the flux traveling downward through a plant canopy (as a function of distance)
dz	infinitesimal change in the distance traveled through a plant canopy
$I^\uparrow[z]$	flux traveling upward from layer z in a plant canopy
$I^\uparrow[z']$	flux traveling upward from the soil substrate at the base of a plant canopy
$I^\downarrow[z]$	flux traveling downward from layer z in a plant canopy
k	extinction (or attenuation) coefficient (Bouguer's Law)
K	absorption coefficient (two-stream model)
LAI	leaf area index
$\Phi(0)$	flux density at the top of a plant canopy
$\Phi(z)$	flux density at distance/depth z into a plant canopy
R_C	canopy spectral reflectance
R_L	leaf spectral reflectance
R_S	soil spectral reflectance
T_L	leaf spectral transmittance
z	distance/depth into a plant canopy

Chapter 8 Population Dynamics

α	competition on a member of species 1 caused by members of species 2
a	success rate with which predators capture prey
β	competition on a member of species 2 caused by members of species 1
b	number of predator births related to number of prey consumed
B	average number of offspring produced per individual per time step
d	death rate of predators
dN	change in population size (or density) over an infinitesimal interval of time
dt	infinitesimal interval of time

$\frac{dN}{dt}$	derivative of N with respect to t
ΔN	change in population size (or density) over a finite interval of time
Δt	finite interval of time
e	exponential function
D	probability an individual will die during a given time step
k_n	parameters of the Runge-Kutta method for numerical integration
K	carrying capacity
K_n	carrying capacity of species n
λ	discrete (or finite) rate of population growth
N	population size (or density)
$N(t)$	population size (or density) at time t (continuous model)
N_t	population size (or density) at time t (discrete model)
r	intrinsic (or instantaneous) rate of population growth
R	difference between the per capita birth and death rates
t	time

Chapter 9 Biospheric Feedback on Daisyworld

a_{black}	fractional area of planet covered by black daisies
a_{soil}	fractional area of planet covered by exposed soil
a_{suit}	fractional area of planet suitable for daisy growth
a_{white}	fractional area of planet covered by white daisies
A_{global}	global average albedo
A_{black}	albedo of black daisies
A_{soil}	albedo of soil
A_{white}	albedo of white daisies
β_{black}	intrinsic growth rate of black daisies
β_{white}	intrinsic growth rate of white daisies
$\frac{da_x}{dt}$	instantaneous rate of change in fractional area of planet covered by daisy species x

γ instantaneous death rate of black and white daisies

L solar luminosity

M radiant exitance ($\mathrm{W \cdot m^{-2}}$)

q' factor describing thermal conduction

σ Stefan-Boltzmann constant ($5.67 \times 10^{-8}\ \mathrm{W \cdot m^{-2} \cdot K^{-4}}$)

S solar constant ($917\ \mathrm{W \cdot m^{-2}}$)

T temperature (°C)

T_{global} global average temperature (°C)

T_{black} temperature of black daisies (°C)

T_{white} temperature (K or °C)

Chapter 10 Modeling Incident Solar Radiation and Hydrological Networks over Natural Terrain

Δ distance between spatial samples in DEM (i.e., DEM cell size)

ζ angle between the surface normal (inclined) and the sun

θ gradient of the surface slope

Ψ solar zenith angle

ϕ aspect (azimuth) of the surface slope

ϕ_s solar azimuth angle

δ solar declination angle

θ_L latitude of site

E_0 total exo-atmospheric solar irradiance

τ atmospheric transmittance

m atmospheric air mass ratio

REFERENCES

Abbott, M. B. and Refsgaard, J. C., 1996, *Distributed Hydrological Modelling* (Dordrecht: Kluwer Academic Publishers).

Aho, A. V., Kernighan, B. W. and Weinberger, P. J., 1988, *The AWK Programming Language* (Reading, MA: Addison-Wesley).

Alstad, D., 2001, *Basic Populus Models of Ecology* (Upper Saddle River, NJ, USA: Prentice Hall).

Atkinson, L., 2000, *Core PHP Programming: Using PHP to Build Dynamic Web Sites*, Second Edition (Upper Sadle River, NJ: Prentice Hall).

Barnsley, M. J., 1999, Digital remotely-sensed data and their characteristics, in *Geographical Information Systems: Principles, Techniques, Management and Applications*, edited by P. A. Longley, M. F. Goodchild, D. J. Maguire, and D. W. Rhind (New York: John Wiley), pp. 451–466.

Beer, J., Mende, W. and Stellmacher, R., 2000, The role of the sun in climate forcing. *Quaternary Science Reviews*, **19**, pp. 403–415.

Betz, A., 1926, *Windenergie und Ihre Ausnutzung durch Windmüllen* (Gottingen, Germany: Vandenhoeck and Ruprecht).

Beven, K. (Editor), 1998, *Distributed Hydrological Modelling: Applications of the TOPMODEL (Advances in Hydrological Processes)* (Chichester: John Wiley & Sons).

Beven, K., 2001, *Rainfall-Runoff Modelling: The Primer* (Chichester: John Wiley & Sons).

Biggs, N. L., 1989, *Discrete Mathematics* (Oxford: Clarendon Press).

Bonham-Carter, G. F., 1994, *Geographic Information Systems for Geoscientists: Modelling with GIS* (New York: Pergamon).

Borse, G. J., 1997, *Numerical Methods with MATLAB: A Resource for Scientists and Engineers* (Boston: PWS Publishing Company).

Bouman, B. A. M., 1992, Linking physical remote sensing models with crop growth simulation models, applied for sugar beet. *International Journal of Remote Sensing*, **13**, pp. 2565–2581.

Bowden, G. J., Barker, P. R., Shestopal, V. O. and Twidell, J. W., 1983, The Weibull distribution and wind power statistics. *Wind Energy*, **7**, pp. 85–98.

Brimicombe, A., 2003, *GIS, Environmental Modelling and Engineering* (London: Taylor & Francis).

British Computer Society, 1998, *A Glossary of Computing Terms*, Ninth Edition (Harlow, Essex: Addison Wesley Longman).

Buchanan, W., 1989, *Mastering Pascal and Delphi Programming* (Basingstoke, UK: Palgrave).

Burrough, P. A. and McDonnell, R. A., 1998, *Principles of Geographical Information Systems*, Second Edition (Oxford: Oxford University Press).

Buttenfield, B. and McMaster, R. B. (Editors), 1991, *Map Generalization: Making Rules for Knowledge Representation* (Harlow, Essex: Longman Scientific and Technical).

Camillo, P., 1987, A canopy reflectance model based on an analytical solution to the multiple scattering equation. *Remote Sensing of Environment*, **23**, pp. 453–477.

Campbell, G. S. and Norman, J. M., 1998, *An Introduction to Environmental Biophysics*, Second Edition (New York: Springer-Verlag).

Cao, M., Prince, S. D., Li, K., Tao, B., Small, J. and Shao, X., 2003, Response of terrestrial carbon uptake to climate interannual variability in China. *Global Change Biology*, **9**, pp. 536–546.

Chernoff, H., 1973, The use of faces to represent points in k-dimensional space graphically. *Journal of the American Statistical Association*, **68**, pp. 361–368.

Clarke, K. C., Parks, B. O. and Crane, M. P., 2002, *Geographic Information Systems and Environmental Modeling* (New York: Prentice Hall).

Cooper, K., Smith, J. A. and Pitts, D., 1982, Reflectance of a vegetation canopy using the Adding method. *Applied Optics*, **21**, pp. 4112–4118.

Cowlishaw, M., 1990, *The REXX Language: A Practical Approach to Programming*, Second Edition (Harlow, UK: Prentice Hall).

Cox, P. M., Huntingford, C. and Harding, R. J., 1998, A canopy conductance and photosynthesis model for use in a GCM land surface scheme. *Journal of Hydrology*, **212–213**, pp. 79–94.

Cramer, W., Bondeau, A., Woodward, F., Prentice, I., Betts, R., Brovkin, V., Cox, P., Fisher, V., Foley, J., Friend, A., Kucharik, C., Lomas, M., Rmankutty, N., Stitch, S., Smith, B., White, A. and Young-Molling, C., 2001, Global response of terrestrial ecosystem structure and function to CO_2 and climate change: Results from six dynamic global vegetation models. *Global Change Biology*, **7**, pp. 357–374.

Crommelynck, D. C. and Dewitte, S., 1997, Solar constant temporal and frequency characteristics. *Solar Physics*, **173**, pp. 177–191.

Dawkins, R., 1982, *The Extended Phenotype* (Oxford: W.H. Freeman).

Deaton, M. L. and Winebrake, J. J., 2000, *Dynamic Modeling of Environmental Systems* (New York: Springer-Verlag).

DiBona, C., Ockman, S. and Stone, M. (Editors), 1999, *Open Sources: Voices from the Open Source Revolution* (Sebastopol, CA: O'Reilly and Associates).

Donovan, T. M. and Welden, C. W., 2001, *Spreadsheet Exercises in Conservation Biology and Landscape Ecology* (Sunderland, MA: Sinauer Associates, Inc).

Donovan, T. M. and Welden, C. W., 2002, *Spreadsheet Exercises in Ecology and Evolution* (Sunderland, MA: Sinauer Associates, Inc).

Doolittle, W. F., 1981, Is nature really motherly? *Coevolution Quarterly*, **29**, pp. 58–65.

Dougherty, D., 1996, *sed and awk*, Second Edition (Sebastopol, CA: O'Reilly and Associates).

Edwards, D. and Hamson, M., 1989, *Guide to Mathematical Modelling* (London: Macmillan).

Fienberg, S. E., 1979, Graphical methods in statisics. *The American Statistician*, **33**, pp. 165–178.

Flanagan, D., 2001, *JavaScript: The Definitive Guide*, Fourth Edition (Sebastopol, CA: O'Reilly & Associates).

Fligge, M., Solanki, S. K., Pap, J. M., Fröhlich, C. and Wehrli, C., 2001, Variations of solar spectral irradiance from near UV to the infrared — measurements and results. *Journal of Atmospheric and Solar-Terrestrial Physics*, **63**, pp. 1479–1487.

Ford, A., 1999, *Modeling the Environment: An Introduction to Systems Dynamics Modeling of Environmental Systems* (Washington, D.C.: Island Press).

Forrester, J. W., 1961, *Industrial Dynamics* (Waltham, MA: Pegasus Communications).

Forrester, J. W., 1969, *Urban Dynamics* (Waltham, MA: Pegasus Communications).

Forrester, J. W., 1973, *World Dynamics*, Second Edition (Waltham, MA: Pegasus Communications).

Freris, L. L., 1990, *Wind Energy Conversion Systems* (London: Prentice Hall).

Friend, A. D., Stevens, A. K., Knox, R. G. and Cannell, M. R. G., 1997, A process-based, terrestrial biosphere model of ecosystem dynamics (Hybrid v3.0). *Ecological Modelling*, **95**, pp. 249–287.

Gao, W., 1993, A simple bidirectional-reflectance model applied to a tallgrass canopy. *Remote Sensing of Environment*, **45**, pp. 209–224.

Gao, W. and Lesht, B. M., 1997, Model inversion of satellite-measured reflectances for obtaining surface biophysical and bidirectional reflectance characteristics of grassland. *Remote Sensing of Environment*, **59**, pp. 461–471.

Gastellu-Etchegorry, J. P., Zagolski, F. and Romier, J., 1996, A simple anisotropic reflectance model for homogeneous multilayer canopies. *Remote Sensing of Environment*, **57**, pp. 22–38.

Gausman, H. W., 1977, Reflectance properties of leaf components. *Remote Sensing of Environment*, **6**, pp. 1–10.

Giordano, F. R., Weir, M. D. and Fox, W. P., 1997, *A First Course in Mathematical Modeling* (Pacific Grove, CA: Brooks/Cole).

Gipe, P., 1995, *Wind Energy Comes of Age* (New York: John Wiley & Sons).

Gipe, P., 1999, *Wind Energy Basics: A Guide to Small and Micro Wind Systems* (White River Junction, VT: Chelsea Green).

Gleick, J., 1987, *Chaos* (London: Heinemann).

Glickman, T. S. (Editor), 2000, *Glossary of Meteorology*, Second Edition (Boston, MA: American Meteorological Society).

Goel, N. S., 1987, Models of vegetation canopy reflectance and their use in the estimation of biophysical parameters from reflectance data. *Remote Sensing Reviews*, **3**, pp. 1–212.

Goel, N. S., 1989, Inversion of canopy reflectance models for estimation of biophysical parameters from reflectance data, in *Theory and Applications of Optical Remote Sensing*, edited by G. Asrar (New York: John Wiley & Sons), pp. 205–251.

Golding, E. W., 1977, *The Generation of Electricity by Wind Power* (London: Spon Press).

Goodchild, M. F., Steyaert, L. T., Parks, B. O., Johnston, C., Maidment, D., Crane, M. and Glendinning, S. (Editors), 1996, *GIS and Environmental Modeling: Progress and Research Issues* (New York: John Wiley & Sons).

Goossens, M., Mittelbach, F. and Samarin, A., 1994, *The LATEXCompanion* (Reading, MA: Addison-Wesley).

Goossens, M., Rahtz, S. and Mittelbach, F., 1997, *The LATEX Graphics Companion: Illustrating Documents with TEX and PostScript* (Reading, MA: Addison-Wesley).

Goudriaan, J. and van Laar, H. H., 1994, *Modelling Potential Crop Growth Processes* (Dordrecht: Kluwer).

Grant, L., 1987, Diffuse and specular characteristics of leaf reflectance. *Remote Sensing of Environment*, **22**, pp. 309–322.

Grossman, P., 1995, *Discrete Mathematics for Computing* (Basingstoke, Hampshire: MacMillan Press Ltd).

Gueymard, C. A., 2003, Direct solar transmittance and irradiance predictions with broadband models. Part I: Detailed theoretical performance assessment. *Solar Energy*, **74**, pp. 355–379.

Gueymard, C. A., Myers, D. and Emery, K., 2002, Proposed reference irradiance spectra for solar energy systems testing. *Solar Energy*, **73**, pp. 443–467.

Haefner, J. W., 1996, *Modeling Biological Systems* (New York: Chapman and Hall).

Hahn, B. D., 1994, *Fortran 90 for Scientists and Engineers* (London: Butterworth-Heinemann).

Hall, N. (Editor), 1991, *The New Scientist Guide to Chaos* (London: Penguin Books).

Hardisty, J., Taylor, D. M. and Metcalfe, S. E., 1993, *Computerised Environmental Modelling: A Practical Introduction Using Excel* (Chichester: John Wiley & Sons).

Harris, J. W. and Stocker, H., 1998, *Handbook of Mathematics and Computational Science* (New York: Springer-Verlag).

Harte, J., 1988, *Consider a Spherical Cow: A Course in Environmental Problem Solving* (Sausalito, CA: University Science Books).

Harte, J., 2001, *Consider a Cylindrical Cow: More Adventures in Environmental Problem Solving* (Sausalito, CA: University Science Books).

Huggett, R., 1980, *Systems Analysis in Geography* (Oxford: Clarendon Press).

Ichoku, C. and Karnieli, A., 1996, A review of mixture modelling techniques for sub-pixel land cover estimation. *Remote Sensing Reviews*, **13**, pp. 161–186.

Inselberg, A., 1985, The plane with parallel coordinates. *The Visual Computer*, **1**, pp. 69–92.

IPCC, 2001, *Climate Change 2001: The Scientific Basis* (Cambridge: Cambridge University Press).

Iqbal, M., 1983, *An Introduction to Solar Radiation* (New York: Academic Press).

Jacquemoud, S. and Baret, F., 1990, PROSPECT: A model of leaf optical properties spectra. *Remote Sensing of Environment*, **34**, pp. 75–92.

Jakeman, A. J., Beck, M. B. and McAleer, M. J. (Editors), 1993, *Modelling Change in Environmental Systems* (Chichester: John Wiley & Sons).

Jones, C. B., 1997, *Geographical Information Systems and Computer Cartography* (Harlow: Longman).

Jones, H. G., 1983, *Plants and Microclimate* (Cambridge: Cambridge University Press).

Kernighan, B. W. and Ritchie, D. M., 1988, *The C Programming Language*, Second Edition (Englewood Cliffs, NJ: Prentice-Hall).

Kirchner, J. W., 1989, The Gaia hypothesis: Can it be tested? *Reviews of Geophysics*, **27**, pp. 223–235.

Kirchner, J. W., 1990, Gaia metaphor unfalsifiable. *Nature*, **345**, pp. 470.

Kirchner, J. W., 2002, The Gaia hypothesis: Fact, theory, and wishful thinking. *Climate Change*, **52**, pp. 391–408.

Kirkby, M. J., Naden, P. S., Burt, T. P. and Butcher, D. P., 1993, *Computer Simulation in Physical Geography*, Second Edition (Chishester: John Wiley & Sons).

Kneizys, F. X., Robertson, D. C., Abreu, L. W., Acharya, P., Anderson, G. P., Rothman, L. S., Chetwynd, J. H., Selby, J. E. A., Shettle, E. P., Gallery, W. O., Berg, A., Clough, S. A. and Bernstein, L. S., 1996, The MODTRAN 2/3 Report and LOWTRAN 7 Model, Technical report, Ontar Corporation, Andover MA.

Kuusk, A., 1995, A fast, invertible canopy reflectance model. *Remote Sensing of Environment*, **51**, pp. 342–350.

Kuusk, A., 1996, A computer-efficient plant canopy reflectance model. *Computers and Geosciences*, **22**, pp. 149–163.

Lal, R., Kimble, J. M. and Stewart, B. A., 1999, *Global Climate Change and Tropical Ecosystems* (Boca Raton, FL: Lewis Publishers).

Lamport, L., 1994, *LaTeX: A Document Preparation System: User's Guide and Reference Manual*, Second Edition (Reading, MA: Addison-Wesley).

Lean, J., 1991, Variations in the sun's radiative output. *Reviews in Geophysics*, **29**, pp. 505–535.

Lee, R. B., Gibson, M. A., Wilson, R. S. and Thomas, S., 1995, Long-term total solar irradiance variability during sunspot cycle 22. *Journal of Geophysical Research*, **100**, pp. 1667–1675.

List, R. J. (Editor), 2000, *Smithsonian Meteorological Tables*, Sixth Edition (Washington, D.C.: Smithsonian Institution Press).

Liu, B. Y. and Jordan, R. C., 1960, The interrelationship and characteristic distribution of direct, diffuse, and total solar radiation. *Solar Energy*, **4**, pp. 1–19.

Lotka, A., 1925, *Elements of Physical Biology* (Baltimore, MD: Williams and Wilkins).

Lovelock, J., 1995a, *The Ages of Gaia*, Second Edition (Oxford: Oxford University Press).

Lovelock, J., 1995b, *Gaia: A New Look at Life on Earth*, Second Edition (Oxford: Oxford University Press).

Lutz, M., 1996, *Programming Python* (Sebastopol, CA: O'Reilly & Associates).

Manwell, J. F., McGowan, J. G. and Rogers, A. L., 2002, *Wind Energy Explained: Theory, Design and Application* (New York: John Wiley & Sons).

May, R. M., 1974, Biological populations with non-overlapping generations: Stable points, stable cycles and chaos. *Science*, **186**, pp. 645–647.

May, R. M., 1976, Simple mathematical models with very complicated dynamics. *Nature*, **261**, pp. 459.

May, R. M., 1991, The chaotic rhythms of life, in *The New Scientist Guide to Chaos*, edited by N. Hall (London: Penguin Books), pp. 82–95.

May, R. M., 2001, *Stability and Complexity in Model Ecosystems* (Princeton, NJ: Princeton University Press).

McGuffie, K. and Henderson-Sellers, A., 1997, *A Climate Modelling Primer*, Second Edition (Chichester: John Wiley & Sons).

McGuffie, K. and Henderson-Sellers, A., 2005, *A Climate Modelling Primer*, Third Edition (Chichester: John Wiley & Sons).

Merz, T., 1996, *PostScript and Acrobat/PDF* (Berlin: Springer-Verlag).

Monteith, J. L. and Unsworth, M., 1995, *Principles of Environmental Physics*, Second Edition (London: Arnold).

Moore, I. D., 1996, Hydrological Modeling and GIS, in *GIS and Environmental Modeling: Progress and Research Issues*, edited by M. F. Goodchild, L. T. Steyaert, B. O. Parks, D. Johnston, D. Maidment, M. Crane, and S. Glendinning (Fort Collins, CO: GIS World Books), pp. 143–148.

Mulligan, M., 2004, Modelling catchment hydrology, in *Environmental Modelling: Finding Simplicity in Complexity*, edited by J. Wainwright, and M. Mulligan (Chichester: John Wiley & Sons), pp. 106–121.

Myneni, R. B., Ross, J. and Asrar, G., 1990, A review of the theory of photon transport in leaf canopies. *Agricultural and Forest Meteorology*, **45**, pp. 1–153.

Niemeyer, P. and Knudsen, J., 2005, *Learning Java*, Third Edition (Sebastopol, CA: O'Reilly and Associates).

North, P. R. J., 1996, 3-dimensional forest light interaction-model using a Monte-Carlo method. *IEEE Transactions on Geoscience and Remote Sensing*, **34**, pp. 946–956.

Oualline, S., 1995, *Practical C++ Programming* (Sebastopol, CA: O'Reilly and Associates).

Oualline, S., 1997, *Practical C Programming*, Third Edition (Sebastopol, CA: O'Reilly and Associates).

Piff, M., 1992, *Discrete Mathematics: An Introduction for Software Engineers* (Cambridge: Cambridge University Press).

Pinty, B. and Verstraete, M. M., 1998, Modeling the scattering of light by homogeneous vegetation in optical remote sensing. *Journal of the Atmospheric Sciences*, **55**, pp. 137–150.

Rasool, S. I. and Schneider, S. H., 1971, Atmospheric carbon dioxide and aerosols: Effect of large increases on global climate. *Science*, **173**, pp. 138–141.

Raymond, E. S., 1999, *The Cathedral and the Bazaar: Musings on Linux and Open Source by an Accidental Revolutionary* (Sebastopol, CA: O'Reilly and Associates).

Robbins, A. D., 2001, *Effective AWK Programming*, Third Edition (Sebastopol, CA: O'Reilly and Associates).

Roughgarden, J., 1998, *Primer of Ecological Theory* (Upper Saddle River, NJ: Prentice Hall).

Russell, G., Jarvis, P. G. and Monteith, J. L., 1989, Absorption of radiation by canopies and stand growth, in *Plant Canopies: Their Growth, Form and Function*, edited by G. Russell, B. Marshall, and P. G. Jarvis (Cambridge: Cambridge University Press), pp. 20–39.

Saltelli, A., Chan, K. and Scott, E. M. (Editors), 2000, *Sensitivity Analysis* (Chichester: John Wiley & Sons).

Saltelli, A., Tarantola, S., Campologno, F. and Ratto, M., 2004, *Sensitivity Analysis in Practice: A Guide to Assessing Scientific Models* (Chichester: John Wiley & Sons).

Saunders, P. T., 1994, Evolution without natural selection: Further implications of the Daisyworld parable. *Journal of Theoretical Biology*, **166**, pp. 365–373.

Schott, R., 1997, *Remote Sensing: The Image Chain Approach* (New York: Oxford).

Schowengerdt, R. J., 1997, *Remote Sensing Models and Methods for Image Processing*, Second Edition (San Diego, CA: Academic Press).

Sellers, P. J., Randall, D., Collatz, C., Berry, J., Field, C., Dazlich, D., Zhang, C. and Collelo, G., 1996, A revised land surface parameterisation (SiB2) for atmospheric GCMs. Part I: Model formation. *Journal of Climate*, **9**, pp. 676–705.

Skidmore, A. K., 1989, A comparison of techniques for calculating gradient and aspect from a gridded digital elevation model. *International Journal of Geographical Information Systems*, **3**, pp. 323–334.

Taylor, B. N., 1995, Guide for the use of the International System of Units (SI), Technical Report Special Publication 811, National Institute of Standards and Technology, Gaitherburg, MD.

Thomas, D. and Hunt, A., 2000, *Programming Ruby: The Pragmatic Programmer's Guide* (New York: Addison-Wesley).

Tucker, C. J. and Garratt, M. W., 1977, Leaf optical system modelled as a stochastic process. *Applied Optics*, **16**, pp. 635–642.

Tufte, E., 2001, *The Visual Display of Quantitative Information*, Second Edition (Cheshire, CT: Graphics Press).

Twidell, J. W. and Weir, A. D., 2006, *Renewable Energy Resources*, Second Edition (Abingdon, UK: Taylor & Francis).

Verhoef, W., 1984, Light scattering by leaf layers with application to canopy reflectance modelling: The SAIL model. *Remote Sensing of Environment*, **16**, pp. 125–141.

Vermote, E. F., Tanré, D., Deuze, J. L., Herman, M. and Morcrette, J. J., 1997, Second simulation of the satellite signal in the solar spectrum, 6S: An overview. *IEEE Transactions on Geoscience and Remote Sensing*, **35**, pp. 675–686.

Verstraete, M. M., Pinty, B. and Dickinson, R. E., 1990, A physical model of the bidirectional reflectance of vegetation canopies. 1: Theory. *Journal of Geophysical Research-Atmospheres*, **95**, pp. 11755–11765.

Verstraete, M. M., Pinty, B. and Myneni, R. B., 1996, Potential and limitations of information extraction on the terrestrial biosphere from satellite remote-sensing. *Remote Sensing of Environment*, **58**, pp. 201–214.

Volterra, V., 1926, Variazioni e fluttuazioni del numero d'individui in specie animali conviventi. *Mem. R. Accad. Naz. dei Lincei*, **2**, pp. 31–113.

Wainwright, J. and Mulligan, M. (Editors), 2004, *Environmental Modelling: Finding Simplicity in Complexity* (Chichester: John Wiley & Sons).

Wall, L. and Schwartz, R., 1993, *Programming in Perl* (Sebastopol, CA: O'Reilly and Associates).

Ward, R. C. and Robinson, M., 1999, *Principles of Hydrology* (London: McGraw-Hill).

Watson, A. and Lovelock, J. E., 1983, Biological homeostasis of the global environment: The parable of the Daisyworld. *Tellus*, **35B**, pp. 284–289.

Watson, D. F., 1992, *Contouring: A Guide to the Analysis and Display of Spatial Data* (Oxford: Pergamon – Elsevier Science).

Welch, B., Jones, K. and Hobbs, J., 2003, *Practical Programming in Tcl and Tk* (New York: Prentice Hall).

Whittaker, R. H., 1975, *Communities and Ecosystems*, Second Edition (New York: Macmillan).

Whittaker, R. H., 1982, *Communities and Ecosystems*, Third Edition (New York: Macmillan).

Wiens, J. A., 1989, Spatial scaling in eclogy. *Functional Ecology*, **3**, pp. 385–397.

Williams, T. and Kelly, C., 1998, *gnuplot: An Interactive Plotting Program*, Dartmouth College, Hannover, NH.

Willis, T. and Newsome, B., 2005, *Beginning Visual Basic 2005* (Indianapolis, IN: Wrox Press).

Wilson, W., 2000, *Simulating Ecological and Evolutionary Systems in C* (Cambridge: Cambridge University Press).

Woodward, F. I., Smith, T. M. and Emanuel, W. R., 1995, A global land primary productivity and phytogeography model. *Global Biogeochemical Cycles*, **9**, pp. 471–490.

Zevenbergen, L. W. and Thorne, C. R., 1987, Quantitative analysis of land surface topography. *Earth Surface Processes and Landforms*, **12**, pp. 47–56.

Index

command-line history, 32
comments
    # symbol, 53, 94
data styles, 36
dummy variable, 102, 198
function fitting, 103, 200
functions
    user-defined, 102
gray-scale palette, 299
handling missing data, 74, 75
hidden-line removal, 49, 50
installing, 323–328
keyword abbreviation, 36
line continuation symbol, 34
mathematical symbols in plot, 42
printing a plot, 55
scripts, advantages of, 32
setting tic-mark interval, 45
terminals, 41, 42
tic-mark labels, 38
time series, 37
    format string, 38
user-defined functions
    arguments, 102
vectors, 313
goodness-of-fit, 18
GPL, 321
gradient, 50, 114, 289, 293
graphics, 56
greenhouse gas, 80, 253

heat diffusion, 261
heuristic model, 252
hidden-line removal, 49–51
homeorhesis, 12
homeostasis, 12, 270
Homo sapiens, 214
human-computer interaction, 32
hydrological
    modeling, 309
    network, 290
    processes, xxv
hydrology, 309
hydrometeorology, 146
hysteresis loop, 248

IDW, 47
immigration, 206
in vitro, 214
infinite series, 175
initial value problem, 230
initialization, 86
instrument failure, 62
integer, 71
integrated modeling environments, 14
inter-specific competition, 238, 239,
    242, 244, 253, 258, 269
international standards
    specifying times and dates, 31, 38,
    60, 72, 343, 344
interpolation, 48, 53
interpolation methods
    inverse-distance weighting, *see*
    IDW
    kriging, 47
intra-specific competition, 218, 238,
    253, 258
inverse functional relation, 10
inverse model, 10
inverse square law, 123
irradiance, 149
    exo-atmospheric, 254
    solar spectral, 123
    spectral, 150
ISO 8610:2004, 31, 38, 60, 72, 343, 344
ISO 9660:1999, 322
isobar, 51
isobath, 51
isohaline, 51
isohyet, 51
isotherm, 51
iteration, 178, 183, 188, 190, 262, 265,
    267

Joule, 121

kelvin, 121, 256
Kernighan, Brian, 16
kinetic energy, 106
Kubelka-Munk equations, 203
Kyoto Protocol, 80, 114